The Poverty of
Green Philosophy

Other Books on Nuclear Energy from Open Universe

Earth Is a Nuclear Planet: The Environmental Case for Nuclear Power, by Mike Conley and Tim Maloney (2024)

The LNT Report: How Bad Science Made the World Afraid of Nuclear Power, by Mike Conley (2025)

Roadmap to Nowhere: A Reality Check on Renewable Energy, by Mike Conley and Tim Maloney (2026)

The Poverty of Green Philosophy

*A Marxist Case for Nuclear Energy
in a Cooperative World*

BILL SACKS

and

GREG MEYERSON

OPEN UNIVERSE
Chicago

To find out more about Open Universe and Carus Books, visit our website at www.carusbooks.com.

Printed and bound in the United States of America. Printed on acid-free paper.

The Poverty of Green Philosophy: A Marxist Case for Nuclear Energy in a Cooperative World

ISBN: 978-1-63770-067-9

This book is also available as an e-book (978-1-63770-068-6).

Library of Congress Control Number: 2025931903

Contents

Preface

We give you instead the great Elon Musk, entrepreneur, inventor, and engineer of SpaceX, PayPal, and Tesla fame. In explaining the rationale behind his latest project, the solar-power venture Solar City, he put it this way: "The sun, that highly convenient and free fusion reactor in the sky, radiates more energy to the Earth in a few hours than the entire human population consumes from all sources in a year. This means that solar panels, paired with batteries to enable power at night, can produce several orders of magnitude more electricity than is consumed by the entirety of human civilization." **It's sort of hard to argue with that kind of logic** [bolding added].

—Michael Mann, *The Madhouse Effect* (2016), pp. 147–48[1]

No, it isn't [bolding added].

—Bill Sacks and Greg Meyerson

Humanity has a problem: anthropogenic global warming (AGW).[2] Not the only problem for sure, but potentially the most threatening to the most people on Earth. Almost every year average global temperature records are broken. Similarly, the atmospheric concentration of CO_2 has been trending upward for more than half a century, amidst annual fluctuations as trees annually green up, lose

---

[1] Mann may be misinformed about energy, but he is an accomplished and highly respected scientist with regard to climate. His latest book (2023) is particularly recommended for the history of the Earth's temperature vagaries.

[2] While this isn't a book about climate, we consider AGW to be scientifically supported for all practical purposes for the time being (FAPPFTTB, the best that any scientific

their leaves, and repeat. As a result of warming effects, people die or suffer problems that make life even harder than before, particularly those worst off. Since AGW is known to be mainly due to emissions and leakage of greenhouse gases (GHGs) from the use of fossil fuels—to generate electricity and heating, facilitate transportation, and underwrite industrial processes—substitutes are sought.

Mass media provide the following story, subscribed to by most governments around the world:

> We must replace coal, natural gas (NG), and oil, which are the main cause of global warming today. We need clean energy from large commercial wind and solar farms. With enough capacity and batteries they could reliably meet instantaneous electricity demand while producing enough surplus energy to maintain and operate themselves, to replace worn-out turbines and PV panels, and to expand themselves to meet growing human needs. Lots of well-paying jobs would be generated and they would still cost consumers less than fossil fuels or nuclear.

Note the one-sided absence of any disadvantages. The media continue:

> Nuclear energy is the most dangerous and expensive energy source, plants take too long to build, and it creates the most toxic and longest-lived waste, which would negatively impact the environment. Moreover, we would quickly run out of uranium and thorium, necessitating wind and solar farms anyway.

Note the one-sided absence of any advantages.

Unfortunately, none of that stuff about wind and solar is true. Fortunately, none of that stuff about nuclear is true either. The little-known origin of our current ideological trap is a 78-year-old massive scientific fraud about the biological effects of radiation, with equally fraudulent reinforcement the following decade. Gunnar Walinder, former chair of the Swedish Radiobiology Society, calls this fraud "the greatest scientific scandal of the [twentieth] century."[3] Chapter 7 discusses the consequences and Chapter 8 the history and essence of this fraud.

---

conclusion can attain, see end of Chapter 10), that it is proceeding apace, and that today it is due mainly to ever-increasing concentrations of greenhouse gases (GHGs) in the atmosphere.

[3] *Has Radiation Protection Become a Health Hazard?* (1995). The present authors, among numerous others, have cowritten several dozen medical and scientific articles in an attempt to expose and expunge this fraud. See, for example, Sacks, Meyerson, and Siegel 2016 at https://tinyurl.com/35sazy5p.

Fossil fuels enjoy many undeniable advantages and have done so for more than a century. However, we hope to persuade readers that nuclear enjoys even more, with none of fossils' disadvantages, and that commercial-scale wind and solar don't even qualify as an alternative. Countries developing these "renewables," even if they also accept nuclear as secondary, are on a path already showing signs of failure. No less a failure, with devastating effects, is the cutting back on fossil fuels in some countries *before* expanding nuclear. Such mindless energy-reducing policies risk innumerable lives and lend credence to fossil fuel proponents.

Two other proposed solutions are intended to circumvent nuclear fission: nuclear fusion (see our Glossary for fission and fusion) and the substitution of NG for coal. We explain the shortcomings of wind and solar in Chapters 2 through 6, fusion's current irrelevance to electricity in the final section of Chapter 7, and the limited effect of switching from coal to NG in Chapter 3. These are all harmful diversions of energy, funds, effort, and attention away from nuclear (unless otherwise stated, we mean fission). As we will show, nuclear is today's sole energy source technically able to halt and reverse AGW.[4]

The chief barrier to the *technical* solution is *geopolitical*, deriving from a fragmented, competitive, profit-seeking global economic system (Chapters 9, 10, and 11). We have no illusions, as will become clear, that either the technical solution will be enacted in its entirety or that the geopolitical obstacles will be overcome in the near future. Instead, we are laying out what we hope to demonstrate are the necessary conditions for halting AGW at some point and enabling its reversal, however long that may take.

We draw on the contributions of many people with regard to both energy and social organization (Chapter 10). We count ourselves among the left, and have throughout most of our lives, tried to improve our use of the methods of analysis first developed and popularized by Karl Marx and Friedrich Engels. While many would call us Marxists, there are tremendous disagreements among Marxists about almost everything, which makes the label less than helpful. A goal we share, however, with all Marxists and many

---

[4] We must also find ways to adapt to AGW, while we seek methods of mitigation. But adaptation will not suffice for long without mitigation. People suffer and/or die due to prodigious rains, floods, and water-soaked sinkholes, to droughts, wildfires, and heat waves, to paradoxical-appearing cold snaps, and to hurricanes, tornadoes, sea level rise, and amplified storm surges. While each of these events had occurred long before our species emerged, they now frequently break records year after year—the hallmark of a relentless trend, despite occasional short-term reversals.

other leftists is to see a cooperative global society replacing the fragmented competitive societies of capitalism. We see that goal as humanity's hope—politically, socially, and economically. As a result, we aim our sharpest criticisms, particularly regarding energy, against most other Marxists and leftists in general, though not omitting warranted criticisms of non-leftists. For reasons discussed in Chapters 9 and 11, most Marxists and other leftists simply refuse to study and inform themselves about energy, uncritically rejecting nuclear and favoring either wind and solar or an overall reduction of energy usage.

While energy is a complex subject (or we wouldn't need a book), we try to clarify each element for a readership with a variety of experiences who are as eager to learn as we are. We necessarily use a small amount of simple arithmetic. For some this will be a breeze, while for others less so. Antinuclear propaganda tends to avoid numerical comparisons, which facilitates the promotion of fear. No one should fall for such arguments.[5]

There are also sections, particularly in Chapter 11, exploring philosophy and the history of ideas that some may find difficult, while others may feel at home. All of this is indispensable for understanding energy (the lifeline of society and human progress) and the social changes that we try to show are required to halt AGW. Ignoring either aspect will almost certainly result in failure.

We return repeatedly to the relative *incompleteness* of certain accounts, meaning only that we provide a more complete account of the subject at hand. Despite the fact that more complete views are almost certain sooner or later to be proven incomplete in other ways, we strive to adduce previously omitted relevant evidence and/or other relevant aspects of the argument that often lead to the opposite conclusion.

We come to this book from opposite directions. Sacks is a former physics professor turned radiologist, now retired, and Meyerson, soon to retire, is an English professor specializing in critical theory, involving philosophy and the history of ideas. Each is less experienced in the principal interests of the other, and each has exerted significant effort to understand the other's contributions to the book, though in many areas we are on level ground. However, in the end everything we say is comprehended and agreed to by both. Moreover, we have been learning and writing articles together for a decade and a half.

---

[5] Neither should anyone fall for the prolific use of misleading numbers that effectively snow the reader/listener and imply that the authors are unassailably knowledgeable and impeccably trustworthy (Chapter 3).

The process of putting this book together has given us an appreciation for the difficulties readers may have in grasping one or another part of the book, but it is vital that everyone understand at some level every aspect of the problem. Only thus can everyone be prepared to see through and reject narratives resting on fantasy, regardless of their appearance of plausibility. It would be well to put aside long-held reluctance to entertain explanations in this book about either the technical or geopolitical aspects. We may be wrong, but at least engaging with our contentions will help advance the discussion.

Finally, when there are two or more authors they have to be listed in some order. Often the first author is the main inspiration, organizer, and perhaps theoretician of the piece, and in some cases the one who has drafted the manuscript. In the present case no such distinction can be made. The drafting of any section by either of us was always preceded and accompanied by mutual discussion. Proofreading and editing has been a cooperative venture. So we decided that the first to be listed would simply be the first to have arrived on Earth.

* * * * *

A note about our title: It is borrowed from Marx's play on words in the title of his short book *The Poverty of Philosophy*. This was a friendship-ending response to a book subtitled *The Philosophy of Poverty* by contemporary French anarchist philosopher Pierre-Joseph Proudhon (to whom we allude again in Chapter 9).[6] Proudhon is an intellectual ancestor of German-British economist E.F. Schumacher, who wrote a book titled *Small Is Beautiful* (1973). As we will see, much of the Green narrative that we critique incorporates that sentiment in the form of locavorism (preference for locally grown food), decentralization, and lowered energy consumption. As we will further see, Green philosophy is conceptually impoverished and leads directly to energy poverty, which leads in turn to unintended economic poverty and decimation of our environment.

* * * * *

All figures are shown in gray scale to avoid prohibitive costs. We provide references to borrowed graphs that allow the reader to access the originals in color when needed to clarify.

---

[6] Both Proudhon and Marx used the French word "misère," which translates as either poverty or misery.

I

Context

1

Controversy

The crisis consists precisely in the fact that the old is dying and the new cannot be born; in this interregnum a great variety of morbid symptoms appear.

—ANTONIO GRAMSCI (1930)

This book is part energy analysis and part economics/geopolitics. The former is pronuclear and the latter is written from a Marxist perspective, which makes it a controversial book on both counts. And it is being written in a period described by US (non-Marxist) historian Richard Slotkin (2024) as "A Great Disorder," a period in which the possibility of culture wars turning into a civil war has become a normal part of our "common" discourse.

Our book has been shaped by this rhetorical context insofar as it owes a great debt to writers and thinkers from antagonistic camps, such that we would be only sort of joking to say that half of them might want to kill the other half—and only sort of joking to fear that both halves might want to kill us. Of course, we're only joking. Sort of.

We return to this rhetorical context below but first we summarize the book's complex argument, which attempts to stitch together the insightful components from narratives belonging to these largely incompatible and mutually hostile outlooks.

Outline of the Argument

In this book we argue for abundant and reliable (continuously available) electrical power, accessible to everyone in the world.[1]

---

[1] Power is the rate of energy production or consumption—energy per time.

This would enable electrification, or electrically-produced non-GHG-emitting synthetic fuels, for all modes of transportation, all industrial processes, and all domestic applications of cooking, heating, and air conditioning. In short, eventually we will need to electrify almost all energy-requiring applications, with the likely exception of heating, which can often be supplied more efficiently (with fewer losses) directly from source to target site. For reasons we will see, electricity will be much less costly in real terms when generated by nuclear than by NG or coal, and still less costly than by wind and solar.

It is crucial that the required energy transition be phased in according to a rational plan. Currently there is no such plan, anywhere. Source substitutions are being mandated in a careless order, determined more by political considerations rather than rational energy policy and often with no consideration of the energy implications of the decisions. California (and until recently the US government) has barred the sale or use of internal combustion engine (ICE) vehicles after a certain date in favor of electric vehicles (EVs), without timely planning for a clean source of energy to charge EV batteries or to manufacture the batteries and the vehicles they power. The US Environmental Protection Agency (EPA) calls for the shuttering of coal plants without ensuring the availability of an alternative source, and where it is being substituted it is generally in favor of NG, another fossil fuel that may offer no improvement because of leakage (Chapter 3).

Insufficient energy immiserates lives and livelihoods. The rapid growth of data processing, AI, and other energy-rapacious computer functions is already taxing the US grid and exposing the insufficiency of our electrical energy sources. An adequate but clean alternative source is urgently needed, particularly since the worsening of AGW waits for no one.

Green proponents of an all-renewable electricity system (all-RE)—consisting mainly of wind and solar—claim that such a system could completely replace fossil fuels and nuclear and produce *all* of society's energy needs reliably. They contend that this system could be completed within a certain schedule, could be sustained in perpetuity (continually reproducing itself), could expand itself as needed, could provide adequate electricity even after all other forms of societal energy were electrified, and would be the best way to eradicate further particulate and greenhouse gas (GHG) emissions and thereby eliminate the negative impacts of fossil fuel combustion on human health and the environment.

To emphasize the supposed eradication of GHGs, the code word "clean" is generally applied to wind/solar energy in the media and by government officials, though it is misapplied, as is the code word "renewable" (Chapter 2). They further assert that such a system would be the cheapest for consumers (Chapter 5), would provide plentiful well-paying jobs (Chapter 6), and would be the safest (Chapter 7).

The common Green rejection of nuclear rests on the belief that it is uniquely hazardous in both operation and waste, that nuclear power plants are too expensive and take too long to construct, and that nuclear energy encourages the proliferation of nuclear weapons (Chapter 7). Many of those with left-leaning politics inextricably link nuclear to capitalism, which they want to abolish. Encouraging all these beliefs, firms profiting from fossil fuels and renewables keep a low profile and instead fund willing environmental organizations to promote antinuclear pro-renewables propaganda. As we will see, renewables are complementary to fossil fuels rather than a threat to their continuation (Chapter 3).

By employing the rhetoric of fear, this propaganda impedes rational assessment of nuclear energy and, by narrowing the choice to renewables alone, also thwarts rational assessment of the all-RE electrical system favored by most Greens (and many governments). Some leftists appeal to the romanticized wisdom of indigenous peoples and First Nations, from whom we may have much to learn but who have never *as a group* faced the emergent problems of a large and technically complex society.

In contrast, we argue that if an all-RE system were hypothetically to be constructed, necessarily aided by the fossil and nuclear matrix, once that support were eliminated the system could neither provide adequate net electricity to run a modern society nor reproduce itself in perpetuity, much less expand itself. Its ratio of energy return over energy invested (EROEI, or conventionally, **EROI**, Chapter 5) is simply too low, and unless it retains nuclear backup, its uncontrollable intermittency makes it unreliable, even with massive amounts of storage. We further show that if some nuclear were retained to provide the necessary backup, it would become apparent that wind and solar are unnecessary and would only diminish nuclear's ability to power the society. In short, production of wind/solar energy at low penetration cannot, without passing a point of diminishing returns, be extrapolated to penetration approaching 100 percent.

We demonstrate that renewables are the most energetically *in*efficient way to produce net energy and are intrinsically the most expensive energy source for consumers (Chapter 5). And while an all-RE system would provide, indeed necessitate, plentiful jobs, the pay would be unlivably low unless subsidized by high consumer prices and/or tax moneys, paid either way by the rest of the working and middle consuming classes, categories that overlap (Chapter 6). Even a non-capitalist (non-profit-based) economic system would find an all-RE system fundamentally unreliable and unsustainable without support from nuclear energy.

A recurrent theme in the book is that proponents of an all-RE electrical system ignore inherent features that result in dyseconomies of scale. These features already produce frequent outages and higher electricity prices in places like California and Germany, where the penetration of wind/solar is among the highest in the world.

Key points include the following: Like any energy source, renewables require **conversion devices**—such as turbines, photovoltaic (PV) panels, or mirrors and fluid heat reservoirs—and receive energy **intermittently**, **uncontrollably,** and **unpredictably** (Chapter 3). Intermittency requires mitigation through **overbuilding/oversizing** commercial wind/solar farms plus plentiful **storage**, with construction of the latter potentionally consuming more energy than construction of the turbines and PV panels. Moreover, no matter how great the storage capacity, the limiting factor is the **timing of arrival** of the *surplus* wind and sunlight to charge the storage apparatus, above and beyond that called for by immediate demand.

Sufficient charging to fill in for episodic shortages is neither predictable nor reliable, leaving occasional **outages** from time to time. Excess capacity and storage apparatus could decrease the frequency but not wholly eliminate outages. The energy needed to produce and install the storage apparatus greatly diminishes the already low EROI (Chapter 5) by amplifying the denominator (EI). Due to their internal chemistry, grid-scale lithium-ion batteries have a tendency to ignite fires that are difficult to extinguish and that damage adjacent batteries and other structures, risking injuries and deaths. Safer batteries, different in design, have been promised but are not yet forthcoming.

The construction of conversion/storage apparatus for renewables, be-cause of their **low energy density** and **low power density** (Chapter 2), consumes **massive quantities** of materials (Chapters 2 and 4), land (Chapter 2), time (Chapter 5), labor (Chapter 6), and energy (Chapter 5). The material requirements challenge world supplies and result in resource **depletion**. Upon their relatively rapid obsolescence, the conversion devices produce largely unrecycled, chemically toxic **waste** in quantities concordant with their inputs, dwarfing that of nuclear "waste" in quantity and relative **hazardousness** (Chapters 7 and 8). The greater material input raises the danger and frequency of **resource wars**.

Additionally, given the current predominance of fossil fuels over nuclear, the energy requirements of renewables construction and backup for intermittency may in fact require multiplication rather than elimination of fossil fuel facilities.

Finally, continuing to build wind/solar farms **diverts massive amounts of public resources** that could otherwise enable construction of an all-nuclear electrical system and electrification of virtually all energy applications, save for some forms of direct heating (though, unlike fossil fuels and nuclear, neither wind nor solar PV operates through heat production).

Figure 1-1 summarizes, in a schematic visual format, comparisons among the three main energy sources with respect to two features: the overall usable net energy and the negative environmental impacts, for a given amount of energy input.

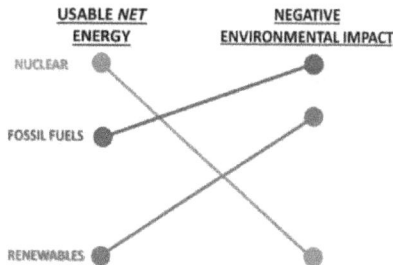

USABLE *NET* ENERGY   NEGATIVE ENVIRONMENTAL IMPACT

NUCLEAR

FOSSIL FUELS

RENEWABLES

FIGURE 1-1

Certain anti-capitalist ecosocialists (Glossary, Chapter 9) call for reducing energy demand to lower the negative impact on the environment. Aside from continuing to deprive millions of people of much-needed energy, this would deplete the energy required to halt and reverse AGW, to repair the damage to human-built structures from AGW's consequences, to desalinate vast amounts of seawater in the face of AGW-enhanced droughts, and to maintain and reproduce the conversion devices that turn energy that comes free from nature—whether in the form of wind, sunlight, fossil fuels, or uranium/thorium—into usable forms, including electricity and heat. The above schematic reflects our ability to lower our impact on the environment while preserving the benefits of centuries of scientific advances in the energy arena. The best solution today to this dual requirement is nuclear fission.[2]

Aversion to nuclear and favoring of renewables is underwritten by a concerted campaign of **disinformation** purveyed through **misinformation**, and by **price manipulations**—lowering the price of renewables (and fossils) and raising the price of nuclear (Chapter 5). The latter is achieved through **regulatory requirements** loosely analogous to requiring vehicle owners to pay to have their brakes inspected and headlights aimed weekly at official locations, and through insufficient funding of research and development of advanced Generation IV reactors.

In addition, **litigations** and **demonstrations** by numerous well-funded environmental organizations delay and sometimes entirely block planned construction of nuclear power plants. Most of the members of these organizations may very well be unaware that their NGOs are funded by the fossil fuel industry and more recently by significant wind/solar money. Helping to favor renewables, electric utilities have been incentivized to build more renewables farms, storage, and NG plants by regulatory guarantee of a 5–10 percent profit on investment. Instead of investing past profits in the anticipation of future profits, utilities are granted the funds to build new commercial renewables farms by the government, and the governmental guarantee of a 5–10 percent profit on investment encourages utilities to build capacity without limit, whether or not it produces energy.

The disinformation/misinformation axis rests firmly on the promotion of nucleophobia that conflates energy with weapons, focuses major attention on, while distorting the truth about, nuclear waste, wildly exaggerates and/or fabricates the negative effects of nuclear accidents, and keeps the public confused as to which information to believe about how nuclear reactors operate. This disinformation is particularly aimed at blocking general recognition of the demonstrable fact that the **safest energy source** of all has been, and for reasons of physics and design will continue to be, nuclear energy (Chapters 7 and 8).

Many countries face geopolitical obstacles to the acquisition of required raw materials that are concentrated and more easily accessed in only certain locations on Earth and not widely available. For that reason and others enumerated below, instituting an all-nuclear electricity system worldwide would likely require a **cooperative global economic and political system** that

---

[2] Nuclear fusion (see Glossary), as opposed to fission, is most likely irrelevant as a sustainable energy source, and may be impossible to achieve on Earth (Chapter 7).

would eliminate profit interests, domestic competition, and international rivalry, and would permit fair and universal distribution of nuclear energy's bounty. Also likely requiring global cooperation would be the planned **phase-in** of nuclear energy as fossil fuels are phased out. As mentioned, some nations have begun the phase-out without adequate nuclear energy ready to take the place of fossil fuels.

Furthermore, the sharing of technological advances to enable the whole world to move forward together would require global cooperation. Unfortunately, an all-nuclear electrical system is rejected by certain ecosocialists, who favor such a cooperative system while also favoring a regime of energy scarcity that is distributed "fairly"—equitable impoverishment. In short, the main obstacle to abundant energy for all is not technical—it is political, economic, social, and epistemological.

The competitive nature of capitalism, with its fragmentation into competing units—whether corporations, political parties, or what have you—tends to drive people into opposing groups with antagonistic world views that the antagonisms distort. The distortions, the result of what we refer to as "epistemic dysfunction," include blind spots recognized, if at all, by the opposing camp. There's no way to reconcile the socioeconomic and geopolitical differences between a Marxian and libertarian outlook. But what is an avoidable and highly consequential error is to conflate technology and socioeconomic outlook (Chapter 9).

The Lack of Critical Evaluation by Certain Public Voices

The epigraph that begins this chapter comes from the early twentieth-century Italian Marxist Antonio Gramsci. While Gramsci's observation was made near the start of the Great Depression (1930), the aforementioned Richard Slotkin repurposes it as part of his argument that US culture (with implications for the global order) is caught in a "no man's land" or purgatory or tower of Babel, where problems ("morbid symptoms") proliferate but no workable solutions are available and where members of opposing camps are unable to talk to each other, producing a communication breakdown.[3]

---

[3] Slotkin (2024) situates the culture wars in a socioeconomic context. As he puts it (p. 383), "American history has been shaped by two political orders: the New Deal order which fell apart in the 1970s, and the neoliberal order, which lost most of its authority after the economic implosion of 2008."

In writing this book we have had to navigate between the serpentine banks of this communication breakdown. The "Blue/Red" divide is unsurprisingly entangled with the discussions around energy—for instance between Green New Deals and "Drill, Baby, drill!" and between renewables and nuclear. The divide is also kaleidoscopically entangled with different views of capitalism. The Blue attraction to Green capitalism appears through Red lenses as industry-destroying totalitarianism, while to the Blue Greens the AGW-denying MAGA Reds are leading us to species extinction—bilateral demonization that proscribes discussion.

We're intervening in this discussion to offer what might at first glance appear to be an impossible position given the current ideological and political atmosphere: a critique of the Green narrative in its various iterations, including ecosocialism, with its fierce and dogmatic antinuclear commitment to an all-RE future,[4] and a critique of the ecomodernist rebuttal, with its unquestioning and dogmatic, but pronuclear, commitment to capitalism. Some ecomodernists also exhibit a tendency to downplay or discount AGW.

Anathematizing nuclear leads most Greens to drop their critical faculties and embrace the work of renewable energy guru Mark Z. Jacobson, a Stanford civil and environmental engineering professor. Jacobson and UC Davis research scientist Mark A. Delucchi's 2009 all-RE proposal, modestly named the "Roadmap," focuses on the US but aspires to global scale. The Roadmap relies on wind, water, and solar (abbreviated WWS), but water (hydroelectricity) is projected to provide only 3 percent of power capacity, as virtually the sole form of storage (Jacobson et al. 2022).[5]

Trust in Jacobson's analysis survives despite fatal criticisms by other scientists and energy analysts.[6] The criticisms have remained

---

[4] Responding angrily to the antinuclear dogma and its stable of experts, well-known US climatologist James Hansen asks rhetorically (2009, p. 203): "Do these people have the right to, in effect, make a decision that may determine the fate of my grandchildren? The antinuke advocates are so certain of their righteousness that they would eliminate the availability of an alternative to fossil fuels, should efficiency and renewables prove inadequate to provide all electricity."

[5] Plus a trivial contribution from concentrated solar power (CSP) of less than half a percent and an even more trivial 30 TWh (terawatt-hours, or trillion watt-hours) of energy storage in batteries.

[6] Instead of successful refutation, Jacobson sued for defamation the lead author of one critical article, Christopher Clack (the only author without an institutional affiliation to back him). Thus, Jacobson pretended that Clack et al.'s scientific rebuttal was just an ad hominem. While he withdrew the suit just as the judge was about to rule against him, the judge charged Jacobson with illegally pursuing a SLAPP (strategic lawsuit against public participation) and ordered him to pay $75,000 to cover Clack's legal expenses. Jacobson's appeal was rejected, and he was ordered to pay even more to cover Clack's additional

largely unpublicized, while the Jacobson proposal, or similar ones, have been popularized by a number of politicians and public intellectuals—including Al Gore, Bernie Sanders, Alexandria Ocasio-Cortez, Van Jones, Bill Nye (the Science Guy), Bill McKibben, Naomi Klein, and, among Green Marxists, Ian Angus. These all-RE popularizers are either ignorant of the critiques or perhaps believe they're not worth the effort of rebuttal—which would be a manifestation of groupthink (Chapter 11). Evaluating everything that Jacobson and his colleagues have written on the Roadmap since 2009 is a daunting task, and the present authors, while having reviewed portions, owe a great debt to the Roadmap's critics, some of whom we name and cite below.

That said, we quote two of the popularizers, Klein and Angus, because they have written relatively comprehensive, non-technical, and widely read books on the causes and consequences of AGW and its presumed solutions. Klein says (2014, p. 101),

Sorting out what mechanisms have the best chance of pulling off a dramatic, and enormously high-stakes energy transition has become particularly pressing of late. That's because it is now clear that—at least from a technical perspective—it is entirely possible to rapidly switch our energy systems to 100 percent renewables. In 2009, Mark Z. Jacobson, a professor of civil and environmental engineering at Stanford University, and Mark A. Delucchi, a research scientist at the Institute of Transportation Studies at the University of California, Davis, authored a groundbreaking, detailed roadmap for "how 100 percent of the world's energy, for *all* purposes, could be supplied by wind, water and solar resources, by as early as 2030." The plan includes not only power generation but also transportation as well as heating and cooling. Later published in the journal *Energy Policy*, the roadmap is one of several credible studies that have come out in recent years that show how wealthy countries and regions can shift all, or almost all, of their energy infrastructure to renewables within a twenty-to-forty-year time frame [italics in the original].

Angus notes (2016, p. 172),

Credible studies from a wide range of environmental groups argue that a full transition to renewable, non-carbon fuels is physically

---

appeal-related legal fees. He has also been ordered to pay $428,723 to cover the legal expenses of the publisher of Clack's article, the National Academy of Sciences (NAS). Jacobson appealed that order and lost in a lower appellate court, but he plans to take it to a higher court, and the current status is in limbo at the time of this writing.

possible. Perhaps the most comprehensive are those done by Mark Delucchi and Mark Jacobsen [*sic*], who presented "A Plan to Power 100 Percent of the Planet with Renewables" [*sic*], in *Scientific American* in 2009, and have followed up with highly detailed technical papers in peer-reviewed journals.[7]

As we will show, this trust in an all-RE electrical system is inseparable from the demonization not only of nuclear energy but also of other technologies that are regarded as hubristic (Chapter 9). Additionally, the Green left's view of nuclear energy as intrinsically capitalist is another basis for their rejection. While most nuclear advocates brook no criticism of capitalism, nuclear energy is no more intrinsically capitalist than agriculture is intrinsically feudal.

*   *   *   *   *

The current Trumpist moment in the US has correlates around the world and fosters profound epistemic dysfunction (Chapter 11). The resulting mix is a chaotic blend of social democratic impulses, best represented by a variety of Green New Deals (GNDs), side by side with a host of impulses that range from libertarian and secular to ethnonationalist, theocratic, and even fascist. The impasse among these antagonistic groups lies at the heart of Slotkin's "Great Disorder." For us, while we find nothing of value and only danger in fascist discourse, the libertarian critique of Green energy and the Green critique of the libertarian commitment to "free market capitalism" hold important insights combined with blindness and dogma.

Authors published by small presses must exert great effort to promote their work. We have been asking writers from these two warring camps—writers whose insights we most value—to read our book and to promote it. One energy analyst, whom we cite repeatedly—pronuclear, a searing critic of renewables, especially their resource intensity, and a libertarian pro-capitalist—upon reading our Preface and an earlier version of this first chapter, lauded the effort we had exerted but refused to endorse the work, even the portion with which he agreed, on the grounds that our promised critique of capitalism was a "deal killer"

---

[7] Our book implicitly exposes the soft underbelly of "peer review," on which Angus in part rests his approval. Indeed, as we mention in Chapter 8, according to a recent report in *The Washington Post* (June 14, 2024, p. A17), over 10,000 scientific papers in 2023 alone passed peer review and were published, only to be retracted later by the publishing journal when fraud was discovered.

(Chapter 10). Another writer, an international relations professor, refused his endorsement, pronouncing our critique of capitalism irrelevant.

As it turns out, in Chapter 10, which explains how intrinsic features of the capitalist system constitute barriers to the global cooperation needed to address AGW, we refer to an argument in this professor's own book: "Solving global warming does not require us to 'tear down capitalism.' The world just needs to be a bit more like Sweden." After all, he and his coauthor reason, if all that is needed is for the world to be a bit more like Sweden—in both its social democracy and its reliance on hydroelectricity and nuclear energy, a nuclear version of a Green New Deal—why would anyone opt for the "great disorder" of a disruptive alternative to capitalism? You would have to be crazy.

In answer, our Marxian narrative argues that the global generalization of Swedish social democracy, with its profound reliance on nuclear energy (plus the tapping of enormous numbers of raging rivers peculiar to only a few lucky countries), would require the kind of global cooperation and regulation that has repeatedly failed to materialize, a failure due to barriers intrinsic to the operation of capitalism. Thus, the world is forced to contend with two dyseconomies, very different in operation, one a technical barrier that, as we will see, disables the all-RE Green story and the other, a geopolitical and economic barrier that disables the seemingly pragmatic nuclear alternative that would be "a bit more like Sweden." In Chapter 10 we argue that a mythical view of "happy competition"—either a libertarian free-market version or one based on Keynesian regulation of the market[8]—has not and cannot work globally and seems to derail the spread of even the best technologies.

Richard Slotkin points to the failures of Keynesianism and neoliberalism (see footnote above). Yet in the introduction to his book, he rules out any anti-capitalist internationalism because twentieth-century attempts to build an alternative to capitalism were short-lived. So he apparently feels he is forced by default to subscribe to both nationalism ("good nationalism") and Keynesianism, whose failure led to neoliberalism and a more recent resurgence of rightwing populism or fascism. Even the best non-Marxist analysts find themselves trapped in a vicious circle.

---

[8] What US Marxist economist Andrew Kliman (2012, p. 198) calls "trickle up" economics.

Contrary to the approach of writers like Slotkin, who are reluctant to propose solutions that are extremely difficult to actualize but that may be the only pathway out of the trap, US legal expert (and Dean of the UC Berkeley School of Law) Erwin Chemerinsky, in his new book *No Democracy Lasts Forever* (2024), shows that adherence to the US Constitution necessarily leads to a lack of democracy, an outcome that the founders (compromising between pro- and anti-slavery activism) desired. He says (p. 183), "Our government is broken and our democracy is at grave risk. But I don't see any easy solutions . . ." He grants the difficulty of changing the Constitution but, nonetheless, does not shy away from urging the necessity of doing so. We follow a similar approach in this book, devoid of illusions concerning the ease of reversing AGW.

However, pronuclear libertarians, liberals, and left liberals, also refusing to propose difficult pathways, feel compelled to regard contingencies, like the nuclear successes of Sweden and France, as generalizable—requiring nothing more than political will and the right policies. They treat the toggling between regulation and deregulation as independent policies that are simply contingent on developing conditions. Whereas in our view these antithetical policies are far from independent of each other. Instead, regulation and deregulation generate each other by virtue of their inevitable failures—failures imposed by capitalism's imperative fragmentation and competition that burst the bonds of governmental attempts to rein them in.

Many believe that the Keynesian New Deal, rather than World War II, rescued capitalism from the depression and fascism (Chapter 10). In the aftermath of 1970s stagflation, marking the end of the Keynesian golden age, opponents of the New Deal (neoliberals) believed that, in the words of Andrew Kliman, "To save the system, the gains of the 1930s had to be rolled back" (p. 201).

Kliman continues:

> Of course, Keynesians and their supporters never fail to place the blame for this [rolling back the gains] on Reagan, Thatcher, and neoliberalism, *but they themselves bear most of the responsibility*. The policies they advocated and implemented failed in the end and, because they failed, new people and new ideas naturally came along to replace them and fix the mess [italics in the original].

Faith in this Keynesianism "helped to demobilize working people. . . . As a result, the new people and ideas that came along were reactionary ones."

Just "a bit more like Sweden" proclaims that what's good for workers—affordable, reliable, clean energy—is good for the system, and that "redistribution of income [and energy] toward the bottom allows more goods and services to be sold, and this *boosts* profitability" [italics in the original]. But, explains Kliman, serious downward redistribution in a capitalist system "could lead to a deep recession, even a depression" (p. 202).

Kliman is writing in the aftermath of the 2008 Great Recession. Recent Green New Deals, with their emphasis on renewables, have been intended to add to downward redistribution. However, the GND repudiation of a high-EROI energy system has so far only amplified the failure, which has amplified the reaction. And along comes Trump, emerging on the scene proclaiming, "Only I can fix it."

A commonsense reading of the Gramsci quote—written in 1930 and resurrected by Slotkin nearly a century later—suggests that we still live in a period with many problems and no workable solutions. As a result, a myriad of fantastic solutions fill the vacuum. We as leftists do not exempt ourselves from this situation. As Kliman points out, "It is one thing to recognize the instability of capitalism, but another to show that an alternative to it is possible" (p. 203). While this book argues that nuclear energy can solve the technical problem, we also insist that "the instability of capitalism," which we explore in terms of its self-negating contradictions is a serious problem for those who, like us, want to see nuclear energy globally scaled. Quoting Michael Shapinker of the *Financial Times*, Kliman agrees that while the 2008 crisis may have weakened confidence in the Reagan-Thatcher TINAism (There Is No Alternative), the left has "'not got a clue' about what might replace it" (p. 203)—other than terms that have yet to be given detailed definition. And we would add that the left in general has not got a clue about which energy source is able to halt and possibly reverse AGW.

It is our hope that this book, which attempts to bring together "ignorant armies that clash by night" (Matthew Arnold, *Dover Beach*), might help us get a clue at least on the latter.

II

Wind
and Solar:
Why We
Shouldn't

2

Energy Density and Power Density

The Nature of Energy and Its Central Role in Society

Energy lies at the foundation of social existence and progress, indeed of life itself. Humankind has progressed from small hunting/gathering bands to complex societies of millions to billions of people. Hunters/gatherers had little leisure time to enable a significant variety of activities and insufficient cumulative experience to produce complex energy technology. That initial phase occupied 95 percent of humanity's roughly 300,000 years on Earth, while large and complex societies have occupied only the most recent 0.07 percent or so, about 200 years.

Along the way, our main sources of energy have progressed from **sunlight** (that powers plant growth) to **human muscle** (using plants and other animals as fuel), to nonhuman animal muscle (for agriculture, lumbering, and building), to **wind** (used to drive grinding mills, to pump water uphill, and to move ships), to the natural downhill flow of **water**, to **fire** (first using harvested wood and other biomass, and later mined fossil fuels—**coal, oil, NG**, the solid, liquid, and gaseous forms of decayed organic matter), to nuclear (also using mined fuel, but fuel that is self-contained, without the need to combine with oxygen from the air).[1] These are all sources of energy that have been stored by nature

---

[1] The independence of nuclear fuel from oxygen contributes to making nuclear safer than any other energy source, even as nuclear weapons appear to be among the most destructive, though no more than conventional weapons (Chapters 7 and 8).

(unless stated otherwise we mean nonhuman nature) and that humans can convert to usable form, such as electricity or heating. "Convert" and "usable" are key words here.

Over a century ago, electricity was captured and put to use. Electricity in wired circuits may be considered to be a source of energy whenever we plug in an appliance or flip a switch to on, but for purposes of this book we distinguish between *sources* of energy when provided by nature and *forms* of energy when converted from nature to a useful application through human effort. Electricity is useful in no small part because it can be transmitted through wires over thousands of miles between source and end-use application.

Almost all sources and forms of energy are subject to risk-benefit considerations, though when it comes to nuclear energy the focus becomes a one-sided emphasis on risk alone, often imaginary and/or deliberately fabricated, while downplaying or completely neglecting benefit. Equally one-sided is the failure to compare dangers among the various energy sources. When comparison is made, it's typically made inaccurately. Nuclear, as we will show, has proven to be the safest source *per unit of energy produced* (Chapter 7).

Fuels for nuclear fission include uranium and thorium,[2] both abundant in the ground and in the case of uranium even more so in the oceans. Controlled continual *fusion*, as opposed to explosive *fusion* in thermonuclear bombs, has not yet been achieved on Earth, though it is the source of the sun's ongoing energy production.[3] Solar energy arrives mainly as electromagnetic radiation, called photons (sunlight). Following convention we use the term

---

[2] To fission is to split a heavy nucleus into two roughly equal parts, along with the release of some energy of motion of the parts and/or electromagnetic energy (like light). We call thorium a fission fuel because it can be mined and carries stored energy, but it does not directly fission easily. Rather Th-232 is the fuel from which U-233, a fissile (fissionable) isotope, is bred in a nuclear reactor—by the capture of a neutron and the emission of two electrons in succession (see "Uranium" in the Glossary for the meaning of symbols like Th-232 and U-233). Similarly U-238 (99.3 percent of natural uranium) can also be considered a fuel, but it is not readily fissile either. Rather it's the fuel from which fissile Pu-239 (plutonium) is bred—by the same process of neutron capture and electron emissions. Fuels like Th-232 and U-238 are called fertile, as opposed to fissile. As a general rule, only the isotopes with an odd number in their symbol are easily fissile. The numerical part of the symbol indicates the total number of protons plus neutrons. It's approximately equal to the atomic mass of the isotope since neutrons and protons have about the same mass and account for almost all the mass of an atom.

[3] Fusion is the opposite of fission. It's the joining together of two or more small nuclei into a bigger one, also with the release of energy of motion and/or photons. Fusion can only occur with small nuclei, while fission can only occur with large nuclei, but energy is released in both cases.

"nuclear" to stand for both the form of energy and the source. There's a sharp distinction between its use as a source of electricity (controlled) and its use as a weapon (uncontrolled, once initiated).

In conversion from one form of energy to another, and in every transfer of energy from one parcel of matter to another, energy is never lost (see $E = mc^2$ in the Glossary). However, some or all of the energy may not be in a usable form, or some or all may not be captured and put to use. Heat is a form of energy, and during every energy conversion or transfer, a portion is unavoidably converted to heat. Heat may or may not be captured, but when it is, heat can be usable to keep a building warm, cook food, or generate steam to power an electric turbine/generator. When not captured, it is energy lost to use. Since total energy is always conserved in any conversion or transfer, the term energy "loss" means only lost to immediate use.

About 18 percent of energy consumed in the US is electrical. In 2023 this amounted to 13.2 quads [4] of usable electrical energy out of 74.7 quads of total end-use energy—Figure 2-1, from the US Energy Information Administration (EIA). Changes are slow from year to year.

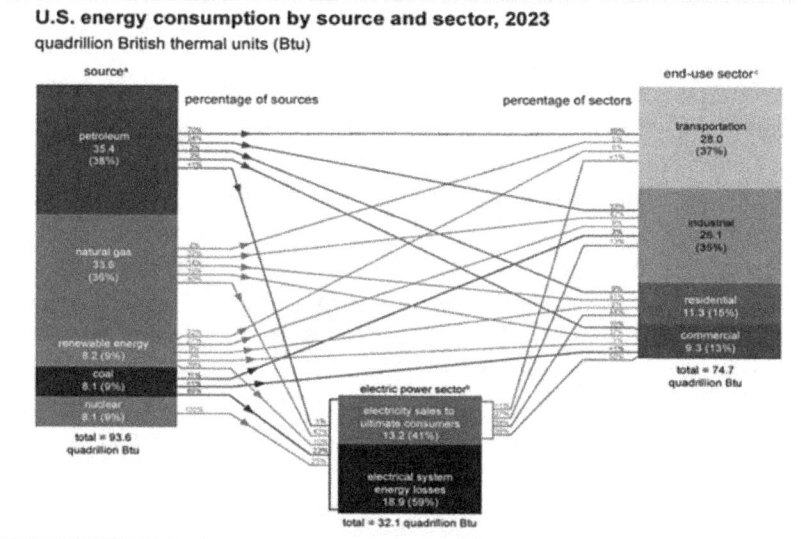

U.S. energy consumption by source and sector, 2023
quadrillion British thermal units (Btu)

FIGURE 2-1

---

[4] One **quad**rillion British Thermal Units (BTU) or roughly 3×10^{11} kWh (kilowatt-hours), or about 300 TWh (terawatt-hours = trillion watt-hours), or 1.055 EJ (exajoules = 10^{18} J)—all commonly used units, even if mind-boggling.

Electrical power requires sources for production and a grid for distribution. The US grid contains 240,000 miles of long-distance (high-voltage) transmission wires (https://tinyurl.com/38jah8h3), transformers, control stations, and over 6 million miles of end-use (low-voltage) distribution lines (https://tinyurl.com/3yru8bbw), along with a few hundred thousand workers who manage the input and output of the grid on a minute-by-minute basis.[5] The US Department of Energy (DOE) estimates that increased data processing alone will necessitate a grid expansion by more than 50 percent, and soon. This requires a rapid increase in miles added per year.

In addition to the needs of data processing, increased penetration of wind/solar in the US will require yet another significant expansion of the grid. DOE estimates of grid expansion to accommodate wind/solar run as high as another doubling of the long-distance lines (http://tinyurl.com/38jah8h3) and a concomitant increase in the short-distance end-distribution wiring (https://tinyurl.com/3yru8bbw), as well as additional transformers, transmission towers, and other equipment. New wind/solar farms also necessitate expansion of the few thousand miles of high-voltage direct current (HVDC) lines that join and synchronize separate networks of AC lines. HVDC lines suffer lower heat losses than AC lines over distances longer than around 375 miles, the break-even point, and are advantageous for carrying electricity underwater from remote offshore wind farms.

Today, most governments put their efforts behind "clean energy," often code words for wind and solar. Secretary General of the UN, António Guterres, calls for the rapid buildup of wind and solar to replace fossil fuels (https://tinyurl.com/9d9h3mdy). He omits mention of nuclear, as though reasons to reject it were self-evident. In this he is far from atypical.

While many countries expand wind/solar, some also accept and some reject nuclear. Those who reject nuclear earn applause from most environmental organizations. In the US, proposed Green New Deals (GNDs, plural) generally reject nuclear and aim to halt AGW mainly through wind/solar—misleadingly referred to as "renewables" (see Sidebar).

---

[5] For readable books on the grid see Angwin 2020, Bakke 2016, or Thompson 2016.

┌─ **Sidebar: What's Meant by "Renewable"?** ─┐

The term "renewable" is ambiguous, meaning either reinfusing with new stock or refreshing by cleansing. Either matter or energy can be reinfused, but only matter can be refreshed. Used energy is partially degraded to heat and cannot be refreshed. Depending on level of consideration, recharging a battery both reinfuses chemical energy and refreshes its charged state.

Wind and sunlight are forms of energy unassociated with fuel (see next Sidebar). They can only be reinfused. They are more accurately termed "indefinitely replaceable." Fossil fuels, while a form of matter, once burned cannot be refreshed, and are renewable only in the sense of reinfusing. Nuclear fuel can be renewed in both senses, either by reinfusing newly mined uranium or thorium into a reactor, or by removing it from the reactor and cleansing it of built-up fission products to refresh it. Yet the term "renewable" functions to exclude nuclear.

Wind and solar are thought to be renewable (reinfusable) because their ultimate source, the sun, is for all practical purposes limitless. But so is the store of uranium and thorium on Earth sufficient to last as long as the sun (Till and Chang 2011).

Nuclear opponents justify their exclusion by arguing that obtaining nuclear fuels, like fossil fuels, requires human effort, whereas nature continues to "renew" (reinfuse) the supply of wind, water, and sunlight free of charge. However, as with fuel-based energy, wind and solar are not usable without conversion devices, each specifically designed for the particular source—wind turbines, solar photovoltaic (PV) panels, or hydroelectric dams. Similarly, nuclear reactors, fossil fuel combustion chambers, steam turbines, electric generators, and power plants. Each requires extraction of materials, refining, transportation, manufacturing, construction, installation, operation, maintenance, and eventual decommissioning with replacement of the apparatus.

They all produce waste that is permanently toxic chemically, requiring adequate containment. Residual nuclear fuel, while no less permanently toxic chemically, is only temporarily toxic radioactively. The very nature of radioactivity is that it decays away and eventually becomes undetectable, though it too requires appropriate containment for extended periods (Chapter 7).

Wind and solar conversion devices are mostly not renewed, since recycling is more expensive than new mining and manufacturing and would diminish profits. Only the energy is renewed, in the reinfusion sense, but only sporadically, with wind gusts or sunrise or parting of clouds, and beyond human control or predictability. Fossil and nuclear fuels do not suffer from such uncontrollable intermittency.

With these provisos, we bow to convention and use the term "renewable" to refer to wind and solar, as well as hydroelectric dams and the less readily available geothermal, wave, and tidal energy.

What Do We Need in an Energy Source?

There are five especially desirable features of an electrical energy source—**relatively clean**, **indefinitely replaceable**, **relatively safe**, **reliable**, with conversion apparatus **capable of generating the energy for its own replacement**. "Clean" energy means clean of GHG emissions and of respiratory toxicity, two concepts that some climate deniers and fossil champions elide to argue irrelevantly (and falsely) that CO_2 does not increase heat trapping nor can it be toxic since we exhale CO_2 with every breath. They correctly point out that CO_2 is necessary for plant life, but many things are both necessary and detrimental, depending on dose (Chapter 8). Like everything else, CO_2 is detrimental in high enough concentrations, both with regard to AGW and respiration. Climate denialists often focus only on respiratory effects and ridicule the idea that CO_2 can be toxic. But since it is heavier than air, at high enough concentrations CO_2 can displace oxygen and asphyxiate.[6] However, its impact on AGW is our focus in this book.

"Indefinitely replaceable" requires that the energy source be inexhaustible for all practical purposes and that a usable form converted from the source be widely accessible. "Relatively safe" means relatively free of hazards under normal conditions of use when compared with its alternatives. "Reliable" means avoidance of outages—supply that meets varying demand. And apparatus "capable of generating the energy for its own replacement" means that the conversion devices must be able to generate enough energy to manufacture their own replacements before they become obsolete, even as they produce a generous portion of energy to run the society. As we show (Chapter 5), the capability of self-replacement is essentially determined by the ratio of the energy return over the energy invested (EROI) of the energy source and its conversion devices.

So which energy sources, if any, satisfy these desirable criteria to a sufficient extent? Our book aims to answer this question for each of the three main sources: wind/solar (renewables), fossil fuels, and nuclear. We place far less emphasis on secondary sources like hydroelectric, geothermal, wave, and tidal energy, because each

---

[6] Tragically, in 1986 Lake Nyos in Cameroon, Africa, burped a giant bubble of CO_2 (https://tinyurl.com/2jzktvbp). Occurring four months after the Chernobyl accident, it hardly made the headlines, though it caused roughly 20–35 times as many human deaths and twice that number of livestock deaths.

suffers from severe limitations, including constrained geographic location and inadequate quantities.

The Inputs to Energy Systems

To convert energy from nature to useful forms, humans must manufacture and emplace conversion devices into power plants or farms. Intrinsic inputs to these facilities include **materials**, **land**, **time**, **labor,** and **energy**, aside from extrinsic challenges to their optimization. Monetary costs are the most susceptible to extrinsic manipulation and are therefore the least relevant index for assessing energy sources. The minimum requirements for intrinsic inputs are mainly determined by physics and engineering, while monetary requirements are largely determined by geopolitical exigencies. The less of the intrinsic inputs required, the more efficient and less costly the conversion infrastructure. Putting it the other way around, US energy analyst Robert Bryce says, "The lower the power [or energy] density, the greater the resource intensity" (https://tinyurl.com/5xen78hf). Densities are ratios of a certain type, which we define below.

Wind and solar farms, hydroelectric dams, fossil fuel power plants, and nuclear power plants all entail serial processes, including raw material extraction, manufacturing, emplacement, operation, maintenance, decommissioning, and replacement. Intrinsic inputs are involved at every step.

Materials require mining and refining minerals, transportation between the different intermediate stages, fashioning the conversion apparatus (wind turbines, PV panels, reactors, furnaces, electric generators, and plant buildings), and installing them. Additional materials are needed to maintain the apparatus. To avoid less immediately relevant and more variable contributions, we neglect secondary materials that are part of the equipment used in each of these stages, such as highway maintenance, gasoline, and the nationwide electric transmission grid.

For reasons that will become clear below, we generalize the definition of **energy density** to be the ratio of energy output over the lifetime of a conversion device (plus farm or power plant) to materials input, including fuel if any. Energy density varies from one type of source to another and is conventionally measured as megajoules per kilograms (MJ/kg).

Power density, as we define it (more below), is the ratio of power output to occupied land. With respect to **land** requirements, we neglect the ever-expanding mines, even though their destructive effects can be severe. Aside from strip mining for coal and tar sands mining for oil, both to be eliminated, rare earth mining (see Glossary, also discussed in Chapter 4) for wind turbines and PV panels in Inner Mongolia, for example, is highly destructive (http://tinyurl.com/yjarken4). While the impact on land use by mining is a greater disadvantage for renewables, we confine our accounting to the land occupied by the wind or solar farms and fossil or nuclear plants. We also overlook the land overrun by the vast artificial lakes behind new hydroelectric dams, particularly since in some places there is now a net removal of dams. The contribution of hydro power is, and will remain, limited, accounting for only a few percent of the total electricity supply in the US, though in some places it accounts for the majority—for instance 90 percent in Norway and almost 100 percent in Paraguay.

In the race to halt AGW, the relevant **time** includes all those processes from mining to installation for each type of conversion device and farm or power plant until they begin to produce electricity. The intrinsic portion of time for such processes, as we will see, can be greatly overshadowed by delays from protests, litigation, licensing, and other extrinsic influences. The relevant **labor**, measured in person-hours, includes that involved in the entire set of relevant processes. **Energy** is required at each stage. We neglect the muscular energy by the workers and the food and oxygen that supply it, since the energy required by equipment is immeasurably greater. The most significant contribution to renewables, to which we return more than once, comes from the fossil/nuclear matrix necessary for the construction and operation of wind/solar farms, but which an all-RE system is intended to eliminate (Chapter 5).

Efficiency

Calls for increased efficiency often focus on end-use, or consumer, appliances. For example, a refrigerator that keeps food at 40°F (4–5°C) is more efficient if it consumes less electricity over, say, a year. Or a washing machine that gets the same amount of laundry clean over a year is more efficient if it does so using less electricity than one of a different brand. The comparison only makes sense between appliances of the same type. In other words, it makes no sense to say some brand of refrigerator is more efficient than some brand of washing machine. The output must be the same qualitative result for efficiency comparisons among appliances, and the comparisons refer to the relative amounts of *energy input*.

For energy *production*, the inputs include all the items listed above, and not just the energy input. Efficiency on the production side, therefore, compares energy output to inputs of various kinds, and is greater with less input per energy output. Efficiency of energy production with respect to *energy* input can be expressed as a pure number—the ratio of energy output to energy input. For efficiency with respect to inputs other than energy, the ratio is not a pure number, but must include the units, such as kWh per ton of material, or per area of land, or per time, or per person-hour of labor, or per dollar. For an energy production method, energy output must exceed energy input, so efficiency in terms of energy input is always greater than 1 and is measured as EROI.

Consumers focus on end-use efficiency, and manufacturers are responding by reducing the energy consumed by their products. But if there are major inefficiencies in energy production, improvements in end-use efficiency will have a limited impact on overall energy usage by the society.[7] What's more, increased end-use efficiency often collides with the Jevons paradox.[8]

There are also severe limits on end-use efficiency improvements. The maximum fuel efficiency of internal combustion engine (ICE) cars has varied from less than 10 mpg to roughly 50 mpg. But further improvement is limited by the energy embodied in the gasoline (when it chemically combines with oxygen). Similarly for appliances like washing machines or refrigerators. In contrast, the efficiency of energy production can be magnified to a far greater degree by switching from wind/solar to fossils, or from fossils to nuclear. Production efficiency, as we will see, can vary by factors of a hundred or more. So, even if end-use devices were maximally efficient, the efficiency of energy production would carry the main burden. Conversely, maximal efficiency of energy

---

[7] A family unable to afford an adequate diet would still be malnourished even if they were to save every crumb from the table.

[8] Nineteenth-century British economist William Stanley Jevons pointed out that increased end-use efficiency generally encourages greater usage, which tends to negate the decrease in overall input. For example, a more gas-efficient car tends to encourage increased driving and greater gas consumption. Thus, attempts to save resources through efficiency at the consumption end are often futile, if not counterproductive. Efficiency improvement at the production end is the most, if not the only, effective way to reduce environmental impact. We return to Jevons in Chapter 9.

production would render end-use efficiency of little concern. Therefore, this chapter focuses on production.[9]

An additional attempt at conserving resources lies in reducing use of energy altogether. This is the recommendation of US ecosocialist Stan Cox (2020) among others (Chapter 9). A recent op-ed in a major newspaper recommended both end-use efficiency and reduced usage: "Eat less meat, buy more efficient appliances, limit food waste, drive electric or hybrid cars, eliminate bottled water, lower thermostats (in winter), try composting and use long-lasting lightbulbs."[10] Fortunately, as we will see, conservation of resources can be achieved alongside abundant energy production, as long as the EROI is high enough—something only nuclear provides (as reflected in Figure 1-1).

Energy Density

The inefficiencies of an all-RE electrical system are embodied in its low energy and power densities. As we said above, energy density is defined in terms of materials input. We define power density (discussed in detail in the final section of this chapter) in terms of land, but for RE it also implicates greater use of materials. In other words, greater output from wind/solar farms requires both more land and more materials. Conversely, high densities mean low requirements of materials and land, respectively. In each case, the focus is on *usable* energy or power.

The material component of energy density is conventionally confined to fuel. However, wind/solar are not fuel-based (see the Sidebar below). Instead, wind turbines and solar PV panels, respectively, receive kinetic and electromagnetic energy directly, without the mediation of fuel. So, to compare the energy density of renewables with fossils and nuclear we need to extend the materials to include conversion devices, storage apparatus, and power plants. Fuel, as it turns out, while being the major material component over a fossil plant's lifetime, is the least of the

---

[9] It's vital to note, however, that efficiency does not stand by itself as a desirable criterion. Reliability requires that it be combined with a measure of redundancy in order to avoid the type of worldwide software outage experienced on July 19, 2024, and the occasional complete blackouts of grid output over widespread geographical areas. But redundancy lessens efficiency, so the trade-off has to be optimized. Therefore, efficiency and redundancy, individually, also have their optimal, or Goldilocks, zones—a concept we discuss more fully in Chapter 8.

[10] Michele L. Norris, *The Washington Post*, August 1, 2021, p. A25.

material requirements for a nuclear plant, both by several orders of magnitude (see following section). So in comparing the energy densities of wind/solar and nuclear (fossils to be eliminated), we can focus solely on the conversion devices and storage apparatus for wind/solar, and power plants for nuclear. Renewables require far more material per net energy output than nuclear, so their energy densities are far lower.

Sidebar: Are Wind and Solar (and Hydro) Fuel-based?

If we consider the Sun-Earth system as a whole, solar and wind energy are indeed fuel-based, but the fuel is inside the sun, namely hydrogen (bare individual protons). Solar energy (sunlight) is carried to Earth as photons (packets of electromagnetic energy—note difference between protons and photons). Under conditions of extreme pressure and temperature in the sun's core, hydrogen nuclei collide at high speed and fuse to form a helium nucleus (two protons, two neutrons). The process releases energy in the form of photons, each with its own frequency. The collection of frequencies makes up the electromagnetic spectrum, which includes (from high frequency to low) x-rays, light, microwaves, radio waves, and so forth. Solar energy then originates in nuclear fusion in the sun's core, and hydrogen is the fuel for sunlight. Fusion on Earth is discussed at the end of Chapter 7.

Wind energy is simply the kinetic energy of moving air on Earth. Masses of air are forced to move when sunlight heats a pocket of air that then expands and, because of its diminishing density, is pushed upward by surrounding, denser air masses that rush in to fill the vacated space, much like a piece of wood under water. Wind is largely guided by the local terrain. The inrushing air masses near the ground are accessible to turbines. Thus, the ultimate fuel for wind energy is also hydrogen in the sun's core, though earth-bound sources of heat, such as urban heat islands or wildfires, contribute secondarily.

Hydroelectric power comes from either a dammed or undammed river whose flow originates in precipitation, either rain or snow, that falls at higher elevations. The kinetic energy of falling water, like wind, is used to turn turbines and generate electricity. If dammed, the timing of the drop in altitude of the water, unlike wind, is under human control, released when needed. Rain and snow originate in evaporation from land or ocean, enhanced by solar heating. Thus, hydro is also a secondary product of solar energy and earth's gravity. Since dammed water can be held back indefinitely (unless it overflows), its energy can be turned from potential to kinetic when needed, independent of nature's short-term timing. In other words, it's dispatchable.

So solar, wind, and even hydro energy do ultimately originate in fuel, but only in the sun's core. While earth-bound water could be con-

sidered fuel for hydro, its readying for use requires no human investment of energy, unlike extraction of earth-bound fuels—coal, oil, NG,
uranium/thorium, and biomass. Of course, construction of the conversion devices in all cases requires human energy, but wind and solar
(and hydro) energy do not employ earth-bound fuel.

Energy density for wind/solar is the ratio of actual energy
output to the materials in the turbines and PV panels as well as
in the necessary storage apparatus, such as batteries. At low penetration, when fossils and nuclear are available to fill in, batteries
may be unnecessary, but storage would become crucial in a hypothetical all-RE system. For nuclear, the materials are essentially
the power plant, since the mass of fuel is trivial in comparison, as
we explain below.

The energy output for renewables is reduced by their intermittency and the efficiency of their conversion devices. Intermittency
is measured by capacity factor (CF), the ratio of actual average
power over some time interval (conventionally taken to be one
year) to the maximum possible power if there were no intermittency. CF over shorter time intervals varies with season and time of
day. Instantaneous CF falls below 1 whenever there is nonoptimal
wind or when cloudiness or night blocks sunlight. Each turbine or
panel is labeled with the maximum power it could produce if it
could operate optimally, as if the CF remained a steady 1. This
label is called the nameplate.

The efficiency of wind or solar conversion devices depends on
their physics and is at best about 20–30 percent, though the efficiency varies with the instantaneous received wind or solar energy
and declines with age. Solar panel efficiency also varies slightly with
ambient temperature. Conversion efficiency is internal to the
device and is independent of CF, which depends on the wind and
sunlight received. AGW is changing the CFs of wind and solar
through the changing weather. Again, both internal and external
features reduce energy output below maximum energy received.

In more detail, the energy output of a wind turbine depends on
both the size of the turbine blades (internal, conversion efficiency)
and the air density and wind speed (external, CF). The instantaneous conversion efficiency is the fraction of the incident kinetic
energy of the portion of the moving air mass impinging on the circle described by the rotating blades that is converted to the turbine's electrical power output. This efficiency declines as the

moving parts wear out. The instantaneous CF is some fraction of the optimum power output at that moment.

Figure 2-2 shows the relationship between wind speed and power output for a particular turbine. There are four relevant ranges of wind speed. Below the "cut-in speed," there is no power produced by the turbine (in this example, 3.5 meters per second, ~8 miles per hour). Between the cut-in speed and the "rated output speed," when the turbine first reaches its maximum (14 m/s, ~31 mph), the power output rises from zero to 100 percent of the maximum. Between the "rated output speed" and the "cut-out speed" (25 m/s, ~56 mph) the output is maximal (nameplate). Beyond the "cut-out speed" the output drops to zero again, as the turbine blades are automatically feathered to protect the device from excessive turning force (torque) that might damage the device, either mechanically or through overheating. So wind downtimes include not only too little wind speed, but also too much. [11]

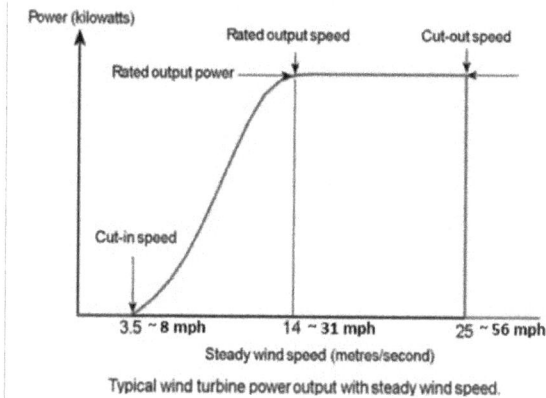

Typical wind turbine power output with steady wind speed.

Source: https://tinyurl.com/333fyhp4

FIGURE 2-2

Just as the instantaneous conversion efficiency of a wind turbine is the ratio of electrical energy produced by the device to the kinetic energy in the wind impinging on the circle described by the rotating blades, so is the instantaneous conversion efficiency of a solar

---

[11] Too little and too much, with an optimal (Goldilocks) range between them, is a recurrent theme in this book and will find its greatest relevance in our discussion of radiation in Chapter 8.

panel the ratio of electrical energy produced by the device to the energy in the sunlight impinging on the panel. Both vary around an average of about 20 percent.

Summarizing so far, the energy density of wind and solar (in an all-RE context) is the total energy produced by their conversion devices over the course of their lifetimes (15–30 years) divided by the materials that go into the turbines, panels, and required storage apparatus. The energy density of nuclear is the total energy produced by a power plant over the course of its lifetime (80–100 years) divided by the materials that go into the power plant (neglecting the trivial mass of fuel).

In all cases, the actual total energy output is the initial maximum power output (nameplate) of the device or plant multiplied by the number of years in its lifespan, reduced by the average CF over its lifespan (and slightly reduced by the decline in conversion efficiency with age and other factors). The CF of a nuclear plant can be well over 90 percent and to the extent that it falls below 100 percent it's due to periods of elective maintenance, under human control, or to intermittent usage to back up the intermittency of wind/solar. In the latter case, of course, its CF is beyond human control and can be far less than 90 percent. The average CFs of wind and solar are discussed further in Chapter 3.

The Energy Density Embodied in Fuel

While fuel is the least significant contributor to nuclear's overall energy density, the reverse is true for coal and NG. In particular, the mass of the nuclear fuel consumed over the 80- to 100-year lifespan of a nuclear plant is very small compared to the mass of materials to construct the plant plus reactor. In contrast, the mass of either coal or NG consumed over their plants' 50 to 60-year lifespans dwarfs the materials that go into building the plants.

Specifically, uranium's energy yield in a light water reactor (LWR, see Glossary and Chapter 5) is approximately 0.04 TWh per ton of uranium ore (before refining and enriching, see Glossary) or, inversely, 25 tons of ore per TWh. This is to be compared with 1,012 tons (equivalent to 920 tonnes,[12] see Figure 2-4) of the materials that go into building the nuclear power plant

---

[12] A tonne is a metric ton (1000 kg or 2,200 pounds), which is 10 percent larger than a ton (2,000 pounds), sometimes called a short ton.

per TWh produced during the plant's lifetime. A fast breeder reactor (see Glossary and Chapter 7), with fuel recycling, would permit a reduction of ore by a factor of around 100, dropping the ore to 0.25 tons (500 pounds) per TWh. Thus, to compare the energy density of wind/solar versus nuclear we need only consider the materials for the commercial farms and the power plants, neglecting nuclear fuel.

The reverse is true for coal and NG, for which the main material contributor is the fuel. The mass of coal per TWh is roughly 550,000 tons and that of NG about 222,000 tons, some 10,000 times the 25 tons of uranium ore for an LWR, and 1,000,000 times the 0.25 tons for a fast breeder. The mass of materials in a coal or NG plant is somewhat less than in a nuclear plant—a combustion chamber is less massive than a reactor—so a coal or NG plant would require somewhat less than 1,012 tons per TWh, rounded off to about 1,000 tons or less per TWh. Thus, for fossil fuels, the predominant contributor to total mass are the fuels, by more than two orders of magnitude—222,000-550,000 tons/TWh of fuel versus around 1,000 tons/TWh for the plant, or less. So, for fossils it is the fuel that mainly determines the energy density and not the power plant.

A major advantage of nuclear lies in the energy density of its fuel. Fossil fuels store chemical energy, which involves the loosely-packed outer electrons, while nuclear energy, as the name implies, is embodied in the densely packed and relatively minute atomic nucleus at the center of the electron cloud (analogous to the sun at the center of the planetary orbits). The much greater amount of energy stored in a nucleus is related to its much smaller size and much greater mass. The close packing of multiple positively charged protons that tend to repel each other requires far more energy input to create and retain the configuration than does the loose packing of electrons.

The electrons are attracted by the protons, even though electrons also repel each other. But the greater separation of electrons entails a weaker mutual repulsion. The close packing of protons is enabled by an equal or greater number of uncharged neutrons that act like glue. Many nuclei are thus relatively stable. Unstable nuclei (with either too few or too many neutrons), seeking stability, spontaneously emit particles and energy from time to time, some very quickly after formation, others much more slowly—a process called "radioactivity."

If an initiating perturbation can cause relatively stable nuclei to break up, they are called "fissile" (meaning able to fission, see Glossary). Radioactivity involves the emission of small particles that hardly change the mass of the parent nucleus, while fission involves a breakup into two roughly half-size nuclei. Under deliberately engineered conditions, perturbations can be applied to fissile material at rates that are controlled (in a nuclear reactor) or uncontrolled (in a bomb). In either case, fissile nuclei are prompted to break apart, releasing much of the stored energy, either constructively or destructively.

Both the electron-associated chemical energy of fossil fuels and the nucleus-associated energy in heavy elements like uranium and thorium come to us already stored by processes in the interior of massive stars that later explode and make the elements available for further formation of stars and planets. The storing of this energy requires no human effort any more than the effort-free arrival of energy in wind or sunlight. What does require human effort is the creation of conversion devices, farms, and power plants, as well as extracting and processing of fuel. But such effort is greatly exceeded by the stored and releasable energy in the fuels, either nuclear or fossil.

As US international relations professor Joshua Goldstein and Swedish engineer Staffan Qvist (2019) put it, one pound of nuclear fuel "produces the same energy as more than 2 million pounds of coal." They continue (p. 22, italics in the original),

> The big difference between kärnkraft [Swedish for nuclear energy] and fossil fuels (or renewables) is the tremendous concentration of energy in kärnkraft fuel. The fuel to run a kärnkraft unit for a year fits onto a truck. The fuel to run a similar size coal plant for a year fills 25,000 railroad cars. The energy released by kärnkraft fuel weighing as much as a single penny equals that released by burning 5 tons of coal.

Their parenthetic reference to renewables requires the expanded meaning of energy density to include the conversion devices and power plants/farms.

The following chart (Figure 2-3), by US environmental scientist Jesse Ausubel (https://tinyurl.com/27v8jwv2), explains that the ratio of mass to energy releasable by burning fossil fuels ranges from about 50 kg/GJ for NG to 100 kg/GJ for brown coal. One

GJ (gigajoule or one billion joules, roughly 278 kWh) is a unit of energy comparable to that used by a 60-watt bulb every night for about a year (roughly 4,400 hours), and a kg, or kilogram, is 2.2 pounds. So it takes about 110 pounds of NG or 220 pounds of brown coal to light your lamp for a year. Even hydrogen—a clean transportation fuel that is manufactured rather than mined or extracted—requires roughly 10 kg (22 pounds) to light your living room for a year.

Figure 2-3 uses a logarithmic scale to fit it into a reasonable size image. This compresses and visually understates the differences, which cover several orders of magnitude (factors of ten).

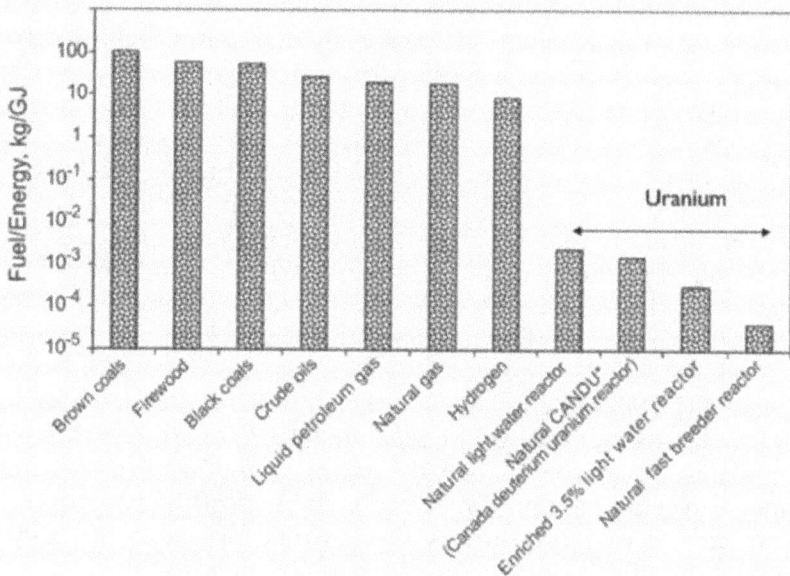

FIGURE 2-3

Nuclear energy ranges from a maximum of 0.002 kg/GJ for LWRs down to about 0.00002 kg/GJ for fast breeder reactors—requiring about 4 to 6 orders of magnitude less (tens of thousands to millions of times less) mass of material for the same energy and far less material mined. That is, lighting your living room for a year takes about 2 grams of uranium in an LWR (the mass of a couple of paper clips) or 20 milligrams in a breeder reactor (the mass of a female housefly).

There are currently 94 working nuclear reactors in the US (down from 104 ten years ago—12 shuttered and 2 new ones operating). They provide close to 20 percent of US electricity, more than any other country in absolute energy terms, though 15 countries use nuclear for more than 20 percent of their electricity (https://tinyurl.com/2wu7rywk). To provide all US electricity, given current demand, would then require increasing the number of 1 GW reactors by a factor of 5. To provide all US energy through electricity (save heating) would require multiplying 1 GW reactors by another factor of 6. That is, we would need 30 times as many comparable reactors (5 x 6) for a total of about 3,000 in the US alone, and some 15,000–18,000 globally. There are now only about 440 operating in the world.

Stored in the US, from some 60 years of operation, there are currently almost 100,000 tons of uranium fuel that have been passed through a reactor just once. This still usable fuel is located onsite, first in cooling pools for a few years and then in concrete/steel casks, and is retrievable. A 1 GW LWR uses around 27 tons of enriched uranium per year on average (some 250 tons of natural or raw uranium), while a 1 GW fast breeder reactor uses only around 1 ton per year. Yet they can both produce 1 GW of power the entire year. The reason for the difference in the amount of required uranium is that, unlike an LWR, a fast breeder can fission not only the U-235 but also the Pu-239 and other transuranics (isotopes heavier than uranium) that are bred by fast neutrons from U-238.

So if the 94 working LWRs in the US were instead fast breeder reactors, together they could generate 94 GW using only 94 tons of uranium each year. Then the nearly 100,000 tons of stored once-through uranium, if repeatedly cleansed of the fission products and recycled into those reactors, would last more than 1,000 years (100,000 tons/94 ton per year = 1,064 years), though a shorter time if more fast breeders were constructed.

But the US also has some 750,000 tonnes (825,000 tons) of depleted uranium (DU, mostly U-238, see Glossary), which is the mostly non-fissile byproduct of the fuel enrichment process (https://tinyurl.com/mpbt4tws). That is, there is 8.25 times as much DU as once-through uranium fuel, much of it having been used for weapons. The DU utilized in a fast breeder would last almost 9,000 years at current usage rates (8.25 x 1,064 years = 8,778 years), for a total of almost 10,000 years (1,064 + 8,778 = 9,842), assuming no increase in the number of reactors. In addition, there is raw uranium in tailings where other minerals are mined, such as rare earth metals (see Glossary, also discussed in Chapter 4). During these millennia no further uranium would need to be mined. Moreover, the US has purchased uranium from Russia that was once in now-decommissioned nuclear warheads. This stock currently provides half the US nuclear contribution to the electricity system, or 10 percent of our electricity.

Nuclear energy, as it turns out, rather than promoting the proliferation of nuclear weapons, is the best way of disposing of the uranium, or for that matter plutonium, from decommissioned nuclear warheads. One application's waste is sometimes fuel for another. And proliferation of weapons is a political decision based on conditions far removed from energy production. This is discussed in more detail in Chapter 7.

The Energy Density Embodied in Conversion Devices

While wind/solar energy can produce electricity without earth-bound fuels, their low spatial concentration (dilution) requires tremendous masses of material (and land) to construct immense fields of turbines and PV panels (or mirrors). And their shorter lifespans require that over time still greater masses of material are required for their replacement, particularly since their constituents for the most part are currently being buried in landfills rather than recycled.[13] Figure 2–4 (from the US DOE's 2015 Quadrennial Technology Review, https://tinyurl.com/bdfvxxn9, and borrowed from Environmental Progress)[14] shows the ratios of materials for the conversion devices, farms, and power plants, per unit of energy output.

The greater tonnage of materials needed for solar and wind farms compared to a nuclear power plant for the same total energy actually produced—almost 18 times as much for solar (16,447/ 920) and a little more than 11 times as much for wind (10,260/ 920)—partly rests on the unpredictable and uncontrollable (temporal) intermittency as well as the (spatial) diluteness of sunlight and wind, both of which require excess capacity to compensate.[15]

---

[13] The National Renewable Energy Laboratory (NREL)—funded and overseen by the US DOE—reports that it costs $20–30 to recycle a PV panel but only $1–2 to dump it into a landfill (https://tinyurl.com/msa8bhbu). As a result, a recycle executive estimates that only one in ten panels are, in fact, recycled.

[14] A typo in the caption of Figure 2-4 refers to its source as Table 10 from the US Department of Energy (DOE) report. This should read Table 10.4.

[15] Many renewables advocates, including Jacobson, claim that wind and solar would eliminate the mining required by fossils and nuclear. They are apparently thinking only of fuel and not of the conversion apparatus. Their incomplete accounting represents a stunning thought gap.

Materials throughput by type of energy source

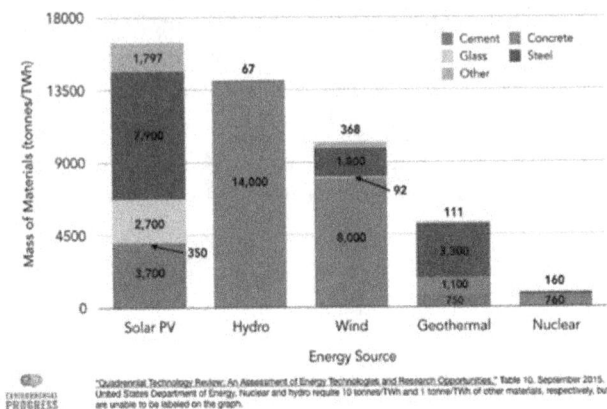

Mass of Materials (tonnes/TWh)

Solar PV — 1,797 / 7,900 / 2,700 / 350 / 3,700
Hydro — 67 / 14,000
Wind — 368 / 1,800 / 92 / 8,000
Geothermal — 111 / 3,300 / 1,100 / 750
Nuclear — 160 / 760

Legend: Cement, Concrete, Glass, Steel, Other

Energy Source

"Quadrennial Technology Review: An Assessment of Energy Technologies and Research Opportunities." Table 10. September 2015. United States Department of Energy. Nuclear and hydro require 10 tonnes/TWh and 1 tonne/TWh of other materials, respectively, but are unable to be labeled on the graph.

FIGURE 2-4

And the shorter lifespans of their conversion devices call for still greater amounts of material (and energy) to replace them, which the graph purports to take into account by providing material input for equivalent energy output. Recycling would lessen the material demands somewhat but would require even more energy—as well as greater monetary costs. Moreover, this graph assumes only partial penetration of wind and solar. Were fossils and nuclear eliminated, the amount of storage equipment (mainly pumped hydro[16] and batteries) would greatly increase these ratios.[17]

Apart from obstacles to accessibility, the materials required for the all-RE plan—for the US, and more so for the world—in some cases exceed known global reserves, and far exceed current annual global production (see Figure 4-3 for the materials required by EV batteries alone). Known global reserves might expand with new discoveries, but late discoveries—such as undersea oil fields, shale oil, or tar sands—often involve extraction more costly than from easily accessible land-based oil fields.

Despite the uncontrollable intermittency of wind and sunlight, Jacobson proclaimed that compensatory excess capacity could provide reliable uninterrupted baseload without significant expensive and energy-wasting storage. This was uncritically accepted by

[16] Discussed further in Chapter 3 and defined in the Glossary.

[17] Indeed, for storage adequate to compensate for downtimes of a couple of weeks (which requires that the batteries be fully charged), the mass of materials in the batteries would be comparable to the mass of materials in the turbines and PV panels (Chapter 4).

many of his supporters. Jacobson (2022) subsequently conceded the need for some storage, but not much.

Low Energy Density Means More Waste

Returning to Goldstein and Qvist (2019, p. 21):

> The amount of toxic waste produced in the electricity generation process [by nuclear] is also many thousands of times less than using coal and even much less than methane [meaning NG].

Considering waste from renewables, obsolete wind turbine blades and solar panels are each projected to reach several tens of millions of tons per year over the next few decades (https://tinyurl.com/2s4y3v46 and https://tinyurl.com/y7e5vxzy). While there are efforts to recycle the fiberglass in blades more cheaply than replacement, this technology is in its infancy (https://tinyurl.com/mvswyar8). And the current efforts aim at a different use for the fiberglass, such as in cement, not in turbine blades. Recycling of solar panels is likewise waiting in the wings.

Waste from renewables is many thousand times greater than the annual 2,200 tons or so of nuclear fuel that passes just once through LWRs, the US's only type of reactor. Stored onsite in 40-foot-deep water-cooling pools, this fuel is wrongly termed "waste" (Chapter 7). If the US would permit recycling, this once-through fuel could be reused repeatedly in a fast breeder reactor after removing the fission products.[18] Since only about 1–3 percent of the energy in uranium is used in the US's current 94 LWRs, by repeatedly recycling the fuel into fast breeder reactors and extracting almost 100 percent of the energy over time, and by thereby spreading over many years the generation of actual "waste," the annual "waste" production would diminish by a factor of 30 to 100, some 22 to 70 tons per year—compared to tens of millions of tons for renewables.

The waste from nuclear plant materials would also be far less, since they last perhaps as much as a century compared to the two

---

[18] While there are thorium breeder reactors under development, in addition to operating uranium breeder reactors, for simplicity we generally refer to the uranium reactors unless otherwise specified. As an aside, the thorium breeder reactors moderate the initial high neutron speeds down to lower, thermal, speeds, while uranium breeder reactors utilize the high neutron speeds for their breeding (hence the term fast breeder).

decades or so for turbines and solar panels and far fewer nuclear plants than wind turbines or solar panels are required to produce the same energy. Furthermore, it is not yet clear how much of the material in a nuclear plant would need to be replaced. But it's the fuel that features in antinuclear propaganda (Chapter 7), and it's the radioactivity rather than the chemical toxicity (common to waste from both renewables and nuclear) that is used to promote fear.

Chemical toxicity of all waste is relatively permanent and must be shielded, though chemical waste often ends up in leaky landfills. In contrast, radioactivity decays continuously and is easily contained and shielded (Chapters 7 and 8). When it comes to mass per unit energy produced (the inverse of energy density), less is better—less fuel and other materials to extract and process, less waste to store or simply release back into the environment, and less energy input (EI) required to extract, refine, and produce the apparatus and fuel.

The relatively safe[19] storage of nuclear "waste" is discussed in Chapter 7, with emphasis on the far more unsafe disposal of waste from fossils and renewables, around which nothing comparable to the often waste-centered antinuclear movement has been organized.

Power Density

Along with many other authors, we define power density as the ratio of power generated by the conversion devices to the required land area housing the devices, neglecting the area of land from which the materials are extracted (mined).[20] Doing so understates the disadvantage of renewables.[21] Ausubel gives a striking example, in which he assumes a 30 percent average annual CF: "To replace the kilowatt hours of the 1½ square mile, 2200 MW Diablo Canyon nuclear

---

[19] We use the term "relatively safe" not to hedge, but because nothing under the sun is perfectly safe; so the best we can ever ask for is relative safety (Chapter 7).

[20] The term "power density" is applied by wind engineers and turbine designers to the power per unit area of the circle described by the turbine's blades. Other energy analysts use the term to mean power output per mass or volume of materials or per average energy input over some span of time. However, we use the term strictly to mean per land area taken up by the wind or solar farm, dammed lake, or power plant.

[21] Ironically, US physicist and leading antinuclear advocate Amory Lovins, as one of his arguments against nuclear energy, focuses precisely on the land for mining, but only for uranium, as though renewables required no mining (an error shared by Jacobson). In fact, the geographic extent of mining is orders of magnitude greater for renewables (and fossil fuels) than for nuclear.

power plant[22] would require a 250 square mile wind farm" (https://
tinyurl.com/27v8jwv2, p.8), with power densities of 566 W/m^2
(watts per square meter) versus 3.40 W/m^2, respectively. The table
below summarizes these figures.

Closer packing of wind turbines can either reduce the area of
the wind farm or increase the number of turbines for the same
area, thereby increasing the nameplate power. But beyond a certain point, the turbulence from closer packing would increase their
mutual interference and further diminish the average power output, rendering the nameplate even more misleading.

The turbine spacing has generally been greater from front to
back than from side to side, for farms with a steady wind direction
(for instance in a mountain pass). This arrangement minimizes the
mutual interference. The recommended distances have until
recently ranged from roughly 2 to 5 times the blade diameter side
to side and 8 to 12 from front to back, with an average spacing of
about 7 blade diameters. For farms with variable wind direction (in
open fields), with turbines that turn to face the wind, the spacing
has to match the greater distance in all directions.

Blade circle diameters can range from under 100 feet to
roughly 800 feet. The larger the diameter, the greater the maximum power output of the turbine, or in other words its nameplate.
The chosen spacing compromises between efficiency and nameplate power of the entire farm. More turbines with closer spacing
lowers the efficiency and the CF.

Recent recommendations to expand the average spacing
from 7 blade diameters to 15 more than doubles the average
spacing and more than quadruples the area, by a factor of 15/7
x 15/7 = 4.59 (https://tinyurl.com/mebwyzf5 or https://
tinyurl.com/2mfk5u66). This would lower the power density of
Ausubel's Diablo Canyon-matching wind farm from 3.40 W/m^2
to about 0.74 W/m^2 and increase the area from 250 to 1,147
square miles (still assuming an average annual CF of 30%, well
inside the usual 25% to 45%). If square, the wind farm would
cover some 34 miles on a side. For CF 25%, the farm would

---

[22] Diablo Canyon, with two reactors, is California's one remaining nuclear plant. It
had been slated to be shuttered by 2025, though the recent inflation in energy prices due
to the Russian invasion of Ukraine has prompted a change of plans. And Diablo Canyon's
power density (566 W/m^2) is far smaller than the 6,500 W/m^2 of the San Onofre nuclear
plant, which California foolishly closed in 2013 and which occupied only about one-eighth
of a square mile (84 acres).

cover 1,376 square miles, a square 37 miles on a side. If rectangular, one dimension would be even larger. This extreme value is used only to give an idea of the land use of an onshore wind farm matching, on average, the power output of a typical nuclear power plant.[23]

Goldstein and Qvist (2019) mention the Swedish nuclear power complex Ringhals, capable of producing up to 4 GW of electrical power 24/7 (almost twice that of Diablo Canyon), and which would yield 35 TWh per year if it ran full out.[24] In fact, it does not run full out and only averages 24 TWh of electrical energy annually, roughly 2/3 of its nameplate annual energy capacity. It does this in a space of about ¼ square mile or 150 acres, for a power density of 4,230 W/m², making Ringhals roughly 7½ times more compact than Diablo Canyon, with its power density of 566 W/m². Unlike a wind farm, for which minimum required areas are dictated by the physics of wind, the amount of land occupied by a nuclear power plant is largely subject, within limits, only to the choice of plant designs, and these, as we see, can vary widely.

They calculate that a wind farm with power output equivalent to that of Ringhals would "entail 2500 wind turbines, 650 feet high, spread over 400 square miles," for a power density 2.65 W/m² (see table below). This figure is comparable (same order of

	DIABLO CANYON	RINGHALS
NUCLEAR POWER PLANT	2,200 MW/1.5 MI² = 566 W/m²	2,740 MW/0.25 MI² = 4,230 W/m²
WIND FARM EQUIVALENT	2,200 MW/250 MI² = 3.40 W/m²	2,740 MW/400 MI² = 2.65 W/m²

POWER DENSITIES OF NUCLEAR VERSUS WIND

[23] Aside from the possible appropriation of land set aside for other uses—such as agriculture, forests, or recreation and leisure—the hundreds of wind turbines can be offensive to eye and ear and, due to rhythmic turbine noise, have a deleterious effect on various aspects of health (see, for example, https://tinyurl.com/6hbmyp5e). These drawbacks have also occasioned frequent protests by neighboring populations (https://tinyurl.com/bdzcb39w and https://tinyurl.com/y5t78za9).

[24] Two of Ringhals's four reactors have been shut down since Goldstein and Qvist published in 2019, leaving the plant's output at 2.2 GW, the same as Diablo Canyon, but we refer to the earlier quantities in their example for illustrative purposes.

magnitude) to Ausubel's figure above (3.40 W/m²). And taking into account the recommended increase in area by a factor of 4.59 (see previous page) would lower the 2.65 to roughly 0.58 W/m². Thus, the power densities of wind farms can be roughly two to four orders of magnitude less—in these examples, roughly 160 to 1,600 times less, or with expanded spacing roughly 770 to 7,300 times less[25]— than that of a nuclear power plant, not because the power densities of wind vary all that much, but because those of nuclear do.

Goldstein and Qvist also note that while their hypothetical 400-square-mile wind farm would provide the same average power (over a year) as Ringhals, the power would be variable and non-dispatchable, sometimes far higher than average and sometimes far lower (possibly zero at times, whenever the wind speed is either too low or too high). And they also conclude that the wind farm would require backup from either fossils or nuclear—or storage. However, as we will show in Chapter 4, even immense storage capacity would not prevent occasional outages without unrealistic oversizing (bigger farms) or overbuilding (more farms).[26]

The following photo from Google Earth, reproduced in Devanney (2023), gives a sense of the relative land area for nuclear versus wind, even though only a small portion of the wind farm is included.

---

[25] The power densities cited above yield the following ratios: 566/3.40 = 166 and 4,230/2.65 = 1,596, or with expanded spacing, 566/0.74 = 765 and 4,230/0.58 = 7,290. To compare costs in Chapter 5, we take an average ratio of land requirements to be on the order of 1,000 to 1 for current spacing (between 160 and 1,600), but some 4,000 to 1 for newly recommended spacing (between 760 and 7,400). These variations do not affect order of magnitude.

[26] Terminology borrowed from US energy analysts Mike Conley and Tim Maloney (2017).

The group of buildings in the middle of the photo, just the other side of the Rhine River, is a French nuclear power plant, which is currently not in operation. The total average power produced by the surrounding wind farm is roughly one-third that of the nuclear plant's former output.

In sum, even assuming equivalent average power output, the nuclear plant would yield reliable power 24/7, while the wind farm's production would be intermittent, sometimes greater, sometimes smaller than demand, and sometimes zero.

Turning to solar farms, the extent of land usage is barely hinted at by this photo (https://tinyurl.com/ytwrj5a3):

The spacing between rows has to accommodate panel tilt that maximizes the received sunlight. The lower the noon sun (farther from the equator) the more vertical the tilt, the longer the shadow, and the greater the required spacing. Even in low latitudes with the sun directly overhead at noon (but not at other times), and the panels horizontal, field workers need enough space to install and service the panels. Total land for a solar farm therefore is two to three times the total area of the panels (Chapter 4). The power density is approximately 20–40 W/m² for solar, an order of magnitude greater than for wind, or one-tenth the land for an equivalent wind farm.[27] And if the recent recommendations for doubling turbine spacing (quadrupling the area) were put into effect, wind's

---

[27] Miller and Keith (2018). Since the power density of a wind farm is roughly 160 to 1,600 times lower than that of a nuclear power plant (with the old spacing), the power density of a solar farm is roughly 16 to 160 times lower.

power density would drop to approximately 0.5–1 W/m², giving solar an advantage of 40 to 1 in power density.

Czech-Canadian energy expert Vaclav Smil (2016) notes that power density for nuclear can be "up to an order of magnitude higher than the power densities for boilers burning fossil fuels." In other words, nuclear plants require roughly ten times less land than equivalent fossil plants. Furthermore, nuclear reactors "have no need at all for pollution controls and for land set aside for the storage of captured waste products," and "radioactive wastes stored temporarily on-site occupy only small areas."

These advantages of nuclear are ignored by opponents like Amory Lovins. Lovins was among those who, following the 1973 oil crisis, began advocating voluntary energy conservation and more efficient electric appliances, largely to avoid expanding the capacity of nuclear energy. Yet nuclear would entail far greater energy savings on the production side than voluntary abstinence and more efficient appliances on the consumption side.

In a striking understatement, Smil refers to Lovins's "questionable procedure," pointing out that, by including mining for uranium ore and assuming the most expensive and least used method of uranium enrichment (see the Glossary) and other cherry-picked factors, Lovins arrives at a deceptive overestimate of the annual land requirements of nuclear. As we have said, mining for uranium can be delayed by decades, and far less is required than, say, coal because of the high energy density of uranium. Smil goes on to say that Lovins's estimates of the continually expanding land occupied by mines (loc. 1864 in Kindle edition)

> would add about 150 ha/year to land claims of a 1 GWe nuclear station.[28] In contrast, Fthenakis and Kim (2009) [estimate that using the most efficient means of uranium enrichment would add only] an additional 2.6 ha/year, less than 2% of Lovins' huge total.

---

[28] The "e" in "1 GWe" indicates the electrical portion of the total power generated in a reactor, which is designated as "GWt" with "t" meaning thermal. Generally GWe is roughly one-third of GWt, a measure of a reactor's efficiency at converting heat to electricity. The abbreviation "ha" is for hectares, a unit of area used by the EU, equal to approximately 2.5 acres.

While Smil notes that even with LWRs nuclear enjoys a huge power-density advantage over renewables, new nuclear reactors, ready this decade (given funding), will have even higher power densities—especially small modular reactors (SMRs) and Generation IV fast breeder reactors.[29]

Smil explicitly discusses the impact of mining and milling on nuclear's power density, but nowhere in our reading of his book *Power Density* does he account for the impact of mining and milling of rare earth metals on the power densities of wind turbines or solar PV panels, which would only strengthen his argument. Taking the latter into account more clearly reveals one of the key disadvantages of wind and solar.

In discussing the dependency of wind's power density on turbine spacing, Smil notes that if the area between turbines is excluded—that is, if most of the land between them could be put to alternative use and not wasted—the power density could be as high as 50 W/m², rather than the 2–4 W/m² mentioned above, or the 0.5–1 W/m² with expanded spacing. Jacobson focuses on the higher power density for wind, arguing that the land between the turbines can be used for other things, such as crops or grazing. He ignores the possibility of negative effects on livestock and humans from the rhythmic noise and turbulence. Nor does he investigate the possible effects of turbulence on crops, such as increased evaporation. Exaggerating alleged hazards and costs of nuclear, renewables proponents ignore objectionable features of wind and solar, which, as mentioned above, neighboring populations do not.

In a world without land restrictions, power density would be of little concern. But we already use 40–50 percent of the world's habitable land area for crops or livestock grazing, even as carbon-absorbing forests are cleared for more. There is nothing Green about taking up more land area for putatively non-carbon-emitting renewables while forests are being destroyed.[30]

---

[29] The new GE Hitachi SMR, the BWRX-300 (boiling water reactor, nameplate 300 MW), boasts a reduction in plant layout by 90 percent (10 percent as large). Since the nameplate is 30 percent of conventional 1GW LWRs, both with nearly 100 percent CF, this would raise the power density by a factor of 3 (30 percent/10 percent) (https://tinyurl.com/yc4968up).

[30] Incidentally, while many Greens consider the burning of biomass for energy to be carbon neutral, this is not necessarily true. Burning trees, for example, contributes more GHGs to the atmosphere in the short run than fossil fuels. Why? Because in contrast to dead organic matter that constitutes fossil fuels, or to, say, the detritus from picked corn fields (termed "stover") that may be destined to be buried in landfills, a living tree absorbs CO_2 from the atmosphere. When a tree is cut down and burned, not only does the burning

Smil further points out that the EROI for wind/solar is well below social requirements and that, as their penetration increases, monetary costs increase dramatically, in no small part due to integrating intermittent sources into the already fragile electric transmission grids that would have to be greatly expanded. He also indicates that the levelized cost of energy (LCOE, discussed fully in Chapter 5) for renewables misleadingly appears low today, largely because of its enormous dependence on the far more energy-efficient fossils and nuclear. Nor does the LCOE for renewables include the costs to health from the air and water pollution from fossil backup for their intermittency—costs including both monetizable healthcare and non-monetizable suffering and lives lost.

Onshore wind power output is greatest in optimal locations, which, as Ausubel emphasizes, are therefore chosen first. So the continuing development of onshore wind becomes progressively less efficient. And since output depends on turbine spacing, wind's contribution to an all-RE system requires further expansion of the total area occupied, whether considering the nameplate power density (called by Miller and Keith the "capacity density") or the actual average power density.

When reading boastful accounts of wind/solar, consider whether the predicted power output represents 1) nameplate capacity, 2) a figure reduced by CF, 3) a figure further reduced by curtailment of wind or solar energy by grid managers when there is an oversupply, or 4) a figure still further reduced by shedding of electricity at the distribution end of the grid, also by the grid managers in response to reduced instantaneous demand.[31]

As you can see, predicted power output is a complicated quantity, and rarely do websites provide enough information for the

---

release CO_2 in comparable amounts to the burning of fossil fuels or corn detritus per unit of energy released, but in addition the removal of the tree prevents removal of CO_2 from the atmosphere that would otherwise have been absorbed. Moreover, while grass regrows in days to weeks during warm seasons, the time it takes to replace most adult trees is at least a couple of human generations. So on human timescales and on the timescale of AGW's progression, the increase in atmospheric carbon from burning recently felled living trees takes a long time to be neutralized—too long to be of use in mitigating AGW in timely fashion. Thus, in practical terms, burning this kind of biomass cannot really be considered carbon neutral.

[31] Renewables advertising commonly uses nameplate in a misleading way. When we're told that a wind or solar farm can power "up to" x number of homes, the estimate is always based on nameplate, even though the nameplate capacity is hardly ever reached, if at all, nor for any significant duration (Chapter 3).

reader to know which of these figures is the one advertised. While pro-renewables ads or articles generally use nameplate, more objective sources, such as the US Energy Information Administration (EIA), might estimate actual energy consumed (the fourth figure above). In general, the greatest reduction from nameplate to consumed energy comes from the CF. Engineering features are generally incorporated into the nameplate, leaving only the external features of nature to determine the CF.[32]

So aside from engineering improvements, because Jacobson (along with other renewables proponents) simply extrapolates a fixed power density to a greater number of wind or solar farms— as though power output remains proportional to total area regardless of quality of location—he overestimates the power density of the renewables farms as a whole. And by overestimating power density, he underestimates total required land area for a given actual average power output.

Were onshore wind farms to be the sole energy source and were they to cover all US energy needs (not just electricity), they would cover almost 43 percent of the land area of the contiguous 48 states of the US. This is based on an estimated power density of 0.9 MW/km² (equivalent to 0.9 W/m²), as suggested by Miller and Keith. If we were to use the estimated power density of 2–4 W/m² from above, using unrealistic assumption of uniformly advantageous wind farm sitings, this would come to roughly 10– 20 percent of the contiguous US, rather than 43 percent.

The foregoing land estimates would diminish for wind combined with solar (with its up to ten-times greater potential power density) and with offshore wind. If, for example, half of energy production were from onshore wind and half from solar, and using wind's lowest land estimate of 10 percent, the land use could be reduced to as little as 5.5 percent [½ x (10 percent + 1 percent] of the land area, and still less if offshore wind were employed. Furthermore, considering that roughly 44 percent of the land area is used for crops or grazing and a little over 3 percent for cities,

---

[32] In contrast, however, Miller and Keith (2018) consider the contribution of certain engineering features to be part of the CF. Thus, they consider that over the last few decades, the CF for wind turbines has risen slightly based on improvement in those engineering features, but at the same time, the need to incorporate less advantageous sites has caused the maximum obtainable power to drop. The result is that the actual power density has remained relatively unchanged. As we've said, the bookkeeping seems easier if CF is defined to include only features external to the conversion devices themselves, mainly weather, leaving the device's conversion efficiency as an independent factor.

leaving 53 percent of the total, 5.5 percent for wind/solar represents a little over 10 percent of what's left, or about 172,000 mi^2 (or 444,000 km^2, greater than the area of California), much of it poorly suited for wind or solar farms. Either way, this indicates that the projected extent of land use for an all-RE system is not trivial. Nor does this account for the land that would have to be covered by additional electric transmission lines required to send energy from far-flung renewable power farms to where it's needed, and possibly additional land for pumped hydro or enormous banks of batteries.

Adding to the foregoing, as AGW progresses and continent-size ice sheets or more localized glaciers continue to melt, to flow down rivers to the sea, and to add to oceanic volume, thereby encroaching on coastlines, the available area of land above sea level will continue to decline, and at an accelerating pace—faster in certain lower-altitude locations (such as Netherlands, Florida, Bangladesh, and a number of small islands). Of course, the disappearance of a glacier exposes land, but in locations not optimal at least for solar farms.

Furthermore, if AGW-associated changes in local conditions cause sites that were once optimal to become suboptimal or almost completely useless, newly built wind or solar farms may become obsolete before their lifespan is complete. This would be a case of stranded assets—apparatus that has to be abandoned before it pays for itself—as well as progressive loss of energy sources.[33] Stranded assets are discussed further in Chapter 10.

Neither fossil fuels nor nuclear require wind's or solar's extensive land use, nor are they dependent on nature's whims—except for the siting of facilities to avoid underground faults, volcanoes, and threatened shorelines. But these are generally known or predictable over lengthy timescales, unlike the generally fickle hourly to daily schedules of wind and sunlight—even if less so for sunlight in desert locations during non-rainy seasons.

In the following chapter we discuss the inescapable intermittency that weather-dependent wind and solar farms introduce into an electricity system and why, in addition to their energy profligacy, their consumer prices are higher.

---

[33] Granting that some previously suboptimal sites might become more optimal does not change the point.

3

Intermittency: The Matrix and Storage

The Implications of Intermittency

Steady baseload electricity and smooth, and reliable midload and peakload must meet fluctuating demand, second by second. Meeting demand merely on average inevitably includes outages. Comparing electrical supply to food availability, even if average caloric intake meets requirements over time—say, a year—food schedules may fail to promote health or survival. The body's ability to store food allows intermittent eating, but only within limits under our control. Erratic eating schedules beyond our control can lead to malnutrition or obesity and even death.

Analogously, too much electricity at certain times can cause power surges or overheating that damage the grid. Built-in protections—fuses or circuit breakers—produce outages. Too little electricity can cause outages (blackouts) or equally unacceptable brownouts.[1] As it is, somewhere in the US an outage occurs every 2–3 hours, usually of short duration, mostly due to grid malfunction—such as gnawing squirrels, waving treetops, or extreme weather events, in which case they can last for days (Thompson 2016).

Intermittent electrical supply, unless rapidly compensated by operators, destabilizes long-distance lines, unsteadying the AC frequency (60 Hz in US, 50 Hz in Europe and other countries[2]), manifested as loss of synchronization between regularly

---

[1] A brownout is a deliberate reduction of electricity by grid operators when supply is less than demand. In contrast, blackouts are generally unexpected and beyond grid operators' control.

[2] Hz means Hertz, or cycles per second.

51

varying voltage and current, causing departures from average voltage.

Whenever supply exceeds immediate demand, the surplus is either transferred to neighboring areas, stored, or curtailed. Transference across national boundaries requires international agreements to accept surpluses on an unpredictable schedule. Storage is discussed in the following section. When neither of these is available, the grid operators simply throw away (curtail) the surplus.

Crucially, wind and solar, being weather dependent, are irregularly, uncontrollably, and unpredictably intermittent. This necessitates immediate availability of either stored energy or backup from a baseload source, either fossil fuels or nuclear. The planned elimination of the latter puts the burden on storage.

An Electricity System's Need for Energy Storage

As Australian energy analysts Graham Palmer and Josh Floyd (2020) explain, energy storage has played a vital role in the advance of social formations throughout human history. Whether converted from food to muscle in humans or domesticated animals, or from wood, coal, oil, or natural gas (NG) to fire, humans could choose when and where to convert energy to usable form.

Other sources, such as flowing water, wind, or sunlight, could be captured and used, but at first the timing and location were determined by nature, beyond human control. Ways were soon found to store flowing water behind dams for subsequent release at will. Ways to control the timed use of wind or sunlight through storage came later. Building storage apparatus requires energy (part of EI), and during charging and discharging some usable energy is lost to heat (reducing ER). Thus, storage diminishes EROI and diverts from net energy to run the society.

Electricity from wind or solar is subject to their uncontrollable, and therefore unpredictable, intensity and time of arrival, necessitating storage to turn non-dispatchable energy into dispatchable. Moreover, to charge storage apparatus, a sufficient supply of surplus energy must arrive before it is needed. The timing does not always oblige, regardless of how much storage apparatus is available (Figure 3–4).

Fuels, in contrast, whether for combustion (fossils, biomass) or fission (uranium, thorium), are their own storage—achieved by natural processes and costing us no EI. The energy from fuels can be converted and released at will, though again, input energy is needed to obtain the fuels and build the conversion devices, and extracting the energy entails some loss to heat. Heat losses are rarely trivial and can amount to more than half of the original energy destined to be stored. While some heat can be captured and used for warming indoor spaces, cooking, or other tasks, that is not always done. Furthermore, capturing and using heat requires still more input of energy, as well as materials, labor, and time.

For electrical systems, storage (whether by fuel, batteries, or other means) is indispensable to smooth a switch between grid contributions from different energy sources (within seconds to minutes), to smooth out midload and peakload during each day (hours), or to fill in for extended periods of wind/solar downtimes (days to weeks or longer). The Roadmap originally included a small amount of potential storage in the form of concentrated solar power (CSP) to cover the first two functions, but nothing close to the requirement for the third. Nor does Jacobson's more recent update (2022) include adequate storage, relying mainly on a small amount of hydroelectricity (of which some unspecified portion would be pumped hydro[3]), accounting for less than 3 percent of total demand. However, in Jacobson's projections, the hydro requirement can sometimes exceed ten times the current available capacity (Clack et al. 2017).[4]

Can Excess Wind and Solar Capacity Alone Solve the Problem of Intermittency?

Two features of any variable phenomenon are averages and departures from average. For electricity, as with food intake, departures from average can be deleterious.

To compensate for intermittency Jacobson relies almost wholly on excess capacity, mainly through oversizing commercial wind/solar farms. He postulates that if in one place the wind is

---

[3] Pumped hydro uses two water reservoirs at different levels. Water is deliberately pumped from the lower to the higher reservoir (using electricity) to be released at will through an electric turbine/generator that also serves as the pump. In 80 percent one reservoir is a dammed lake in the US; in 20 percent both reservoirs are independently created (or discovered) for the purpose, in which case water may require transportation. Pumped hydro now accounts for 96 percent of electrical energy storage in the US (https://tinyurl.com/5t266enf), but it's limited by geography and topography. And in the US and other countries dams are being dismantled. While batteries contribute little so far, being portable they are likely to proliferate.

[4] Currently in the US, hydro produces 6–7 percent of electricity, though in Norway, an extreme case, it accounts for 90 percent.

blowing too slowly or too fast, and/or the sun isn't shining, an excess is almost certainly occurring elsewhere. And that the surplus can always be transmitted from latter to former. This may sound plausible, but it is magical thinking.

For one thing, even if the wind is blowing optimally in one place when not in another (similarly for sunlight), the windy/sunny location may not have power to spare. When surplus does exist, the grid must be able to switch rapidly to link the two locations, the distance between which varies randomly as well. And the heat loss is proportional to that distance. Plus, broad geographical areas tend to harbor similar weather, so oppositely supplied regions tend to be farther apart. Lastly, even if some temporary deficits may link to a momentary surplus, not every pairing will involve matched amounts of surplus and deficit.

There are three possible correlations between location pairings—direct, inverse, and absent. Direct correlation means simultaneous surplus or simultaneous deficit. Inverse correlation means deficit here surplus there. Absent correlation means randomness of deficit and surplus. Reliable electricity purporting to depend on correlation requires not just coincidence, but causal and inverse correlation, with all deficient locations simultaneously paired with surplus locations.

Since the Roadmap rejects significant storage and since there is no reason to expect a causal inverse correlation, Jacobson's plan relies on an absence of correlation (randomness).[5] In other words, that during a deficit in one location, the odds are that a surplus will be available somewhere else that can be transmitted by the electric grid. Not only is the Roadmap's optimism based on an absence of correlation, its feasibility requires that compensation be reliable. Moreover, the grid must be capable of handling rapidly changing paths, without losing frequency or phase synchrony between voltage and current, in addition to its everyday second-by-second distribution of energy from the various sources of supply to the many sites of consumption.

Even if inverse correlation magically were to arise, such that

---

[5] This is not to be confused with Jacobson's implication (in 2009, see Sidebar) or explicit statement (in 2022) that wind and solar reliably complement each other at the same geographical location. In his 2022 paper he says (p. 439), "The second correlation plot in Figs. 3 and S2 suggest [*sic*] that, in U.S. regions, solar and wind are anticorrelated, thus complementary in nature. In other words, when the sun isn't shining during the day, the wind is blowing and vice versa." It's important to note that in 2022 he was referring not to "real data," but to his modeled projection 28 years into the future, hence perhaps his softening to the term "suggest."

timing for mutual compensation were reliable, success would require that the amount of surplus power always match the simultaneous deficit. Meeting this requirement is prevented by the fact that local demand is partly random, and it is relative to local demand that surplus and deficit are defined.

Likewise, absent correlation would prevent reliable mutual compensation, while direct correlation would prevent mutual compensation altogether. In fact, direct correlation is common, i.e., weather systems can cover vast areas for days or longer.

To obviate the problems of distance between farms of the same type, Jacobson intimates that solar and wind can reliably compensate for each other's deficits and claims that this is demonstrated by "real data." We examine this claim in the following Sidebar.

Sidebar: Can Wind and Solar Compensate for Each Other?

In both the November 2009 *Scientific American* article coauthored with Mark Delucchi and during a February 2010 TED debate with ecologist and pronuclear convert Stewart Brand (https://tinyurl.com/26fbynh3), Jacobson displays a graph showing wind, solar, and hydro (and a trivial amount of geothermal) mutually compensating for each other's hourly shortfalls (Figure 3-1, taken from the magazine article). The graph shown during the debate is a smoothed version of the magazine figure, published three months earlier. Hydro and geothermal, while capable of providing reliable baseload, contribute little, with wind and solar doing the lion's share and largely compensating for each other—bolstering the claimed continuity of supply.

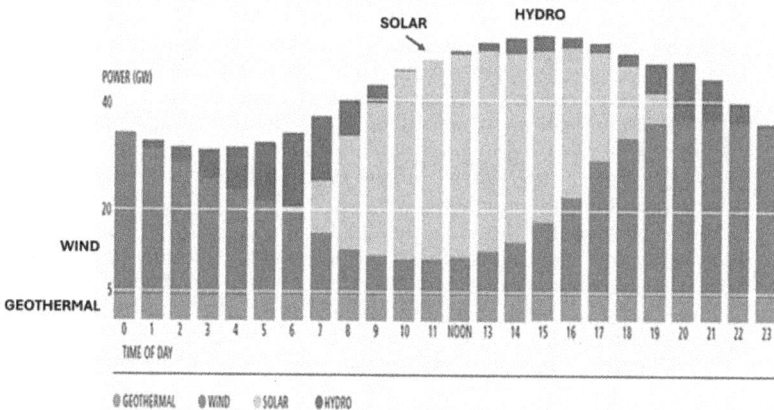

FIGURE 3-1

The magazine caption reads, "CALIFORNIA CASE STUDY: to show the power of combining resources, Graeme Hoste of Stanford University recently calculated how a mix of four renewable sources, in 2020, could generate 100 percent of California's electricity around the clock, on a typical July day. The hydroelectric capacity needed is already in place."

Twice during the debate (at 13:14 and 15:58) Jacobson emphasizes that this graph is not hypothetical but "is from real data in California." Several points about this "real data": First, as the caption states, it's data from a particular day, taken to be typical. In July. In sunny California. The graph was projected to be still valid in 2020, ten years after its publication. What does "real data" have to do with such a far-off projected outcome, restricted to a particularly sunny location, during a specific summer day, cherry-picked for its appearance of compensation? Jacobson's clear implication is that, being taken from "real data," this graph is a trustworthy basis for his assumption of reliable availability of electric power. And since the Roadmap purports to cover the entire US, and eventually the world, the implication is that continuous availability is reliable everywhere all the time. Otherwise, the graph would be irrelevant.

Second, the smoothly varying appearance is an artifact of averaging over an interval longer than seconds or minutes, namely hours—an interval dictated not by "real data," but chosen to produce a smooth, appealing graph. The graph obliterates any display of the second-by-second or minute-by-minute fluctuations that can cause electronic devices—in the absence of their all-but-shunned storage or reliable baseload backup—to fail at unpredictable times. For a more realistic set of data, see Figure 3-2 below, though the scale there is in weeks, and it would be even more spikey if the scale were shorter.

Third, the question is not whether wind and sunlight happen to fluctuate in opposite directions, but rather how their combined output compares to instantaneous demand as it varies over that day. So this is incomplete data, "real" or not. Jacobson's emphasis on "real data" misleads by turning an exceptional day into a typical one, by turning typical into universal (for example, not seasonally dependent), and by making instantaneous output today relevant to local energy demand ten years in the future.

According to a show of hands before and after the debate, Jacobson swayed what appeared to be half a dozen people, out of an audience of several hundred, away from nuclear toward wind and solar, with none swayed in the opposite direction. But then this particular 2010 audience favored nuclear energy by a ratio of some 3 to 1 (even in the moderator's estimation). Jacobson's small measure of success represents a soft echo of the way that he and his colleagues have successfully impressed certain public figures and a significant portion of the environmental movement.

Finally, Jacobson and his colleagues are so focused on oversizing each farm that in their original version of the Roadmap they called for over-building the number of farms by only a little more than 4 percent of the total average output, and they relegated a small amount of storage to a portion of the 7.3 percent of total average power output provided by CSP.

Jacobson's 2022 update still includes a pittance of hydro storage, roughly 3 percent of total power capacity, compared with 44 percent for wind (offshore plus onshore) and 52 percent for solar PV (farm plus rooftop), and less than half a percent for CSP (2022, Table 3, p. 433).

The current US electrical system—even with its reliable baseload from fossils and nuclear and with less than 20 percent coming from wind and solar (greater in some locations)—is overbuilt by 150 percent, not just 4 percent, making it 2½ times as large as its average actual output—almost 1,200 GW capacity versus roughly 480 GW average demand. The excess mainly buffers predictable peaks in demand and occasionally unpredictable local extreme events. Yet the grid still experiences local blackouts, sometimes over wide areas. These are due, however, not to persistent intermittency of supply but rather to chance occurrences like wildfires, storms, or operator error, which may be impossible, or at least infeasible, to prevent.

CF Is Not a Fixed Quantity for Wind or Solar

Jacobson's analysis suggests that that to obtain a desired electrical output over the course of a year, each farm must be oversized by the approximate inverse of its CF. Thus, a farm planned to produce 100 MW on average, but which provides power only, say, one fifth (20 percent) of the time, or 20 MW on average, would produce the desired 100 MW if it were five times as large, with a nameplate of 500 MW. Even so, the instantaneous power output would generally be higher or lower than 20 percent of its nameplate.

The CF of a farm is not a fixed quantity. Rather, CF varies with both the selection and duration of the time interval. If the average CF is 20 percent over one year, it may differ slightly over other years (inter-annually) and be much smaller or larger over shorter timescales (intra-annually). Figure 3-2 illustrates this for the case of wind during a particular year in Germany (borrowed from Conley and Maloney, 2017). Thus, a farm's nominal CF involves a tacit choice of interval. Jacobson's tacit choice, which is conventional, is a one-year duration, the so-called "annual" CF, which is generally averaged over several years. Both specific time interval and its duration are arbitrary and ungeneralizable, all the more so as climate changes.

Departures from average have their biggest impact at shorter timescales, since rapid fluctuations tend to average out over longer intervals. Figure 3-2 shows wind's contribution to Germany's electrical energy in 2014. The black represents actual wind energy and the light gray the shortfall relative to the maximum possible, total installed capacity (nameplate) as it slowly grew that year through additional installations. Note how the summed output from numerous wind farms fails to smooth out the variability, but

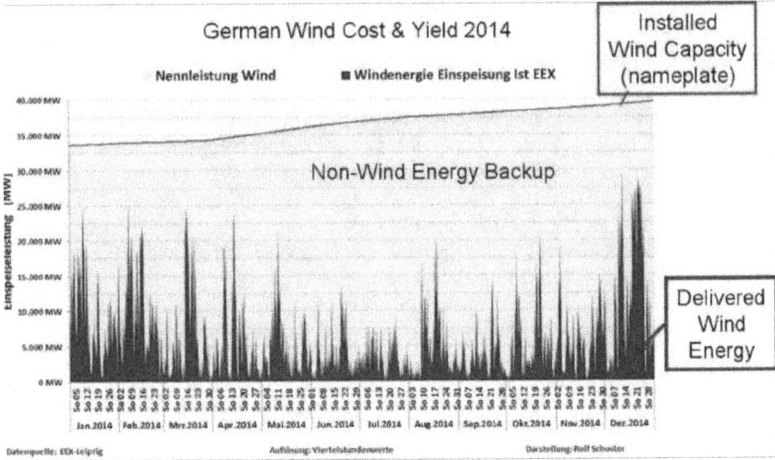

German Wind Cost & Yield 2014

Nennleistung Wind   ■ Windenergie Einspeisung Ist EEX

Installed Wind Capacity (nameplate)

Non-Wind Energy Backup

Delivered Wind Energy

Datenquelle: EEX-Leipzig   Auflösung: Viertelstundenwerte   Darstellung: Rolf Schuster

FIGURE 3-2

instead embodies it. The extreme spikiness is peculiar to wind, which generally arrives in gusts rather than in continual and smooth breezes or steady gales. Solar input, which follows cloud motions and the Earth's rotation, varies more slowly.

Averaged over short time intervals CF varies tremendously—from as low as zero (complete interruption of output) to 80–90 percent, and perhaps even 100 percent on occasion when the time interval is short enough. However, as seen in Figure 3-2, Germany's actual contribution from all its wind farms in 2014 never approached 100 percent of nameplate. The intra-annual CFs can vary significantly by season, with German wind generally much better in December to February than in July to August. Any year would suffice to illustrate the point.

In Germany the light gray portion of the graph is generally filled in mainly by coal, in many countries mainly by NG. The annual CF for 2014 is defined as the fraction of the entire area occupied by the black portion for that year. But the frequency with which supply falls below average is at least as important for understanding the full implications of wind's (or solar's) intermittency. Even more important is the relationship of those peaks and valleys to the instantaneous and varying demand (not shown in the graph). The magnitude of the departures from instantaneous demand, their frequency, and their specific timing are crucial for estimating the amount of storage that would be required for the Roadmap's proposed elimination of reliable baseload. But a good guess is the best that can be done, since the fluctuations are random and will vary from year to year.

The following Sidebar is intended to clarify the dependence of CF on the time interval under consideration, both its duration and its specific choice.

Sidebar: The Dependence of CF on Choice of Time Interval

Figure 3-3 represents the variation in the particular magnitudes of the CF for a set of selected durations—from a second to a day to a year—along with the frequency with which each specific magnitude would be attained over many years, considering many specific intervals of the same duration. For example, year upon year, or day upon day. (This is known in the math business as a probability density function.)

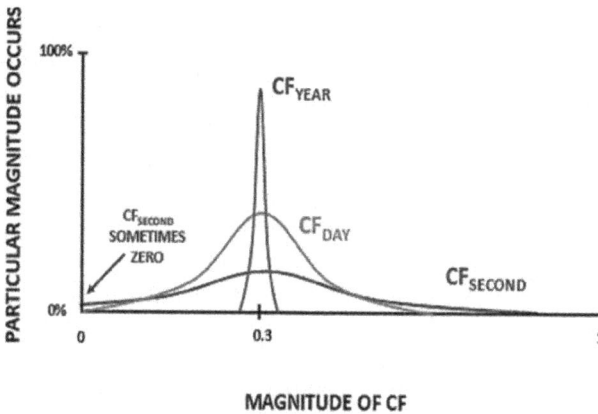

FIGURE 3-3

CF_{YEAR} means the average CF over a year (a particular yearlong interval), CF_{DAY} over a particular day, and so on. It's easy to see the trend so that you can roughly fill in the curves appropriate to other intervals, such as minutes, hours, weeks, and months.

Each of the CFs will average 0.3 over many years, but as you can see, the spread in specific magnitudes from year to year, or day to day, or second to second—the behavior of departures from average—varies systematically from one duration to another, with the spread being the greatest for the shortest time interval (in the graph one second) and the least for the longest time interval (one year). This can also be seen by inspecting Figure 3-2 above and imagining moving a narrow or wider interval across from left to right and observing the rough magnitude of average CF within that moving interval.

We can also extrapolate this beyond both the short end of the spectrum and the long end, considering split seconds and intervals longer than a year, such as a decade. The pattern would extend itself, with still greater spread for split seconds and still narrower spread for a decade.

These drawn curves reflect inevitable tendencies and are representative of what would actually occur. It could be that the output during any one day, for example, might be very steady or might be very variable over the course of the day, but over enough particular choices of day these curves would tend to describe the actual variation.

In contrast to wind and solar, a fossil fuel or nuclear plant operated continuously for baseload, with no unplanned departures from average, has a steady and higher CF, over all timescales. However, the same baseload plant, if used to "back up" or "buffer" intermittent sources (called load following), would have a variable and generally lower CF, since with load following, departures from average or from demand tend to predominate. The departures would be in synchrony with the departures of the renewable for which they are filling in, though in the opposite sense. Buffering wind and solar, therefore, will lower the annual CF of a baseload plant and its EROI. What matters more than the fact of intermittency is its uncontrollability and unpredictability.

Furthermore, since deficits can occur at any moment, the backup plant has to idle continually so it can ramp up almost as suddenly as the wind speed changes or direct sunlight disappears. The energy produced between such calls generally goes unused, since demand from elsewhere is random. The plant's input energy (for instance to obtain the fuel and for operation and maintenance) rises without increasing usable return, thereby lowering the plant's EROI. Since backup is mostly from coal or NG, the idling adds to GHG emissions, which is charged to the coal or NG plant but really belongs on the renewables' bill. Nuclear backup would eliminate such emissions, even if some of its energy were wasted. In the absence of nuclear, wind and solar can at best lower GHG emissions without eliminating them.

Granted, greater oversizing/overbuilding of renewables and/or greater storage capacity would diminish dependence on fossil backup and thereby lower GHG emission, but this would further increase the cost of all inputs, further lowering wind/solar's already low EROI (discussed in detail in Chapter 5).

The Solar Contribution to Germany's Electricity

Examination of actual CFs helps clarify the intermittency problem. We examine Germany's electricity system because of its extreme variation in short-term CF, and next compare it to France's system next door, one of the world's lowest GHG emitters. We begin with solar, which is less intermittent than wind. The worst days occur in winter, when there are 9–10 hours of daylight and 14–15 hours of darkness and the noon angle of incidence is lowest. Back in 2010–11 installed solar capacity totaled 15.5 GW. While average annual CF for German solar is 10–11 percent (wind averages 18–21 percent),[6] it was typical during 2010–11's December–January that the entire solar array had a daily CF of 0.0066, or 0.66 percent.[7] Average daily output was 0.0066 x 15.5 = 0.102 GW (102 MW), 1/150 of its nameplate, illustrating the often deceptive and irrelevant nature of nameplate capacity. What matters most is minimum output, as it has the greatest impact on usefulness, cost, and EROI.

By Jacobson's logic, to reliably obtain 15.5 GW without storage (were solar the only source) would require an oversize/overbuild roughly 150 (the inverse of 0.0066) times the installed capacity, or 2.3 TW, almost the total consumed electrical power in the world (~2.6 TW). The magnitude of this required excess is mainly due to the intermittency and to a far lesser extent to Germany's high latitudes. But even with 2.3 TW capacity, outages would still occur, because of the frequently positive correlation among mutually distant farms (when some underproduce, all are likely to underproduce). To prevent outages with only excess capacity would require oversizing/overbuilding by not just the inverse of the annualized CF (10–11 percent), but by the inverse of the daily CF, and not just the daily but the hourly, and not just the hourly but . . . and so on.

However, continuing with Jacobson's (incomplete) logic, based on Germany's average annual CFs and optimistically using

---

[6] In contrast, the US, with a quarter of Germany's renewables and almost entirely at a lower latitude range, enjoys an average annual CF of roughly 20 percent for solar. And US annual wind CF averages 35 percent. Before it was shuttered, Germany's nuclear CF was upwards of 90 percent at all timescales, and coal ~60 percent, reduced by its frequent filling in for renewables.

[7] One of us cited these figures at the time (Roberto, Meyerson, et al. 2012, p. 68).

the high end of each range, solar and wind installations would require oversizing/overbuilding by factors of at least 9.1 for solar (1/11 percent) and 4.8 for wind (1/21 percent). If Germany's system were to become all solar and wind, say half each, the average overbuilding/oversizing factor would be about 7—½ x (9.1 + 4.8).

Even this optimistic approach would expand land usage and increase materials, labor, time, and energy to produce, install, and repeatedly replace them. And it would still fail at shorter time scales. Just how vast the oversizing/overbuilding would have to be to avoid outages at shorter timescales is entirely uncertain. It would, however, be indeterminately larger than 7.

The question is not how often each wind/solar farm is producing enough energy, but rather how often each is producing too little, including none at all. A glance again at Figure 3-2 suggests just how often that occurs for wind. Indeed, the power output (black spikes) is below average most of the time.[8]

Nor is this example unusual for Germany. According to US energy expert Mark Nelson,[9] daily CFs in 2017 for solar were, on January 7, 8, 9 and 11, respectively, 0.9 percent, 0.5 percent, 0.8 percent, and 0.8 percent, all below 1 percent. Seventeen days in the month had CFs below 2 percent, and the deficits were made up mainly by coal. Without storage (or wind), if coal or any other reliable energy source were also unavailable, as Jacobson's plan proposes, there would have been virtually no electricity available on those days. The availability of wind might compensate slightly, but it has similar, if less extreme, problems with intermittency, as we explore in the next section.[10]

Without significant amounts of reliable baseload, the complete elimination of random outages approaches impossibility. At best, with enough hypothetical oversizing/overbuilding and storage capacity, the frequency of blackouts might be brought to within "acceptable" limits, but the amount of requisite apparatus would be prohibitive and could exceed the world's resources (Chapter 4).

---

[8] Excursions above average are greater and much briefer, excursions below average less extreme and longer. While the relationship to instantaneous demand is the relevant comparison, we compare here to the average because the fluctuating instantaneous demand data is not shown. The point, however, would be the same.

[9] Formerly of the Breakthrough Institute, later Environmental Progress, and now founder and managing director of Radiant Energy Group.

[10] To convince yourself we are not cherry-picking, the following website provides the data in an interactive electricity map of the world: https://app.electricitymaps.com/.

The Contribution to Germany's Electricity from Wind

Nelson has recorded the actual daily, monthly, and annual CFs for German wind for 2011–2017. During this period, daily wind CFs for the entire country vary from a high in January 2012 of 82 percent to a low in February 2013 of 0.9 percent. Average annual CFs range in these years from 18 percent to 22 percent, an expectedly smaller range for the longer interval (see Figure 3-3). We focus on 2016 with annual CF 18.6 percent, an arbitrary choice but representative of that 7-year period, and ask the question: What if wind farms alone had been Germany's sole source of electricity?

For purposes of illustration, and with no loss in generality, we assume an average and steady consumption (demand) of 50 GW, for which production must be adequate. By Jacobson's logic the wind capacity would have required an excess 5.37 (the inverse of 18.6 percent) times that demand. Thus, 5.37 x 50 GW = 269 GW of wind power capacity would have sufficed to avoid outages, says Jacobson. But some days receive more than 50 GW and others less. To supply the latter, we would be forced to store surplus on days that receive more.

In Germany, while winter sees the least sun, it sees the most wind, and in the absence of solar, the wind surplus would have to be large enough in winter to accommodate the other seasons. Not only are excess wind turbines required but also adequate storage capacity. Jacobson's shunning of storage is tantamount to his assuming that the CF every day is 18.6 percent, all day long, with never a surplus of wind received but never a deficit either.

Let's now assume that both the excess wind capacity and adequate storage capacity were both available. The surplus each day is the difference between wind energy received and that day's demand. Beginning on January 1, 2016, the CF that day was 10.1 percent (8.5 percent below the annual average of 18.6 percent), and wind power received was therefore 10.1 percent x 269 GW, or 27.2 GW—less than the needed 50 GW. So without a stored surplus from December, this day would have run short of electricity demand. However, January 2 CF was 19 percent (0.4 percent above the average), and received wind power was 19 percent x 269 GW, or 51.1 GW, and so on. Figure 3-4 should help follow the account.

In 2016 daily CF for the first 52 days (through February 21) was always above the annual average of 18.6 percent (except on January 1) and averaged 31.2 percent (12.6 percent above 18.6

WIND ENERGY IN GERMANY IN 2016

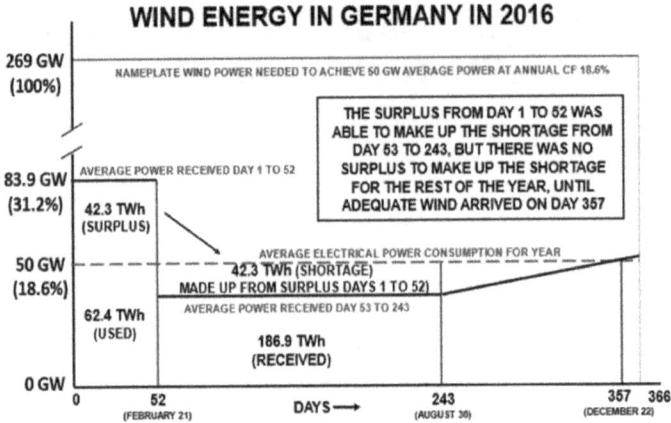

FIGURE 3-4

percent). So the average wind power for those 52 days was 31.2 percent x 269 GW, or 83.9 GW. For those 52 days, then, there was an average surplus of 33.9 GW (83.9-50, or 12.6 percent x 269 GW). From day 53 to the end of the year the wind occasionally exceeded demand, but there was never again a sustained surplus.

Therefore, the surplus energy from the first 52 days would have to be stored to compensate for the coming deficits. This surplus energy was 42.3 TWh (33.9 GW x 52 days x 24 hours/ day). Since the average energy consumed per day is 1.2 TWh (50 GW x 24 hours), the surplus stored in the first 52 days represented 35.3 days-worth (42.3/1.2) had no more wind been received. In other words, the stored surplus alone could have compensated for zero wind for the next 35.3 days, a little over a month. However, since the subsequent days were not without some wind, the stored surplus would have actually compensated for more than a month—as it turned out, more than 6 months, up to day 243 (August 30). But on day 244 (August 31), the stored-plus-received energy fell slightly short of demand. The deficit would not have ended until day 357 (December 22). Which means that in 2016, if wind plus storage alone supplied Germany's electricity, there would have been a blackout or brownout from August 31 to December 22, almost 4 months. And this would be due not to a lack of adequate storage apparatus (assumed to be adequate), but rather to nature's timing, completely out of human control.[11]

---

[11] Prolonged wind deficits are not uncommon in Germany. In reality, the demand was covered in 2016 by several sources—coal and a much smaller amount from NG (both GHG emitters) plus imported energy from some neighboring country. Such imports come largely

Now let's assume the storage apparatus to hold the 52-day surplus consisted of lithium-ion batteries. A 1kg lithium-ion battery can hold 150 Wh, so to store the 42.3 TWh would require 282 billion such batteries (42.3 TWh/150 Wh), at a cost of almost $21.2 trillion (about $500 per kWh storage capacity for grid-scale batteries). Germany's current GDP is $4.3 trillion, so if hypothetically no money were spent on anything else, it would take around 5 years to purchase enough batteries—once.[12] But the batteries only last for a few hundred charging-discharging cycles, perhaps 3–5 years, depending on the frequency of their use. So their replacement would be required every few years, about the time the first round were paid for (and nothing else). Furthermore, disposal of the toxic waste materials would add further costs.

In this example, for simplicity we treated electricity demand as a constant. This allowed us to focus on the intermittency, oversized/overbuilt wind capacity, and storage, but electricity demand fluctuates above and below the annual average. As mentioned on page 57, the US grid, with an average demand of around 480 GW, is built to accommodate almost 1200 GW at peak. To meet fluctuating demand with likewise fluctuating, unreliable, uncontrollable, and unpredictable solar and wind would pose additional challenges to planning an all-RE system. It's one thing to plan for an annual average demand but quite another to plan for fluctuating demand that is unknowable.

Additional considerations include heat losses every time a battery is charged or discharged, spontaneous fires that plague lithium-ion batteries (Chapter 4), and the changing availability of wind and solar energy in light of accelerating AGW, all of which magnify the infeasibility of an all-RE system.

In summary, the example of Germany helps illustrate the challenges posed by intermittency. But the main point is that departures from average, the relative timings of surplus and deficit, and particularly their seasonal character necessitate vast amounts of storage, requiring in turn vast amounts of materials, money, and particularly energy (EI), among other costs. Between vast excess generation capacity and immense storage capacity, both required by the weather dependency of wind and solar, the result is an unworkably low EROI. And an electrical

---

from France's reliable (non-GHG-emitting) nuclear output. In fact, the all-time monthly high that Germany ever imported from France was 2.5 TWh in July 2014, a little over 2 days' worth (https://tinyurl.com/yc5wesb2).

[12] Even the US GDP in 2023 was $25.5 trillion, so the batteries would cost 83 percent of that figure.

system that has too low an EROI cannot sustain a complex modern society (Chapter 5).[13]

GHG Emissions—Comparison of France and Germany

The primary motivation for the turn toward renewables is to eliminate GHG emissions (and respiritory pollutants) from the electricity system, and secondarily to avoid nuclear. Let's see how successful wind and solar have been in reducing emissions so far. The emissions can be expressed as grams of CO_2 per kilowatt-hour of electricity (gm CO_2/kWh), though as we will shortly see, methane leakage is significant but neglected.

Central to our technological argument is that wind and solar do not scale. That is, they would fail at high penetration, most dramatically as they near 100 percent and thereby eliminate the matrix. Yet without nuclear, entirely clean electricity generation would require that very 100 percent penetration—a Catch 22.

At low penetration, while far less efficient than fossils or nuclear, renewables are efficient enough to provide some electricity to the grid, even as they lower the EROI of the system as a whole and render it more expensive in materials, land, labor, time, and energy. Moreover, new wind or solar farms could, under certain circumstances, serve to reduce but not eliminate CO_2 emissions. That their emissions have not declined significantly, if at all, is directly due to their continued reliance on coal and NG to fill in for the intermittancy of theier wind/solar.

US energy expert Meredith Angwin (2020, p. 200) describes a study of twenty-six OECD countries over more than two decades (1990–2013) that found for every 1 percent increase in new wind or solar capacity (not actual delivery, but nameplate) there has been an accompanying 1.14 percent increase in the fossil-fueled backup capacity—primarily NG. Because this additional fossil capacity is intended only for backup, its emissions will not be as continuous as those of a fossil plant that is run 24/7. Nevertheless, in some places a new backup NG plant would be active more than half the time, whenever the wind or solar farms produce less than their assigned share of instantaneous demand (Figure 3-2).

---

[13] Some German business executives are already urging their government to restart some of Germany's 17 shuttered nuclear reactors (https://tinyurl.com/4xc2kf6p)—to no effect so far.

More importantly, wind and solar construction has not kept up with the fast growing energy needs of industrializing economies, mainly China and India, where the rapid opening of new coal mines and plants are swamping the global growth of renewables (https://tinyurl.com/mtexrpvf). In short, despite the hopeful propaganda, we're not seeing a global transition from fossils to renewables; rather, both are increasing in absolute terms.

Germany illustrates the case of moderate penetration. As of December 2023, its solar capacity (nameplate) was ~81 GW (https:// app.electricitymaps.com/), well above the average demand for that month of ~55 GW and constituting 32 percent of its total electrical capacity of ~250 GW (not the hypothetical 269 GW nameplate that would have been required of an all-wind electrical system, as discussed above).[14]

Germany's coal and NG capacities are 37.5 GW and 34.8 GW, respectively, both used to fill in for renewables' deficits, though mainly done by coal. The total capacity of coal plus NG (72.3 GW) is a bit less than the solar capacity (~81 GW). Its wind capacity (~66 GW) is smaller than its solar, but wind is available a much greater pro- portion of the time, in part because wind doesn't suffer from the same interferences as solar. But even though the total capacity of solar plus wind, ~147 GW, is more than half of the total 250 GW (includ- ing biomass and hydro), consumption requires backup from coal and/or NG most of the time. Thus, Germany consistently fails to sat- isfy its promised decrease in CO_2 emissions.

Let's look at the CO_2 emissions. From the electricity map, at the time of this writing Germany is enjoying peak sun, and solar energy is producing nearly 57 percent of German electricity—31.6 GW out of a total instantaneous demand of 55.7 GW. In any interval during which solar is able (and permitted) to do this kind of work, German GHG emissions are on the low side, something like ~150–300 gm CO_2/kWh.[15] But when the sun is not shining, like at night (which averages 12 hours over the year), German emissions reach 650–700 gm CO_2/kWh or more. During such periods 40–50 percent of German electricity comes mainly from coal and NG.

---

[14] Note that total capacity outweighed average demand in December 2023 by more than 4 to 1 (250 versus 55). In contrast, total grid capacity in the US outweighs average demand by only about 2.5 to 1, because the US excess is only to accommodate high demand peaks (needed by every electricity system), while the German excess is largely to fill in for the intermittency of their renewables. Update: Germany's solar and wind capacity has now increased from ~147 GW to ~172 GW—100 GW (solar) plus 72 GW (wind).

[15] The grams of CO_2 produced per kWh by a particular country at any moment in time depend on the specific mixture of sources. Respectively, these are responsible for approxi- mately 820–1,152 (coal), 490–593 (NG), 26–35 (solar), 11–13 (wind), 11–24 (hydro) and

In neighboring France solar is generating (at the time of this writing) nearly 18 percent of French electricity. With this sun, French emissions clock in at 26 gm CO_2/kWh (roughly one-tenth that of Germany's sunny daytime emissions). But even when the sun is not shining in France, emissions only rise from 26 to 46. So solar power can reduce CO_2 emissions in both cases—in Germany from ~700 to ~250, in France from 46 to 26, but only for part of the 24 hours.[16] The difference then is almost entirely due to France's use of nuclear power as baseload electricity, allowing it to virtually eliminate coal and minimize NG (though transportation in France, and everywhere else, is still heavily dependent on oil, with its gm CO_2/kWh roughly 2/3 that of coal).[17]

At the other end of the spectrum, sunny California fares better than Germany. Its electricity is heavily dependent on NG. California uses no coal, but in the Sidebar below we show that, because of methane leakage, a switch from coal to NG offers little to no improvement in GHG emissions for a couple of decades.

Referring to actual production rather than capacities, over the last year the electricity map shows that California got 20 percent of its electricity from solar (with wind providing 11 percent and hydro 8 percent), but it still got 44 percent from NG, and the other 17 percent from a mix of nuclear, geothermal, biomass, and imports. The NG capacity has even been upgraded in the last few years, due in part to the irrational closure of the San Onofre nuclear plant in 2013. This leaves just one nuclear plant with two reactors, at Diablo Canyon, while Germany has shuttered all seventeen of its reactors.[18]

5–12 (nuclear), so that each country's CO_2 emissions will be somewhere between around ~5 and 1,100 gm CO_2/kWh. That the last four involve any CO_2 at all is due to their current dependence on coal or NG for the construction, maintenance, and operation of the conversion devices over their lifetimes. If nuclear were permitted to power everything, including its own construction, its CO_2 emissions would be zero (given new methods of making cement, the only remaining source of CO_2 emissions from a nuclear plant).

Over the last 12 months France's average has been 48, and Germany's 420; Sweden's 22 and Poland's 779. The relatively low figures for France and Sweden are due to their great dependence on nuclear or nuclear and hydro, respectively, while, despite contributions from wind and solar, Germany and Poland depend largely on coal. Again, none of this includes the methane (NG) leakage that inevitably accompanies the use of coal, NG, and oil (see the Sidebar below), which makes Germany and Poland even worse.

[16] On November 30, 2023, during evening hours, Germany's emissions exceeded 800 gm CO_2/kWh.

[17] There was recently a plan to cut back on the fleet of French nuclear plants, but as of this writing President Macron's government has fortunately abandoned that plan.

[18] The seventeen reactors were all built within 20 years, from 1969 to 1989. They were located in twelve power plants, five of which contained two reactors each. The 2011 Japanese earthquake and tsunami, and their effect on the Fukushima Daiichi nuclear plant

Shutting down nuclear plants inevitably increases GHG emissions, despite the growth of wind and solar. Moreover, it inevitably increases the cost of electricity to the consumer (https://tinyurl.com/484ce6bc). The fallacy that wind and solar are cheaper than nuclear comes from cherry-picking only a part of the entire life cycle, as we explore in Chapter 5.

In the 1970s and 1980s France's rapid transition to clean electrical power, through nuclear, was possible because a nuclear plant fully replaces a fossil-fuel plant and can occupy the same site with suitable modification. The French state, lacking any significant domestic sources of fossil fuels, was motivated not by environmental concerns but rather by the desire for "energy independence." They nevertheless incidentally demonstrated that only through a significant contribution from nuclear can electricity be reliably and abundantly supplied without generating GHGs, and without excessive monetary expense. In contrast, paraphrasing Robert Kennedy Jr.—an avid public advocate of renewable energy and former investor in NG—solar and wind farms are in fact gas plants (or coal plants). In short, despite illusions to the contrary, electricity derived from wind and solar is not "clean energy." (We return to Kennedy in Chapter 6.)

Are NG Plants Cleaner than Coal Plants?

Because NG emits less CO_2 than coal (per kWh generated) when each is burned, a switch from coal to NG is intended as a method of avoiding nuclear expansion. There are even pronuclear energy analysts who believe that NG is the right bridge while nuclear is being expanded and phased in to replace fossil fuels and the inherently intermittent renewables altogether. This belief, in effect, mistakes CO_2 emissions for carbon or GHG emissions. But CO_2 is not all of carbon (and carbon is not all of GHG). Based on this mistaken concept, certain countries, including the US, have replaced some coal plants with NG plants, sometimes by adapting the old plant for its new fuel, rather than building a completely new facility.

---

prompted demonstrations by the large German antinuclear movement demanding the shuttering of Germany's nuclear reactors. The politicians acceded, and the last three reactors were shut down on April 15, 2023. It seems doubtful that then Chancellor Merkel, given her physics degree, was ignorant of nuclear's advantages, so deference either to the antinuclear movement or to Germany's coal magnates with their prodigious underground resources, or perhaps both, prevailed over common sense, and Germans (and the rest of the world) are paying for it heavily.

The substitution of NG for coal has led those countries to boast of a reduction in their carbon or GHG emissions. This boast is advertised in government announcements and media reports, and it is widely believed by the public, and perhaps by government officials and journalists. But all they have done, at best, is to reduce their CO_2 emissions, which has done nothing to reduce methane leakage, and methane is the main ingredient of NG. In other words, methane burned is not the same as methane leaked, and leakage accompanies extraction, pipeline transmission, processing, and power plant usage. In fact, methane leaks with the extraction of coal and oil as well as that of NG/methane itself. As the lightest of hydrocarbons (a single carbon atom with four hydrogen atoms, CH_4), methane is the most volatile, and this gas generally accompanies the liquid (oil) and solid (coal) forms.

Therefore, accurate accounting for total carbon emissions from newly converted NG plants requires inclusion of NG/methane leakage along with the combustion that turns CH_4 into CO_2. Given the present rate of leakage, which globally averages roughly 4 percent, it takes some 20 to 30 years before the substitution of a coal plant by an equivalent NG plant results in a significant reduction in total carbon emissions. Longer where the leakage rate is greater than 4 percent. Only decreased or eliminated leakage would offer an immediate gain from a switch, but there are millions of unidentified and uncapped leakage sites.[19] And this is aside from other sources of methane, such as melting permafrost and cattle burps.

Although not acknowledging methane leakage, Robert Bryce correctly indicates an even larger problem—the rapidly increasing coal usage by China and India alone, but also Indonesia and other industrializing countries. Their rising coal usage outweighs the reduction in CO_2 emissions of the other four of the world's six largest economies—US, UK, Germany, and Japan (https://tinyurl .com/mtexrpvf). The reduction in the CO_2 emissions by these four is mere noise, swamped by the signal from the industrializing world, whose nations look first to coal, even as some are also building nuclear plants.

While the growth of coal usage currently outweighs methane's contribution, it is still necessary to understand the role of the leakage. The following Sidebar explains the dilemma. Because it is the gross effect of methane that matters, we present relatively accurate graphics rather than precise numbers and complicated formulas.

---

[19] "Methane Hunters," *Scientific American*, September 2021, p. 62.

Sidebar: Replacing Coal with NG (Methane)—Burning versus Leakage

While it varies somewhat depending on the method of burning NG and the grade of coal used, NG emits approximately 50 percent as much CO_2 as coal *for the same energy return*. However, the main constituent of NG, methane (CH_4), leaks from ground to atmosphere during extraction of any fossil fuel—fracking NG, pumping oil, or mining coal. Methane also leaks during NG's ensuing stages—distribution through pipelines, preparation of fuel, and electricity production by power plants. That is, methane leaks even when burning NG, whether at a power plant or in your gas stove or furnace. Thus, use of NG incurs greater leakage than use of coal (or oil), approximately twice as much as coal.

Methane is a much stronger GHG than an equal weight of CO_2. But over a few decades, a puff of methane in the air slowly turns into CO_2(and water) through chemical reactions, so its greater GWC (global warming contribution) is of shorter duration.[20] In contrast, CO_2 leaves the air mainly through absorption by ground and oceans, a much slower process, perhaps centuries. The dwell time of methane has a half-life of 9.1 years, while that of CO_2 is not accurately known. In other words, if no further methane were added to the atmosphere, half would disappear every 9.1 years—an exponential decrease. But since, as long as we continue to use fossil fuels more methane will leak, its addition and disappearance combine, which we explore momentarily.[21]

The Intergovernmental Panel on Climate Change (IPCC) estimates that a single ejection of methane has 84 times as much GWC over the first 20 years as a single ejection of the same weight of CO_2, dropping to 30 over the first 100 years because of its faster disappearance. These two intervals were chosen arbitrarily, merely to quantify the gradual diminution of methane's GWC. In other words, 20 and 100 have no particular physical significance—any pair of intervals would serve.

Many defenders of NG versus coal claim that because methane spontaneously disappears from the atmosphere much faster than CO_2, its effect on AGW can be neglected. But while this may be true for a single ejection of each gas, with the amount of CO_2 vastly greater than that of methane—an idealized hypothetical—it does not apply to continual emission/leakage of both.

---

[20] We will define the contribution of each gas to AGW over time as its global warming contribution (GWC). This is different than the relative global warming potential (GWP), for which that of CO_2 is defined by the IPCC as 1, regardless of how great its contribution. GWP is useful for relating the AGW impact of another gas relative to that of CO_2 at any point in time, but it precludes description of the changing impact of CO_2 over time—hence the need to define GWC.

[21] If CO_2 emissions and methane leakage were to cease today, to decrease CO_2's atmospheric concentration within a few decades would require active removal, while methane could be allowed to disappear on its own.

To estimate the effect on AGW of continual burning-plus-leakage, we admittedly make an assumption. We assume that the relative contributions to AGW over any time interval following a single pulse of the two gases follows the same time course as the instantaneous contributions at the end of that time interval after the onset of continual emission/leakage. That is, the GWC of a *single pulse* of each gas integrated *over a time interval* versus the *continual addition* of each gas examined at *a point in time at the end of that interval*, with these two approaches assumed to yield roughly equivalent results mathematically. And we use the IPCC's estimates of the relative GWC between the two gases at 20 and 100 years.

While estimates of global average methane leakage tend to converge on 4 percent, some studies estimate leakage in certain locales to be as high as 9 percent (https://tinyurl.com/2r9hma6e). That is, 4 percent of NG molecules leak before they can be burned. If the leakage could be stemmed, the effect of methane could be greatly diminished, but virtually nothing effective is being done at present to control it in most places in the world. Nor is anything effective being done to capture CO_2 after burning methane.

The September 2021 *Scientific American* article cited in footnote 19 mentions leakage from some 1.6 million uncapped oil wells in the US alone, some of which have simply been abandoned ("orphaned") by oil companies. Only about 2,400 such wells were capped in the US a few years ago, a trivial proportion. Recall the infamous 4-month leakage of methane in 2015 from Porter Ranch in the northwest suburbs of Los Angeles, in which 100,000 tons were accidentally released to the atmosphere, constituting the largest methane leak in US history. How many smaller leaks have failed to garner this degree of publicity is unknown and probably unknowable. So for purposes of illustration we settle first on 4 percent and then review the effect of doubling that rate to 8 percent.[22]

Figure 3-5 shows the relative GWCs of CO_2 and methane (CH_4) associated with a continually emitting NG plant and a power-equivalent coal plant. The contribution of the CO_2 associated with the NG plant is roughly half as high (50 percent) as that associated with the coal plant. At 20 years the GWC of each molecule of CH_4 is shown as only 1.27 times that of the CO_2 for the NG curve. The IPCC's ratio 84 applies to an equal weight of the two gases (not equal numbers of molecules). To compare molecule for molecule this factor has to be reduced twice, first, because only 4 percent of the NG (methane) molecules leak, leaving 96 percent to be burned (96/4 = 24), and second, because a methane molecule weighs only 16/44 (= 0.364) that of a CO_2 molecule, so there are more methane molecules than CO_2 molecules in an equal weight (1/0.364 = 2.75 times as many). Thus,

---

[22] The warming due to fossil fuels incurs the amplifying feedback of melting permafrost at high latitudes as well as undersea clathrates (methane trapped in ice), both of which release prodigious amounts of methane to the atmosphere. Neither this source of methane nor that from agriculture and livestock are counted by fossil fuel defenders, but they all add to AGW, making removal of significant amounts of CO_2 even more urgent.

84/(24 x 2.75) = 1.27 per molecule. The same reductions apply at 100 years, reducing 30 to about 0.45, as shown in Figure 3-5.

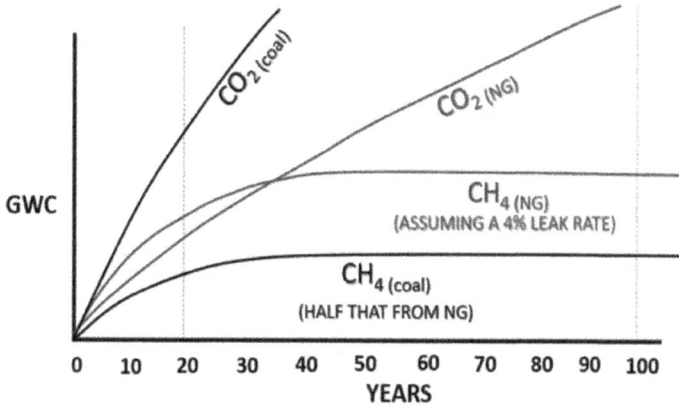

FIGURE 3-5

Rising curves that tend to level off to a plateau are characteristic of a continually injected substance that simultaneously undergoes an exponential disappearance—whether it's a gas emitted into the atmosphere or a medicine injected into a vein. CO_2 behaves the same as methane, but because of its much slower disappearance rate (longer dwell time), it takes much longer to reach its plateau, far outside the range of the graph. As a rule of thumb, the time to reach the plateau is approximately equal to five times the disappearance half-life. Rate of injection and rate of disappearance are independent of each other, the former under human control, the latter a matter of physics and chemistry.

The net GWC associated with the coal plant can be pictured as the sum of the two coal curves and that for the NG plant as the sum of the two NG curves, as shown in Figure 3-6 (reduced in height to fit on the page).

Note how much slower the coal and NG curves depart from each other, such that at 20 years the NG is roughly ¾ that of the coal curve rather than the ½ that would obtain if there were no methane leakage.

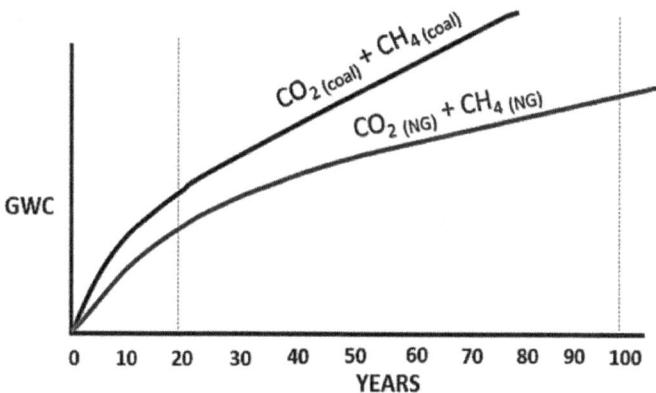

FIGURE 3-6

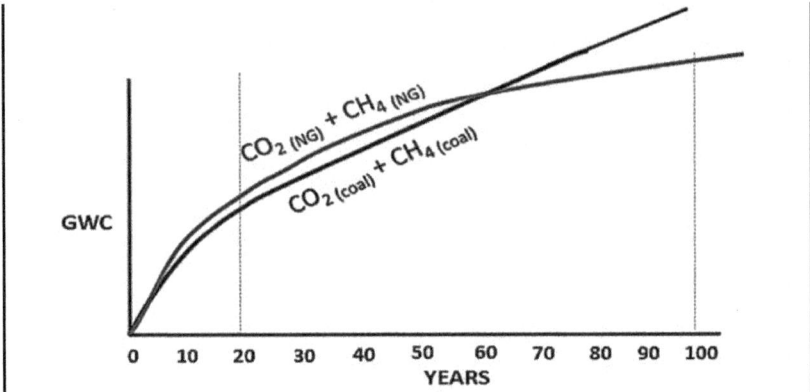

FIGURE 3-7

If the leakage rate were less than 4 percent, there would be more advantage to switching from a coal plant to an NG plant, but if it were more than 4 percent, say 8 percent, the emissions from the NG plant would actually have greater net GWC than the coal plant for the better part of a century, as shown in Figure 3-7.

The Confusion of Possibility with Certainty

To navigate around intermittency, Jacobson claims his computer program can match supply from an all-RE electricity system with instantaneous demand 30 years into the future (it was 10 years in the debate with Brand). His program, LOADMATCH, ostensibly demonstrates that, despite the intermittency, demand and supply can be precisely matched every 30 seconds, decades into the future. Jacobson (PNAS, 2015) explains that they run the program over and over, each time adjusting assumptions about the effects of AGW, population growth, and other factors. Each run takes only a few minutes, and for some runs supply and demand match in every one of the more than two million half-minute intervals over a two-year span. They then select the cheapest successful run and claim to have demonstrated the feasibility of an all-RE electrical system, calling for its construction over 35 years. Jacobson implies that this proves not just the possibility that an all-RE system could work, but the certainty of its success.

He also assumes that businesses will postpone, for up to 8 hours, processes that can be delayed. He leaves open whether this will be voluntary or coerced by the state, or even whether it will be coordinated, or whether a maximum delay of 8 hours will always suffice.

Addressing the kind of management Jacobson calls for in earlier versions, Weißbach et al. (2013) say,

> Adapting the demand to the output at all times . . . is obviously not acceptable, because one becomes dependent on random natural events (wind and PV solar energy). A developed and wealthy economy needs predictably produced energy every time, especially the industry needs a reliable base-load-ready output to produce high quality goods economically.

The credulity of Jacobson's admirers rests in part on the confusion of a model with reality. So let's review the essential features of a model.

What Is a Model and How Is It Validated?

A model is a theoretical construct that purports to mimic some feature(s) of the real world.[23] A model can be quantitative and/or qualitative; it can explain structures (spatial) and/or processes (temporal). Models can help us understand how the real world works and how it has changed over time by making visible certain structures and functions so that their interrelationships can be comprehended. By studying past and present, models can also help us plan for the future. But a model is distinct from the real world, no matter how closely it may mimic it, and the future is distinct from the past and present, no matter how closely it may replicate them.

To judge how closely a model approximates the real world, it has to be validated, which can only be done by comparing its implications with past or continuing events. A model is adjusted as many times as necessary until it reflects as accurately as possible the past or the present, at which point the model is considered valid, at least for all practical purposes for the time being. However, an important feature of the past is that it has already happened, so records of the past are generally available. In contrast, the future has not yet happened, so records are nonexistent.

---

[23] Since dismissive talk about models plays a major role in climate denialism, it is worth noting that we use models all the time, usually without realizing it. Even language, for example, is a model. Words are not the things they describe. They are representations of those things, representations that people come to agree upon when they learn a language. Our very consciousness consists of internally modeling an image of the world surrounding us that integrates the input of all our senses—a capability that is a latecomer in animal evolution (Godfrey-Smith 2016). In short, as with things like air and radiation of which we may be unaware, we live in a sea of models and cannot do without them anymore than we can do without paradigms (discussed further in Chapter 11).

To be useful for planning future events, as Jacobson's model claims to be, validation through comparison with past or current events can at best be a first approximation. A model's relevance to the future makes assumptions about changes in the future and therefore must always be regarded as tentative. Validating its accuracy for the future can only be done when the future arrives. Any claim that a model will represent the future with certainty ignores the possibility of unexpected events.[24] Thus, Jacobson's claim for the accuracy and precision (both defined in the following section) of his modeling of the future, and not just one hour from now, or tomorrow, but decades into the future, is by its very nature unjustified and unjustifiable.

As Palmer and Floyd (2020) put it (p. 120, the two citations within the quote and the emphases are in the original):

> It is possible to demonstrate mathematical and internal consistency within energy-economy models relating to envisaged futures, or to internally *verify* that the model's code is consistent with the modeller's conceptual understanding of the situation under investigation, but it is impossible to *validate* that the model outputs represents [*sic*] the "truth" [by referring only to the model's internal features] (Oreskes et al. 1994). In the scientific literature, the peer-review process generally ensures that there is a lower bound to numerical verification, but the process says nothing about the "truthfulness" of the model. It is only through the process of actually building and operating the proposed infrastructure and plant within the envisaged socio-economic context that the "truth" can emerge. A useful rule of thumb is that claims [of validity] based on conceptual modelling should be treated with a healthy dose of scepticism (Alexander and Floyd 2018).

Uncertainty is inherent in projections of models into the future, no matter how closely they have reflected the past. Jacobson provides many predicted quantities with completely unjustified precision, implying the virtual absence of uncertainty in outcome. This can give an unfounded impression of accurate foreknowledge. Before we examine the precision implied by Jacobson through his number of significant figures, the follow-

---

[24] Analogously, even wind tunnel tests of a new aircraft design do not substitute for test piloting the finished plane. It takes special training and skill to be a test pilot because of the ever-present possibility of unanticipated events that did not occur during wind tunnel and other tests. Similarly for the design of nuclear reactors, computerized vehicle controls, bridges, buildings, medicines, and any other invention.

ing Sidebar explores the use of significant figures and the difference between accuracy and precision.

Sidebar: Accuracy, Precision, Uncertainty, and Significant Figures

Reality exists out there, independent of what we may know of it. We learn about the quantitative aspects of reality through either counting or measurement, depending on whether the quantities are discrete or continuous (at the relevant scales). Either may be performed more or less accurately. Accuracy is a relationship between reality and our counting or measurement.

Precision is something entirely different and independent. With counting we can be very precise, because our estimate is a whole number. But precise or not, it may or may not be accurate. For example, as the joke goes, Margie and Alan are riding in a train and pass a flock of sheep. Alan wonders aloud how many sheep there are, and Margie replies "217." "How did you count them so quickly?" asks Alan, astounded. Margie answers, "I simply counted their legs and divided by four." Now, whether Margie is right or wrong about the number of sheep, a matter of accuracy, her answer is perfectly precise. She doesn't say, "Somewhere between 215 and 219," which would be less precise. In fact, the owner of the flock knows that there are really 220 sheep, which means that Margie's answer is precise but not quite accurate. Had she said, "220, plus or minus 2," she would have been accurate but not entirely precise.

When we measure continuous quantities, on the other hand, we're necessarily imprecise, even if we are fairly accurate. Measurement is ultimately a side-by-side comparison of an arbitrarily dubbed standard quantity, like a length or a mass or a time interval, that is embodied in a material object. For example, by international agreement two objects, among others, have been held since 1875 under tightly controlled temperature and pressure conditions in the International Bureau of Weights and Measures in Paris. One is a platinum ruler with two marks engraved on it a fixed distance apart and the other is a platinum alloy cylinder. By international agreement the distance between the two marks on the ruler is defined as the standard meter and the mass of the cylinder as the standard kilogram. Through a chain of validating events, similar objects, held for convenience in other countries, have been compared to these two standard objects—the rulers held side by side and the cylinders balanced on a scale—to verify that they have the same length and/or mass as the two standard objects.

So to determine the length of our SUV, we hold a measuring tape alongside the vehicle and find that the length is 186 inches, plus or minus half an inch. This measuring tape has been validated for its accuracy through a chain of events that ultimately compared the distances between the various markings directly with the standard meter in Paris. However, each time a comparison is made, there is a possibility, almost a certainty, of a very slight and undetectable error. Therefore, the accuracy of a measurement made with this tape is slightly uncertain. But when we express the measured length of the SUV as 186 inches plus or minus half an inch, whether or not it is accurate the pronounced length is not entirely precise. We can, if necessary, try to improve our chances of obtaining an accurate number by measuring several times, perhaps by different people, and averaging the results, which can lead to both greater accuracy and greater precision.

Uncertainty of a measurement or a count, then, is expressed as the degree of precision. So the greater the imprecision, i.e., the wider the "plus or minus" range, the greater the uncertainty. Conversely, the greater the precision, the smaller the uncertainty. When no such expression of uncertainty appears, the explanation is not quite complete and is misleading, whether intentionally or not.

Imprecision in a count, like in a measurement, can also be expressed as what is called a "confidence interval," such as when a published poll gives an estimate of how many people intend to vote for candidate A and how many for candidate B (assuming only two candidates). The pollsters interview a randomly selected sample of the voting population. By doing so, they can be entirely precise as to the percent who say they intend to vote for A or B. But because this is only a sample of the voting population, they may or may not be representative, though the larger the sample (the closer to surveying the entire population) the more likely the results will be representative of the entire population. This is why the results of such surveys are always given as something like "plus or minus 3 percentage points," with the uncertainty larger if fewer people are interviewed and smaller if more are included. If we could sample the entire voting population, our result could be entirely precise, but, assuming the counting is without error and reported honestly, it may or may not accurately represent the outcome on election day, because people may not be truthful or they may change their minds or they may fail to vote or some other influence may render the actual vote different from the polling estimate.

Engineers and physicists, among others, have a second way to express the degree of precision or imprecision of a measurement, and that is through the number of significant figures. To say "plus or minus 1 inch" is an explicit statement of the degree of precision; to say that the length of the SUV is 184.0 inches is an implicit statement of the degree of precision, because the conventional meaning of 184.0, particularly among engineers and physicists, is a number between 183.95 and 184.04, which range brackets the expressed number. Every number between those two, if rounding off by one decimal place (digit), will yield 184.0. But if all we know is that the length of a car is somewhere between, say, 178 and 182 inches, then it should be expressed as 180 +/- 2 rather than 180, which stands for a number between 179.5 and 180.4. Similarly, 180.0 stands for a number between 179.95 and 180.04. The latter proclaims a tighter uncertainty interval than the former, and for the car length neither degree of precision is warranted in this example.

In contrast, Margie's pronouncement of her count as 217 does not imply merely a number between 216.5 and 217.4 sheep, because the number of live sheep in a flock is the result of counting rather than measuring and is necessarily a discrete whole number.

So when dealing with measurements rather than counts, the number of significant figures serves as an expression of precision. For example, the number 0.02570 is more precise than 0.0257 (without the zero at the end). The first of these has four significant figures, and the second has three. That is, the zeros immediately following the decimal point only establish order of magnitude but not precision. Thus, 0.00257 has the same precision as 0.0257, but is one order of magnitude smaller (the zero before the decimal point is conventional but does not contribute). When the digits all occur before the decimal point, as with 396,000 (the decimal point is understood to be at the end), the zeros establish order of magnitude. The first zero may or may not be a significant figure, but the second and third are less likely to be significant. So 396,000 has three (or possibly four) significant figures and may have the same precision as 396, but it is definitely three orders of magnitude larger.

Now we're prepared to evaluate the validity of Jacobson's measurements, which he often provides with four, and sometimes five, significant figures. For example, in his 2015 paper, he gives the total electrical demand in the year 2050 with five significant figures (1,572.8 GW, Table 1), and in his 2022 paper, he gives the proportion of total supply provided by solar PV farms with four significant figures (30.77 percent, Table 3). And countless other measurable quantities in these complex papers are similarly expressed with unjustifiable precision. There is no way that the expressed precision in these numbers could possibly be valid, given the fact that they are mere predictions rather than actual measurements of existing quantities.

Expressions of such extreme precision can certainly serve to impress those members of his audience who are less experienced in the meaning of significant figures—impress them with the illusory implication of unobtainable accuracy in his estimation of outcomes a few decades in the future. In other words, expressions of extreme precision imply measurements possessing extreme accuracy, with little uncertainty. Jacobson apparently either does not care or has not recognized (is it possible?) that those members of his audience who do have the experience might see through his claims. But the slim likelihood that the lay members of his audience will ever come into contact with the criticisms by those with more experience stands as a layer of protection for his reputation as a trustworthy advocate for an all-RE electrical system.

Nor does LOADMATCH take account of such things as the distinct possibility that by 2050 the large hydroelectric dams will no longer have enough water behind them to provide reliable electric power. Along the Colorado River, Lake Mead (the largest fresh water reservoir in the US) and Lake Powell (the second largest) have already (as of early 2023) dropped more than 180 feet below their peak heights in the 1980s.[25] And because of the trend of annually declining average snowpack in the Rockies and earlier spring melting and runoff, not all of which enters the Colorado River, these levels continue to drop, some 7 to 8 feet per year. Consumption of drinking water plus surface evaporation (the latter accounting for approximately 6 of those feet each year) are not being fully replaced. Since hydroelectric power is based on the height of water behind the dam, Hoover Dam has lost almost one-third of its approximately 2 GW capacity when Lake Mead is full. If the lake loses another 150 feet or so, the dam's ability to generate power will be completely lost. Similarly for Glen Canyon Dam at Lake Powell, upstream from Lake Mead and the Grand Canyon.

A picture tells some stories more clearly than words. The photos show the same northernmost portion of Lake Mead from satellite images as its volume has declined over the 22 years from 2000, when it was near peak, to 2022, including the drop in the last year (https://tinyurl.com/bddyz2dr).

While Lake Mead has lost about 15 percent of its height, it has lost more than 70 percent of its volume.[26] This climate-related diminution affects the lake's ability to provide both electrical power and fresh water for roughly 25 million people in the surrounding seven states and part of Mexico. Water regulators are already suggesting that the tens of millions of people who depend on these two reservoirs for clean water will have to cut back their water usage by 25–30 percent, for a start (https://tinyurl.com/2mpmjwp4).

Yet Jacobson projects that hydro (both hydroelectric from dams and pumped hydro) will remain available to buffer the vagaries of electrical supply from wind and solar farms. He predicts that the available power capacity of hydroelectric in 2050 will be 88.8 GW (2022, Table 3). This exceeds the yearly average power capacity over the last two decades in the US, which never even reached 80 GW—ranging from just under 77 to just under 80 GW—with

---

[25] By the end of 2023 there was slight recovery in each due to a wet winter early in the year, but drought conditions remain and the trend is still downward.

[26] The relationship between height and volume depends on the contour of the lakebed. It is widest at the top where each foot of height loss represents more loss in volume than a foot of height loss at lower levels. The loss in height is more directly relevant to the availability of potential hydroelectric power, while the loss in volume is more directly relevant to the availability of fresh water for drinking and other uses. Continuing drought is diminishing both.

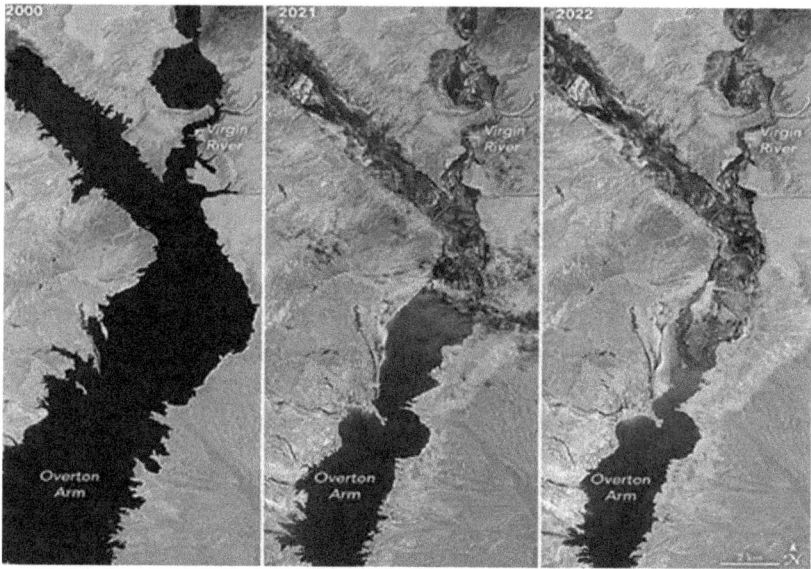

pumped hydro growing slowly and amounting to just under 23 GW of those totals, as of 2020 (https://tinyurl .com/3s8mh9t8). The AGW-driven diminishing availability of hydro alone is invalidating the Roadmap's predictions.

For these reasons and numerous others, Jacobson's modeling of future match of supply and demand, with both successful and unsuccessful computer runs that are wholly dependent on his assumptions, should not be confused with what will actually happen.

An important source of improvements in engineering and design is the discovery of the unintended consequences of previous attempts to develop and deploy new technologies. In daily life this is encountered as trial and error. As US engineer and popular author Henry Petroski (2004) says (paraphrasing), the mother of invention is less necessity than the occurrence of failure.

No one can anticipate every surprise that may arise. Any engineer or scientist, or indeed any prognosticator, who pretends that projections into the future, particularly by several decades, will be free of uncertainty and error, and that therefore their proposed plan shares no features with a trial, is deserving of no one's trust.

Finally, while many environmentalists, would-be energy analysts, and hopeful entrepreneurs fail to "do the math"—a widely used phrase in critical writings—Jacobson and his colleagues certainly do math, and plenty of it. But they do lots of wrong math, and in so doing impress the unwary reader with their appearance of care and attention to detail. This derails many in their audience from approaching this work critically. Whatever Jacobson's intention, the sheer abundance of the math has the effect of snowing much of the audience, and, being charitable, perhaps even the authors.

4

Material Resource Requirements: Technical and Geopolitical Limitations

We touched on the material and land requirements of various energy sources in Chapter 2. Here we present more detail on these resource implications for wind/solar, leaving the requirements of energy and labor for the following chapters.

Materials Used for Wind and Solar

The permanent magnets in wind turbines require a small but critical amount of so-called "rare earth" metals (see Glossary), used also in electric cars, hybrid vehicles, and stereo speakers, among other devices. While not crucial for solar PV panels, rare earths have been found to improve their performance. The multiplicity of applications exerts pressure on world reserves, which amplifies geopolitical conflicts, while these minerals are also indispensable for the weapons often used to gain access to the reserves.

To generate electricity, wind energy is converted through electric turbines driven by propeller-like blades mounted tens of stories high with blade lengths comparable in dimension.

Solar energy must be converted either through PV panels that turn incident sunlight into immediate electricity or through mirrors

83

directed at heat retaining fluids or solids that store the heat for later deployment to generate steam to drive turbines (concentrated solar power or CSP).

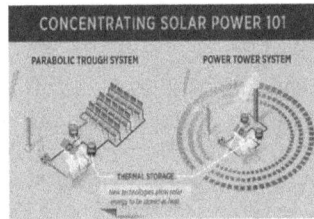

The growth of PV is greatly outpacing that of CSP (https://tinyurl.com/5xhunrsd), with commercial solar farms outpacing rooftop installations, as large aggregations use installation processes more efficiently.

Magnitude of Materials for Wind and Solar and Obstacles to Their Accessibility

The Roadmap calls for over half a million additional 5MW wind turbines, between 2015 and 2050 in the US alone, with a total nameplate power of 2,580 GW. Given an average annual CF of about 30 percent, those added turbines would represent a growth of actual average wind power from about 22 GW in 2015 to a projected 796 GW in 2050—a roughly 36-fold increase in 35 years. The total buildout of wind and solar farms over the last 40 years or more would have to be duplicated each year on average for the next 35 years. As of this writing (2024) this massive acceleration has not even begun, requiring an even greater speed of expansion from the time of this writing to 2050.

An additional 2,580 GW of wind (nameplate) would require something like 464,400 tonnes (510,800 tons) of rare earths alone

(~180 tonnes per GW)[1]—all of which would need replacing by 2050. Mark Smith, CEO of Molycorp (a rare earth supply company), unsurprisingly gives an estimate more than twice that of an executive from a wind turbine manufacturing company. The reality may lie somewhere in between.

Over the 35 years from Jacobson's original projection, and using Stiesdal's more conservative estimate, this would average 13,270 tonnes (14,600 tons) of rare earths per year, but Jacobson envisions building much faster in the early years and slowing only later, when approaching completion. So in the early years this 14,600-ton annual requirement would be greatly exceeded.

Jacobson ignores the fact that China's rulers corner 90 percent of the world market for extraction, preparation, and production (Figure 4-1), and that the US already imports for all uses 11,000 tons of rare earths from China annually (70 percent of US usage). The Roadmap fancifully calls for more—much more—than 14,600 tons in addition.

Where Clean Energy Metals Are Produced

Production of key resources is highly concentrated today. Charts show the top three producers.

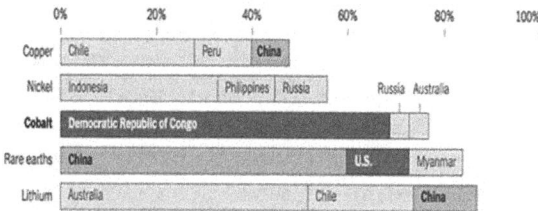

	0%	20%	40%	60%	80%	100%
Copper	Chile		Peru	China		
Nickel	Indonesia		Philippines	Russia	Russia	Australia
Cobalt	Democratic Republic of Congo					
Rare earths	China				U.S.	Myanmar
Lithium	Australia			Chile		China

And Where They Are Processed

China dominates the refining and processing of key metals.

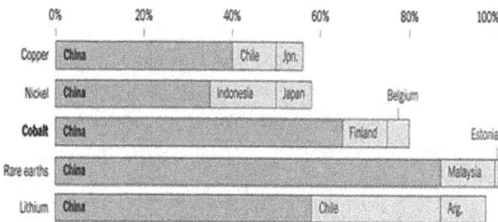

	0%	20%	40%	60%	80%	100%
Copper	China		Chile	Jpn.		
Nickel	China		Indonesia	Japan	Belgium	
Cobalt	China			Finland		Estonia
Rare earths	China					Malaysia
Lithium	China			Chile		Arg.

Source: International Energy Agency · By The New York Times

FIGURE 4-1

---

[1] Kalantzakos 2018, p. 190, footnote 66, quoting Henrik Stiesdal, the Chief Technology Officer of manufacturing giant Siemens Wind Power.

Conley and Maloney (2017) point out that Jacobson's plan for the US alone would consume 33 percent of the world's known copper reserves and 90 percent of the silver reserves. While new deposits might be discovered, world supply imposes severe limitations on build rate, if not on the entire project. Furthermore, Jacobson's all-RE plan would deprive other applications of these minerals, pending discovery of alternative substances. Since the most advantageous deposits are generally exploited first, the ensuing cost increases amplify the problem.

Thus, an all-RE plan, due to its low energy and power densities, magnifies resource depletion and collides with accessibility of world supplies. The inescapable need for energy storage, particularly for lithium-ion batteries, would challenge world supplies that much more. In Chapter 10 we discuss the geopolitical obstacles that, if anything, overshadow these technical considerations.

The Negative Environmental Impact of Waste

The magnitude of waste over time is directly related to the material requirements and lifespans of conversion devices, plants, and storage apparatus, putting aside for the moment the waste from fuels. If the lifespan of a wind turbine or solar PV panel is roughly one quarter that of a nuclear reactor and plant, this adds another factor of 4 to their waste stream—mitigated to the extent that some of the material is reused or recycled. Figure 2-3 provides the material requirements not for each lifespan, but rather for a given amount of energy output (TWh) regardless of how many lifespans that requires, so the shortened lifespans are already accounted for.

The nuclear reactors shown in Figure 2-3 are LWRs. Small modular reactors (SMRs) require even less material per MWh than LWRs by an average factor around 2, turning ratios of 18 and 11 into roughly 36 and 22 times as much mass of waste from solar panels and wind turbines, respectively, versus the potential waste from nuclear plant structures, for the same actual energy produced, though much of the materials in nuclear plants would last longer than a century.

Figure 4-2 shows yet another estimate of relative waste, in terms not of mass but of volume. Volume considerations, more relevant for disposal sites than weight, give even higher ratios for waste disposal—300 to 1 for solar versus nuclear. Moreover, solar panel waste is quite toxic (chemically), never decays, and is found to leach from landfills into soil and water supplies.

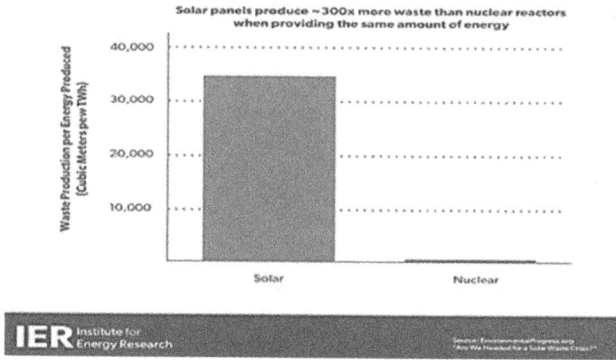

FIGURE 4-2

Turning to fuel, which excludes renewables, the US has accumulated almost 100,000 tons of once-through nuclear fuel over more than five decades.[2] Some is in cooling pools at power plants and some in onsite concrete/steel casks. If the entire 100,000 tons were in casks, they would fill a football field to a height of less than 30 feet (https://tinyurl.com/3tbjemzc) or a small corner of a Walmart warehouse (https://tinyurl.com/4bked2en). The roughly 40,000 cubic meters (m^3) produced something like 20,000–30,000 TWh over the five-plus decades, for a concentration of some 1–2 m^3/TWh (for fuel). Compared to the potential nuclear *plant* waste of approximately 120 m^3/TWh (1/300 x ~35,000, Figure 4-2), fuel adds little (confirming the conclusion in Chapter 2). So 300 to 1 solar to nuclear waste (fuel + plant) volume is the right order of magnitude.

Unlike once-through fuel from LWRs, many countries, including major producers and consumers of solar panels, have no plan to dispose of solar waste safely. China produces twice as many panels as the US and has no such plan (https://tinyurl.com/2p865ztb). Japan has handed the solar farm managers voluntary control over their own waste disposal with unenforceable guidelines (https://tinyurl.com/2p8m2htw). California, with the greatest solar nameplate capacity in the US, has recently ruled that PV panels can be treated as "universal" rather than "hazardous" waste (https://tinyurl.com/2p8mmmue).

---

[2] We deal with the composition and relative safety of once-through nuclear fuel in Chapter 7. Claims of the extreme hazards of nuclear "waste," central to the antinuclear narrative, are simply false and encompass serious misunderstandings fostered by deliberate fabrications.

The Monetary Cost of Materials and Land for Solar Panels

Let's begin with Elon Musk's solar reverie, as celebrated by Michael Mann in our opening epigraph. Musk claims that solar could supply the world's power with just enough batteries to compensate for night-time, assuming every day is sunny. However, his meaning is ambiguous. He refers both to "energy," by which he could mean all primary energy, and "electricity," which is roughly one-sixth to one-fifth of primary energy. So we'll do the calculation both ways, first for electricity alone and next for total primary energy. (Throughout these calculations we often provide one more significant figure than is warranted. We do this to avoid compounding round-off error, but again, what matters is order of magnitude. We discuss significant figures in the final Sidebar in Chapter 3.)

- Given that the radius of the earth is approximately 4,000 miles, the total area that the earth presents to the sun as a disc is πR^2 = **5.02×10^7 mi^2 = 1.29×10^{14} m^2**.

- The areal density of solar power over this disc, arriving at the *top of the atmosphere* is **1,360 W/m^2**.

- So the total solar power received by the earth at top of atmosphere is **17.5×10^4 TW** (1,360 W/m^2 x 1.29×10^{14} m^2). TW means terawatts, or a trillion watts.

- The average *electrical* power used by the world in 2023 was **3.4 TW**, and total *primary* power was **19.6 TW** (which includes electrical power, a factor of 5.8 greater). First let's consider just electrical power.

- So **5.15×10^4 times** (17.5×10^4 TW/3.4 TW) the global electrical power usage arrives at the top of atmosphere, and is captured.

- As an aside, let's check Musk's estimate that the total global electricity consumption for an entire year is exceeded by the sunlight that arrives at the earth in just "a few hours." There are 525,600 minutes in a year, and 525,600/5.15×10^4 = **10.2 minutes**. So Musk missed a chance to strengthen his claim by changing "a few hours" to "a few minutes," and if he meant primary energy it would still have taken less than an hour, **~59 minutes**.

- But this is only the solar power captured by the earth's disc at the top of the atmosphere and does not yet account for the areal density reduced by the earth's spherical surface or by passage through the atmosphere or by the intermittency at ground level or for the power lost in the conversion of solar energy to electricity by a PV panel.

So let's get real and bring this down to Earth.

- The intensity of solar radiation reaching Earth is actually spread over the surface of a sphere, not a disc, and a sphere has an area that is 4 times as large as a disc of the same radius (the fact that only half the earth is illuminated by the sun at any one time is taken care of by the CF, which we get to momentarily). So instead of 1,360 W/m^2, it's actually 340 W/m^2 (1,360/4) on average at the top of the atmosphere.

- At ground level, given the absorption and reflection by the atmosphere, the 340 W/m^2 is reduced by about half to **170 W/m^2**.

- Let's stipulate a generous 20 percent for the average annual CF of solar around the world. This reduces the average available solar power to **34 W/m^2** (170 W/m^2 x 0.2).

- The average efficiency of PV panels is about 20 percent. This further reduces the average power that can be generated to **6.8 W/m^2** (34 W/m^2 x 0.2).[3]

- Thus, after conversion to electricity, the solar power impinging on the top of the atmosphere is reduced on average over the globe by a factor of 0.005 (6.8/1,360). In other words, 1,360 W/m^2 is reduced by a factor of 0.25 x 0.5 x 0.2 x 0.2 = 0.005, or 1/200.

Now let's see what this means in terms of number of PV panels and total solar farm area.

- The **total area** of PV panels required to convert solar energy to global average electrical power is 3.4 TW/6.8 W/m^2 = 5.0 x 10^{11} m^2, or about **1.93 x 10^5 mi^2**.

- The area of a PV panel for commercial solar farms is **21 ft^2**.

- The **number of panels** then is [1.93 x 10^5 mi^2 x (5,280 ft)^2/mi^2]/21 ft^2 = **256 billion panels**.

- The spacing between rows of panels required for installation and servicing, depending on latitude and the consequent angle of incidence of sunlight, roughly doubles or triples the **land**

---

[3] The commercial rating for solar panel output of some 250 to 400 W, for panels measuring about 2 m^2 in area, or 125 to 200 W/m^2, is an artifact of the laboratory conditions under which panels are tested. Their ratings are measured using artificial lighting of intensity 1,000 W/m^2, which greatly exceeds not only the average ground-level solar intensity of 170 W/m^2 but even the maximum solar intensity of roughly 680 W/m^2, when the sun is directly overhead on a cloudless day. In short, no solar panel will ever generate its rated capacity, just as no wind or solar farm will ever generate its nameplate for any significant interval, advertising implications to the contrary notwithstanding.

area occupied by the solar farm from 1.93 x 10^5 mi^2 to somewhere between **4 x 10^5** and **6 x 10^5 mi^2**.

- The land area of the contiguous US is 3.12 x 10^6 mi^2; so to provide the total energy currently used globally each year, solar farms would occupy at least 4 x 10^5 mi^2/3.12 x 10^6 mi^2 = **13 percent of the contiguous US** (and up to 19 percent) — **at least the combined area of California and Texas** or **the combined area of France and Spain**.

And how much would this cost?

- Aside from the cost of land, the **price** per panel is $1,200 x 256 billion panels = **$307 trillion**. This is more than **10** times the **US GDP** and roughly **3 times** the **world GDP**.

- And that's the price of solar panels alone — **without the storage** — though Musk grants that batteries would be needed to balance night and day. But he misses the point that in order to charge the batteries there would have to be twice as many panels to absorb twice as much solar energy during the day. We took care of that with the generous CF, and in any case a factor of 2 doesn't change the orders of magnitude we are talking about. At that, this fails to account for **replacement** on obsolescence — it accounts for only one round of construction — and it assumes that there are enough **raw materials** in the world and assumes an adequate **speed of production**.

- The $307 trillion also neglects the monetary cost of the processes that come after the manufacture and purchase of the solar panels, including installation, maintenance, operation, and decommissioning, plus the expansion of the grid. But there's little point in adding in those figures when the price of panels alone reveals the absurdity of Musk's claim.

- Given the lifespan for solar PV panels, given predictable inflation, given technical improvements, etc., it would cost on the order of another **$307 trillion every 10–20 years**.

- Finally, all of this has dealt mainly with average power as though it were constant, compensating only for the average reduction due to clouds and nighttime.[4] That is, it fails to account for the inherent unpredictability and uncontrollability of the intermittency discussed in Chapter 3, which add incalculable cost.

---

[4] Solar farms located closer to the equator would have better than average insolation. So rather than 170 W/m^2 they would at most be exposed to 680 W/m^2 (approximately 50 percent of the intensity at the top of the atmosphere directly under the sun). Adjusting the foregoing figures by this factor of 4 would hardly change their order of magnitude.

If Musk is referring to total global consumption of *primary* energy, rather than just electricity, we would have to enlarge by a factor of 5.8 (19.6 TW/3.4 TW):

- Total area of panels required to convert solar energy to total global primary power would be 2.9 x 10^{12} m^2, or about **1.12 x 10^6 mi^2.**

- The number of panels required would then be **1.49 trillion** (not billion).

- The land area occupied by solar panels would be 2.3 to 3.4 x 10^6 mi^2, at least **74 percent of the area of the contiguous 48 states** to possibly more than the entire 48 states—or **from 60 percent of the area of Europe outside of Russia to almost its entirety**.

- The panels alone would cost **$1,790 trillion, almost 18 times** the **world GDP.** And all the provisos listed above apply.

Either way, Musk is fantasizing.[5]

The Additional Material Requirements of Storage—Batteries

Unlike Jacobson, Musk does count on storage, but then he sells batteries. Nor is Musk alone. In the face of declining hydropower, estimates of materials for storage apparatus focus on batteries.

Storage materials of any type lower the EROI of renewables (by adding energy input than return) and raise the monetary cost per electrical energy produced. The additional waste stream from batteries greatly compounds the environmental impact of renewables. Pumped hydro in the US today, though far less significant than backup from fossils and nuclear, constitutes over 95 percent of energy storage, with batteries less than 5 percent. But many all-RE proponents seek to greatly increase the number of batteries, largely because of portability and the speed of mass production.

The energy capacity of a battery (or any storage method) is measured in watt-hours (Wh), while the power capacity is measured in watts and is the maximum rate at which the energy can be released without overheating the circuit. The capacities of both energy and

---

[5] In a recent video Musk says he is abandoning solar energy and that he has had approving thoughts about nuclear for many years (https://tinyurl.com/yh42m3km). [Note: we have discovered this video is no longer available].

power are specified for each type of battery. The actual power of any discharge event, which cannot exceed the battery's power capacity, is determined by the amount of energy stored, which diminishes as it discharges, and the resistance in the circuit through which the batteries are discharged, including the resistance in the battery itself. Too great a speed of energy release (too great an electrical power) would risk destroying the battery and/or transmission line or even starting a fire (we discuss battery fires below).

US plant scientist and ecosocialist Stan Cox (2020), to whom we return in Chapter 9, refers to electrical engineering professor William Pickard (Washington University in St. Louis), who estimates the tonnage and volume, rather than monetary cost, of lithium-ion batteries (p. 63). Pickard simply asks the question: If a person needed fully charged batteries for wind and solar downtimes that last two days or two weeks, how much would the batteries weigh in each case?

In 2014 the batteries available to Pickard held less energy when fully charged than ones today. So we use our own calculations but credit him for the conceptual approach. US electrical usage is 33 kWh/day per capita, so to hold two days' worth of charge each person would require on average almost half a ton of lithium-ion batteries hooked to the grid [(2 days x 33,000 Wh/day)/(150 Wh/kg) = 440 kg = 970 lb = 0.49 tons]. To cover two weeks would require 3.4 tons of battery per capita. If the entire country were to experience windless and sunless days for two weeks, the 330,000,000 US residents would require more than a billion tons of batteries (3.4 tons/person x 330,000,000 persons = 1.12 billion tons).

Durations of low (or zero) output between two days and two weeks, and more, are not uncommon (Ruhnau and Qvist, 2021, discussed below) and are rising in likelihood as AGW causes more extreme weather events.[6] But since batteries can't be transported large distances instantaneously to cover local deficits, they must be attached to the grid at a number of geographical centers throughout the US—and fully charged. This would require a greatly expanded, and smart, grid, which would require additional energy and monetary cost to build it.

In addition to tonnage, Pickard also estimates volume. Again, using more recent figures, the volume of two weeks' worth of batteries would occupy a volume 200 times that of the Great Pyramid

---

[6] The February 2021 Texas rolling outages spanned almost 2½ weeks, though each location experienced a shorter spell.

of Giza in Egypt. And 1.12 billion tons is the weight of 170 Hoover Dams (6.6 million tons each). For the US alone.

And if we haven't already reached TMI (not Three Mile Island, but too much information), 1.12 billion tons of batteries contain a little more than 22 million tons of lithium (as the third lightest element, lithium constitutes only about 2 percent of the battery's weight[7]), while the known world reserves of lithium (land-based) are roughly only 19 million tons (https://tinyurl .com/3r5std5h). Recycling lithium is five times more expensive than mining and refining fresh ore (https://tinyurl.com/ 7avccn4y).[8] So the use of enough lithium-ion batteries to fill in for a two-week downtime in the US alone would exhaust the known land-based world reserves, and the batteries would require replacement every few years. There is an estimated additional 250 million tons of lithium in the ocean, but it is more expensive to extract (in dollar and energy terms) because of the extreme dilution.[9]

Moreover, since 85,000 tons of lithium are currently mined annually, 22 million tons would absorb almost 260 years' worth (assuming greater reserves could be located), though that could be sped up somewhat. To hold two days' worth, the US population would require only one-seventh of that, just over 3 million tons of lithium, or a little more than 35 years' worth of current extraction/production. But the known world's supply would be exhausted in a little over 6 years, without recycling. And this is for grid batteries alone, and just for the US, aside from the many other uses for lithium, and other countries.

---

[7] The estimates of this proportion vary widely in our experience, as well as that of Paul Martin, an expert in chemical processing (https://tinyurl.com/5djjb4e4). Many estimates refer not to lithium alone, but rather to lithium carbonate (Li_2CO_3), 5.3 times as heavy as lithium. Such estimates are 10–11 percent of the battery's weight, so 2 percent is close enough.

[8] Competing economic interests, each having to maximize profit, conflict with resource conservation. This leads to faster depletion and earlier peaking of resources—unless competing interests prevail on governments to require each company to recycle and perhaps to subsidize the process. By then, however, the resource may have become an endangered species. More on this in Chapter 10.

[9] Dissolved uranium is also currently more expensive to extract, in terms of energy and money, than land-based mined ore, but an electricity system based on nuclear energy would enable such an effort should land-based reserves be exhausted. However, fast breeder reactors using recycled uranium would obviate the need for any further extraction for many decades, or even centuries (as we saw in Chapter 2). In contrast, an all-RE electricity system would effectively disable the extraction of oceanic lithium, or even its recycling, due again to its low EROI.

The 1.12 tons of batteries for the US, would translate to 24 times as much for the entire 8 billion world population, or 27 billion tons to cover a 2-week downtime. This needn't occur simultaneously, and probably never would, but large commercial-scale batteries are not quickly transportable to places of need. Allowing for some places that might virtually never experience a complete deficit of both wind and solar, such as the Sahara for solar or Northeast Brazil for wind, let's halve this figure, or 13.5 billion tons. Remembering that this is proportional to the duration of the deficit, which is only a rough estimate, all we need for a grasp of the problem is order of magnitude.

Since the 256 billion solar panels (derived above) needed to handle all the world's electrical needs by Musk's reasoning, each weighing some 50 pounds, would total 6.4 billion tons, we see that the mass of materials in batteries (27 billion tons for a 2-week downtime) can be comparable to, or exceed, that in solar panels. Wind turbines require roughly 20 percent of the mass of materials in solar panels for the same nameplate power capacity; so if we use Jacobson's plan for dividing the task approximately equally between wind and solar, this would reduce the mass from 6.4 billion tons of material to 3.8 billion tons ($\frac{1}{2}$ x 6.4 + $\frac{1}{2}$ x 0.2 x 6.4 = 3.8). The batteries would even further outweigh the requisite panels plus turbines.[10]

The US holds less than 10 percent of global land-based lithium reserves. Bolivia, Chile, and Argentina together hold more than 50 percent of those reserves, Australia some 8 percent, and China about 6 percent (https://tinyurl.com/3nka24sy). While new deposits may be found, the required batteries would enhance the stress on world supply and intensify the competition for these dwindling reserves.[11]

The International Energy Agency (IEA) report in May 2021, assessing the material requirements of wind/solar farms plus battery storage (https://tinyurl.com/266c7xa3), is summarized by *The Wall Street Journal* (May 11, 2021, https://tinyurl.com/mvyphf4j):

---

[10] And if the world were to electrify virtually all energy usage (save some direct heating), the 3.8 billion tons of materials for turbines and panels would have to be raised by a factor of 5.8 (derived above) to 22 billion tons, still exceeded by the mass of batteries for a 2-week downtime.

[11] Among other elements, batteries require cobalt, largely mined under slave-like conditions by children in the Democratic Republic of the Congo. Most of the mines are owned by Chinese corporations and bring enormous profits (Kara 2023).

The IEA finds that with a global energy transition like the one President Biden envisions, demand for key minerals such as lithium, graphite, nickel and rare earth metals would explode, rising by 4,200 percent, 2,500 percent, 1,900 percent and 700 percent, respectively, by 2040 . . . The world doesn't have the capacity to meet such demand.

As to the impact on the environment, lithium from expired batteries becomes part of permanent landfill. Replacement on a regular basis further depletes world reserves and amplifies mounting waste. Lithium's (chemical) toxicity is well known since it is also a psychiatric drug with side effects including kidney disease and death, given high enough doses.

Meeting power demand requires enough charged batteries to supplement incoming wind/solar power (which at times may suffice, Chapter 3). Avoiding fires from excessive current requires the discharge of multiple batteries simultaneously, thereby increasing their number even further. Even when idle, lithium batteries can discharge spontaneously. Five years ago (2019) *The Korea Times* reported that spontaneous fires in the previous year had completely destroyed 20 of the country's almost 1,500 battery facilities, called Energy Storage Systems (ESS): "An ESS is a large stack of rechargeable batteries. It is often used for storing cheap off-peak electricity or coupled with solar, wind or other intermittent power generators." The fires are still being investigated by the South Korean Ministry of Trade, Industry, and Energy, who reportedly "have no clue" as to the cause (https://tinyurl.com/mrm6vbeu). Really?

Such fires might be preventable with additional capacity and with differently designed batteries. But two years later, on August 2, 2021, one of the 14-ton Tesla lithium-ion batteries near Melbourne, Australia, caught fire. Containing the fire to prevent its spread to scores of neighboring batteries took 150 firefighters more than 4 days (https://tinyurl.com/5x7enpns).

The fires are difficult to extinguish because one of the battery electrodes is made of $LiCoO_2$ (lithium cobalt oxide), which releases oxygen at higher temperatures. Merely blocking access to ambient oxygen with water or a blanket can therefore be futile. Moreover, a quenched fire can reignite. In both 2022 and 2023 more than 200 e-bicycle batteries caught fire in New York City, some causing building fires, with concomitant injuries and deaths (https://tinyurl .com/3a8xdk5m). Large lithium-ion batteries are a major hazard.

Materials Needed for EVs

Despite the foregoing unsolved battery problems, some governments (including US and California) push full speed ahead to mandate production and sales of EVs to replace ICE vehicles. But the materials may simply be unavailable to complete the transition. Figure 4-3, borrowed from Palmer and Floyd (p. 141), shows material requirements for EV batteries compared to current annual production and known reserves. Manganese is omitted, as its requirement is at least an order of magnitude less than the other four.

Note how the required quantities (righthand column) dwarf current production (tiny boxes) and exceed or are comparable to known global reserves (middle column). Vehicles powered by hydrogen fuel cells, produced with nuclear energy, may be able to succeed, if the associated problems can be overcome. Better and

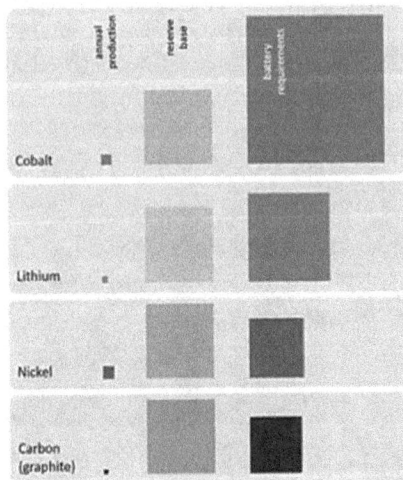

FIGURE 4-3

more accessible public transportation could also alleviate the stress on energy resources.

Bryce provides a bar graph (Figure 4-4) of figures from the IEA, showing the near monopoly held by China on the various constituents of EV production(https://tinyurl.com/ex8rby4w).[12] The elements are lithium (Li), nickel (Ni), cobalt (Co), and graphite (Gr), a form of carbon. Thus, China has virtually complete control over the manufacture of EVs for US roads:

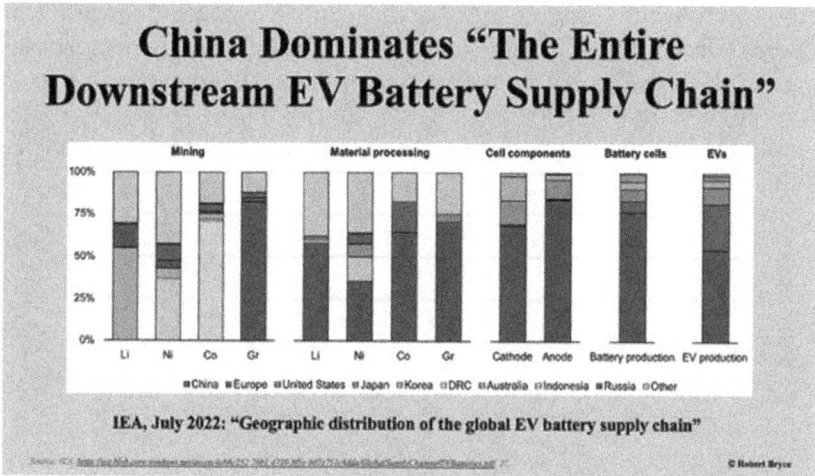

China Dominates "The Entire Downstream EV Battery Supply Chain"

IEA, July 2022: "Geographic distribution of the global EV battery supply chain"

FIGURE 4-4

Aside from material requirements, charging an all-EV vehicle fleet in the US would require roughly 40 percent of the current US electricity supply (https://tinyurl.com/bduxbth9). Just to eliminate ICE vehicles, electricity production plus grid are presently insufficient. And the switch to mitigate AGW would require that the source of electricity be carbon free. If the source remains an NG or coal power station, an EV merely transfers your toxic exhaust from your neighbors to those living near the power station.

One final point concerning the dangers of war entailed by various energy sources: Predictions of the peaking of minerals are fraught with inaccuracy since new geological deposits might be discovered (for example, shale and offshore oil), and there are

---

[12] This figure is similar to Figure 4-1, which shows the geographic distribution for all uses of the constituent minerals. Figure 4-4, in contrast, shows only those involved in EV manufacturing and is broken down into the various stages of the manufacturing process. From the fourth bar on, the lowesr section (the darkest) represents China.

sive processes. But in a fundamentally competitive international political economy, resource wars eventually arise with increasing demand in the face of constant or declining supply, whether or not the Earth's supply of any particular resource has peaked. Rivalry often takes precedence over actual peaking from a worldwide perspective, though peaking can become a problem locally (discussed more fully in Chapter 9).

Suffice it to say that claims by Jacobson (and others) that nuclear energy leads to nuclear weapons and war are simply false, as we show in Chapter 7. The more likely pressure toward war derives from the far greater material requirements of wind turbines and solar panels. Nuclear energy mitigates the pressures toward resource wars due to the immense energy density of nuclear fuel and the far smaller material demands it imposes (as discussed in Chapter 2).

5

EROI: The Key to Energy Efficiency

Historically, newly discovered energy sources and newly invented conversion technologies have generally enabled an increase in output of *usable* energy for the energy *humans had to exert* to obtain it, expressed as the ratio EROI. In other words, the EROI of a device or infrastructure is a measure of the amplification of energy exerted by humans provided by the device/infrastructure.

As the *ratio* of ER to EI has grown larger, so has the *difference* between them, ER − EI. This difference is the net energy produced (EN)—EN = ER − EI. The magnitude of EROI has ranged from not much more than 1 up to hundreds or thousands in modern times, and this magnitude is a measure of the efficiency of energy production. However, the entire goal of energy production is EN, the net energy used to run society—the surplus beyond the input required to obtain the output. EN underwrites the basic modern social needs of transportation, communication, education, health, manufacturing, shelter, food, entertainment, art, aesthetics, relaxation, stress relief, and, in the face of progressive AGW, increasingly needed functions such as desalinization of ocean water, removal of CO_2 and methane from atmosphere and oceans, construction of additional conversion devices as needed, and so on for a long list of necessities and desirables, all of which require energy, and in general, lots of it.

The progressive amplification of EROI, and hence EN, has freed successively larger portions of the population from producing energy (converting a source to useful form) and enabled ever-greater ways for people to put that energy to use. Life expectancy has grown, and the more varied division of labor has improved the potential quality of life. New energy applications have eased many lives and made them far more interesting, even as other applications have been put to destructive use. In Chapter 9 we discuss authors and organizations who, without foundation, claim that certain energy sources inherently dictate their use. However, the triaging of applications is mainly determined by social organization and need, not by the form or source of energy.

After more than a century of productive use, one set of sources has been found to spawn an unanticipated side effect—fossil fuels (decayed biomass). Their burning and leakage (of the gaseous form) fill the atmosphere with gases that trap some of the outgoing heat—heat that would otherwise be radiated back into space to keep an energy balance between incoming sunlight that peaks in the range of yellow and outgoing radiation that peaks in the infrared. A balance would keep the Earth's temperature and energy content relatively constant. The physics of so-called greenhouse gases (GHGs) has been understood for well over a century, but the greenhouse effect of fossil fuels has only become detectable in the last few decades.[1]

The production of abundant energy is needed to preserve life and the proliferation of applications, but sooner or later we have to halt AGW. Some believe that using abundant energy and halting AGW are incompatible goals. They propose reducing energy production to slow or halt AGW (Chapter 9). We show that both objectives are technologically possible, though the latter is geopolitically impeded, if not prevented, without a major transformation of the global economic system.

The Central Role of Net Energy

Having reviewed material and land inputs, we focus in this chapter on the *energy* input for an electricity system, comparing all-RE to all-nuclear and focusing attention on EN. While many all-RE advocates regard EROI as irrelevant, we show its central relevance regarding what is feasible, what is possible, and what is desirable. Only by ignoring, or being unaware of, EROI can Green advocates find anything attractive about wind and solar. EROI focuses strictly on the intrinsic features of energy systems, thereby avoiding the incidental and highly variable impacts of extrinsic factors. Looking ahead, the approximate EROI for wind and solar is 2–4, for fossil fuels 30–40, and for nuclear 100–1,000⁺.

---

[1] The glass in a greenhouse lets in sunlight mainly as visible light but traps both the reradiated, mainly infrared (longer than red), wavelengths and the warmed air, keeping the interior hospitable to plants. It's the trapping of reradiated infrared that the atmosphere shares with a greenhouse, while gravity traps the air.

As mentioned in Chapter 2, the capture of wind represented an early advance in EROI beyond human and animal muscle, first when sailing ships were invented and later windmills, which also use unfurled sails to catch the moving air. Used to grind (mill) grains and pump water uphill, windmills provided another leap in energy yield, though still small. This lasted for centuries until the enormous increase in EROI provided by steam power—extracted from burning wood or, when that ran short, coal—enabled capitalism's industrial revolution, first in England then spread to much of the world. As oppressive as the new economic system was for millions, life expectancy soared. The energy yield from wood or coal through the mediation of steam underlay an increase in EROI by an order of magnitude (approximately tenfold). The subsequent discovery and extraction of petroleum in the mid-nineteenth century, in both the Caucasus and in Pennsylvania, enabled the multiplication of modes of transportation, followed by the twentieth-century discovery of nuclear energy that raised EROI by at least another order of magnitude. While first used destructively in wartime, nuclear was subsequently adapted to generate electricity for domestic economies.[2] What the future holds for EROI remains to be seen.

Again, EROI is the mathematical expression of efficiency, but only for an energy *source*. For an appliance that uses, rather than produces, energy and whose purpose is home heating, cooking, agriculture, transportation, light, entertainment, computation, or some other desired application, EROI is meaningless since there is no ER. And as mentioned in Chapter 2, efficiency of energy production has enormously more impact than efficiency of end-use applications.

Contributing to the low EROI of renewables (in the absence of fossil or nuclear backup) is their need for four things, each of which adds to EI for a given ER and none of which is required by fossils or nuclear, for a given ER. These are excess conversion devices (because of low CF), storage apparatus (because of intermittency), expansion of the grid (including DC transmission lines), and frequent replacement apparatus (because of weather-accelerated obsolescence).[3] In addition, both storage and addi-

---

[2] Like nuclear, numerous inventions for waging war have later found domestic application (https://tinyurl.com/4u725b3b).

[3] EI is required at numerous junctures: mining the minerals, refining the ore, transporting it to the factory, running the factory, transporting the product to the field or plant, installing the devices or constructing the plants, maintaining the devices or plants, operating them, decommissioning the components upon obsolescence, and removing them to make room for replacements, whereupon the cycle begins again. Recycling can save material but calls for additional EI. The total EI combines these component energies, averaged over the lifetime of each type of device and/or power plant.

tional transmission diminish ER through heat losses. With reduced ER and increased EI, EN is squeezed from above and below. The greater the penetration, the greater the reduction of EROI (Figure 5-3).

To clarify how EROI is related to societal need Australian energy analyst John Morgan quotes US energy expert Charles Hall (https://tinyurl.com/mrfrwvsy):

> Think of a society dependent upon one resource: its domestic oil. If the EROI for this oil was 1.1:1 then one could pump the oil out of the ground and look at it. If it were 1.2:1 you could also refine it and look at it, 1.3:1 also distribute it to where you want to use it but all you could do is look at it. Hall et al. 2008 examined the EROI required to actually run a truck and found that if the energy included was enough to build and maintain the truck and the roads and bridges required to use it, one would need at least a 3:1 EROI at the wellhead.
>
> Now if you wanted to put something in the truck, say some grain, and deliver it, that would require an EROI of, say, 5:1 to grow the grain. If you wanted to include depreciation on the oil field worker, the refinery worker, the truck driver and the farmer you would need an EROI of say 7 or 8:1 to support their families. If the children were to be educated you would need perhaps 9 or 10:1, have health care 12:1, have arts in their life maybe 14:1, and so on. Obviously to have a modern civilization one needs not simply surplus energy but lots of it, and that requires either a high EROI or a massive source of moderate EROI fuels.

As considerable research shows, wind and solar have barely adequate EROIs when the energy source is viewed in isolation. Only when it is embedded in a high-EROI fossil fuel and/or nuclear matrix might the overall EROI of the combination be sufficient to permit more than mere admiration of the product. But without such a matrix, the EROI of wind and solar in isolation becomes too low for any useful deployment.

One example for which the EROI is close to 1 is ethanol production. It is added to gasoline because it allows more complete and cleaner burning—cleaner of respiratory toxicity but not of GHG emission per usable energy output. Moreover, as a gasoline additive, ethanol reduces mileage efficiency. Since respiratory cleanliness is the goal, rather than energy efficiency, the low EROI of ethanol is accepted (it may even be less than 1). But for energy to run a society, we seek sources with EROI as much greater than

1 as attainable, consistent with availability, sustainability, relative safety, and relative cleanliness.

If the only non-GHG-emitting source of energy had a lower EROI than fossil fuels, then, were the social organization hospitable, we would collectively decide which feature is the more desirable and might intentionally opt for the cleaner, less efficient source. Fortunately, the cleanest, safest, and most sustainable energy source turns out also to be the most efficient (highest EROI).

Wind/solar lack predictability, reliability, and controllability. Even aside from their impact on EROI, these features would prevent the maintenance of a stable economy that could meet a wide diversity of human needs in an extensively populated world and would obstruct even the economic growth needed to keep pace with population needs, whether in industrialized or "developing" nations. Low EROI returns us to an energy analog of feudalism, mimicking times past (and in some countries representing the present) when the vast majority of the population was (or still is) needed just to generate food. Palmer (2014), in his preface, compares Cambodian rice farming to that in Australia. The former requires 40–80 person-hours to harvest a hectare, while the latter, using a modern harvester, requires 15 minutes of one farmer's time, for a ratio of 160–320 to 1. As a result, agriculture occupies 70 percent of Cambodia's workers and only 2 percent of Australia's workers.

Similarly, mechanization and concentration have reduced US agricultural labor from over 90 percent of the population to about 2 percent. Given hospitable social organization, efficiently produced electricity could free a similar proportion from other basic industries to develop and fill more interesting and less grueling jobs. There would almost certainly be resistance to an attempt to return agriculture to its former inefficient status. Yet most of the resistance to driving energy back toward such a state with renewables is coming from a rightwing backlash to Green policies. Meanwhile much of the Green left applauds such a retrogressive development, presumably without their realizing its retrogressive implications.

Calling for wind and solar to replace fossils and nuclear is like calling for oxen and plows to replace tractors and combines, underwritten by an unwarranted but widely promoted fear of tractors that, after all, have in fact caught fire, run people over, amputated limbs, and caused fatalities. Indeed, in the midst of the "peak oil" debate earlier this century some thinkers advocated the return to a "world made by hand," the title of James Howard Kunstler's 2008 novel.

Because of its inefficiency (low EROI), Jacobson's Roadmap would return humanity to a system simulating a labor-intensive farming economy, this time in high-tech garb. And, as we mentioned in Chapter 3, the plan calls for the subordination of industrial and other work schedules to nature's dictates of fluctuating wind and solar energy production. In his 2015 update, Jacobson called such subordination "flexible response"—inverting the usual response of supply to demand.

The Progressive Cheapening of Renewables

The current progressive cheapening of portions of the manufacture of wind turbines, PV panels, and batteries is touted by RE proponents as evidence that renewable energy will become ever cheaper—eventually cheaper than nuclear (and for that matter fossils). Not acknowledged is the fact that the progressive cheapening has slowed dramatically, as initial cheapening always eventually does. The price decline of new technologies is always steepest at first and later levels off as the learning curve exhausts its benefits and the factory production stage approaches its maximum potential for automation. Furthermore, the cheapening does not come from the labor-intensive costs of mining, transportation, and field installations of the turbines and PV panels, but rather from the factory production of parts where significant productivity gains are possible.[4] Such accounting, moreover, ignores the required excess capacity, batteries, and additional transmission lines.

But more importantly, any net cheapening of wind and solar conversion apparatus is completely dependent on the current existence of the fossil and/or nuclear matrix and government subsidies. Without reliable high-EROI sources to supply the energy for production of conversion and storage apparatus, and to back up remaining deficits, these would become far more costly. The successive replacement of fossils and nuclear by wind and solar represents a self-negating process.

---

[4] And this one-sidedly neglects the similar progressive cheapening of nuclear technology if its development were to be unleashed and allowed to flourish globally. But as we have pointed out in Chapters 2 and 4, the intrinsic cost of nuclear per energy produced is less than that of renewables to begin with, and by a wide margin.

Values of EROI for the Three Types of Energy Sources

Figure 5-1 is a slight modification of a bar chart from Weißbach et al. (2013) showing estimates of EROI for various energy sources and for some unspecified penetration of wind and solar.[5] While their figures come from Germany, they provide some idea of the relative orders of magnitude. They also draw a horizontal dashed line at 7 to represent the minimum EROI needed to run a relatively simple society, which would be substantially higher for a modern complex society as we saw above in Morgan's quotation of Hall.

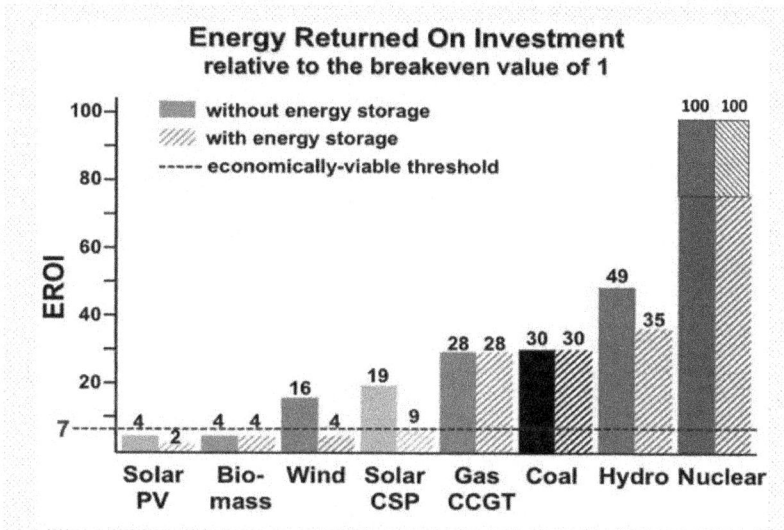

Energy Returned On Investment
relative to the breakeven value of 1

- without energy storage
- with energy storage
- economically-viable threshold

FIGURE 5-1

Note that the EROI values are lower when storage is included, but only for renewables (including hydro), since fossils and nuclear and even biomass, being fuel-based, are their own storage.[6] Given

---

[5] The original bar chart labeled nuclear's EROI for typical LWRs as 75, which is an underestimate, as it was largely predicated on a compromise between different methods of fuel enrichment, some less efficient than others. So for our purposes, we have labeled the bars as 100. Generation IV reactors will have even higher EROI values.

[6] Weißbach et al. show a drop in EROI for hydro when storage is added. While dammed lakes (like fuel-based sources) are their own storage, they only consider "run of the river" (ROR) hydro. ROR means tapping a river's kinetic energy as it flows without a dam, and pumped hydro is the storage method. This requires energy input (EI) to construct and maintain the pumps and to construct the upper reservoir, and the increase in EI lowers the EROI, just as for storage of wind/solar energy.

the current energy mix in the US, the overall EROI is around 40, according to US energy analyst James Conca (https://tinyurl .com/ma5vt7sw). Conca, incidentally, reproduces Weißbach's original bar chart.

Real Cost Ratios for Wind/Solar versus Nuclear to Obtain the Same Net Energy

Using Weißbach's figures (as modified above), Figure 5-2 compares the EIs for the three main energy sources for equivalent net energy produced (EN). For simplicity, we average the respective EROIs of wind and solar as they would be in the absence of the matrix—namely, 4 and 2, respectively—into a single ratio, 3. These include estimates of the necessary storage (buffering) and the average annual CFs so that the energy return (ER) is the actual energy produced per year rather than the nameplate capacity. The relative values of EI assume that each is the sole source of electricity. Any combination would yield intermediate values. Finally, since EN is the objective of an electricity system, we define ENOI (EN/EI), which is EROI -1.[7]

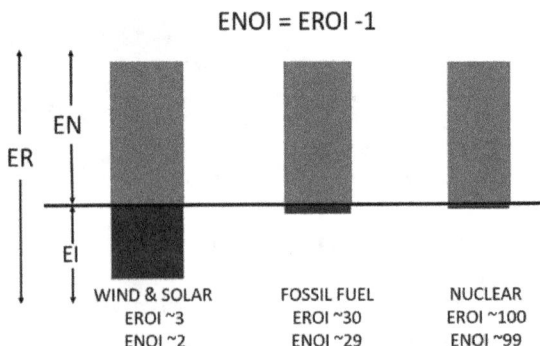

FIGURE 5-2

EI for wind/solar (were an all-wind/solar system possible) is much larger than for either fossils or nuclear—in fact roughly 15 times that of fossil fuels (29/2) and roughly 50 times that of nuclear (99/2), with comparable ratios of environmental damage from resource depletion and waste production.

---

[7] EN = ER – EI, so EN/EI = ER/EI – EI/EI = EROI –1.

A comparison of the requisite energy inputs (EI) turns out to be a reasonable and conservative comparative estimate of the associated inputs of materials, land, labor, and money. Each of these is far greater, and therefore far more costly, for wind/solar than for nuclear. Thus, the comparison of intrinsic costs is quite different from a comparison of prices to consumers. Unlike wind/solar, fossils and nuclear expend a minuscule portion of their respective ERs to obtain more usable energy (EI), nuclear even less than fossils.

Let's review the ratios for each category of input. From Figure 2-4 and the accompanying discussion, and averaging between wind and solar, the ratio of **materials** to build a renewables farm with the same total energy return (ER) as a nuclear plant is approximately 15/1, which becomes roughly **22/1** for the same net energy (EN)—3/2 times as great.

From the same chapter, and the table showing "POWER DENSITIES OF NUCLEAR VERSUS WIND," and using Goldstein and Qvist's conservative numbers, the ratio of **land** is approximately 167/1 for the same total ER and roughly **250/1** for the same EN. The ratios of **labor** hours for the same ER and EN—looking ahead to the table in Chapter 6 and averaging the ratios for wind/solar to nuclear—are on the order of 57/1 for the same ER and on the order of **85/1** for the same EN.

The range of input ratios covers an order of magnitude, 22/1 to 250/1. Since the ratio of the relative **energy** inputs (EIs) for the same EN is roughly **50/1**—on the low side of the range—this can be considered a conservative surrogate for the ratio of the respective monetary costs.

The EI ratio of 50/1 for the same EN is also dependent on the continuance of the fossil and nuclear matrix as the primary, if not sole, source of energy needed for the initial construction of wind/solar farms. If, on the other hand, the EI for their construction were to be provided by wind/solar themselves, more EI would be required, raising the ratio above 50 to 1 (the presence of storage is already included in this ratio, taken from the numbers by Weißbach et al.—Figure 5-1).

Why is this true? After all, it takes the same energy to build a wind/solar farm regardless of the source. But if those farms had to provide the energy for their own replacements, that part of the total energy they produce (ER) would divert energy away from their EN (this is illustrated below in Figures 5-4 and 5-5 on pp. 113–14). With wind/solar farms diverting energy from their EN, more wind/solar farms (or larger ones) would be required to reach the original EN. The construction of more or larger farms, in turn, would require greater EI. Absent a fossil/nuclear matrix, and with wind/solar farms having to rely on themselves for the energy required by their construction, maintenance, decommissioning, and replacement, the EI ratio would be even greater than 50/1.

Thus, we can use the conservative ratio 50/1 to compare the amount of net energy (in TWh) that is ultimately produced by each dollar devoted to building and maintaining a wind or solar farm in an all-RE electricity system to the corresponding amount of net energy produced by the same dollar devoted to building and maintaining a nuclear power plant in an all-nuclear electricity system. The nuclear dollar produces at least on the order of 50 times as many net terawatt-hours (TWh) as the wind/solar dollar. With Generation IV nuclear designs and taking into account the increased material and labor inputs for batteries, the ratio would far exceed 50/1. Conversely, to obtain the same net TWh from an all-RE electricity system would require at least 50 times as many dollars as from an all-nuclear electricity system.

This puts a floor on an estimate of the intrinsic costs (as opposed to consumer prices) of renewables versus nuclear. One might still ask: Isn't a turn toward wind and solar, even with a rejection of nuclear, at least a step in the right direction? The short answer is, No. It's decidedly a step in the wrong direction, in no small part because, in the absence of nuclear energy, wind and solar, due to their intermittency, require the continuation of fossil fuels and cannot replace them, even if under certain scenarios it might diminish their use somewhat (explored in Chapter 3).

What rationale, given a general understanding of all the many dimensions, would lead to spending a dollar on a minuscule quantity of uncontrollably intermittent wind/solar net energy rather than on a prodigious quantity of reliable nuclear net energy? Absent the phobias of nuclear energy and radiation (Chapters 7 and 8), there would be no such rationale.

The Effect of Storage on EROI Varies with the Degree of Penetration of Renewables

Figure 5-3, borrowed from Palmer and Floyd (2020, p. 83), shows the effect of storage on the EROI of wind/solar as their penetration approaches 100 percent. Their estimates of EROI for renewables in Australia differ from Weißbach et al.'s estimates from

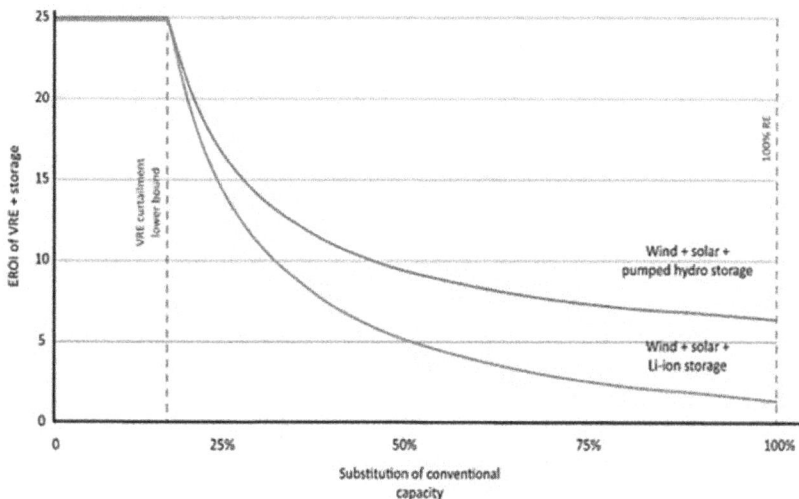

FIGURE 5-3

Germany. But the general sense of inadequacy is the same. In fact, their estimate of an initial EROI of 25 is generous, and were it closer to that of Weißbach, the values near 100 percent penetration (what they call "Substitution of conventional capacity") would be that much lower. They also use the term "VRE," meaning variable renewable energy. Under that label they blend wind and solar together, as we do in Figure 5-2 above.

Note how rapidly EROI drops as the penetration passes a certain point, at which compensation by fossils and nuclear begins to diminish and storage starts to become indispensable. Also note how nearing 100 percent penetration EROI approaches the minimum level required for a society with greatly limited complexity—around 7 for pumped hydro storage and below 1 for batteries. Even at 50 percent penetration, the EROI with either form of storage is far below that needed to satisfy our current electricity requirements.

For battery storage, at 100 percent penetration wind/solar becomes an energy sink rather than a source (EROI < 1). Moreover, for initial values less than 25, the EROI would drop below 1 at even lower penetration. Any system for which EROI even borders on 1 would leave virtually zero net energy to run a society and would therefore not be worth the effort to construct it. Consideration of EROI alone reveals that any attempt to follow Jacobson's Roadmap, or any other plan that relies on wind/solar to any significant extent, would be (and is) sheer folly.

The High EROI of Nuclear

The EROI of nuclear is still evolving with new Generation IV reactors. It's potentially much higher than the 100 of an LWR. Rather than wastefully discarding the non-fissionable but fertile U-238, fast-neutron breeder reactors continually convert U-238 to Pu-239, which is used as additional fuel alongside U-235.[8]

Fast breeder uranium reactors can, with repeated fuel recyclings, extract almost 100 percent of the energy in the uranium (U-235 plus U-238). Since the 94 remaining LWR reactors in the US all have their once-used fuel stored onsite, still containing the vast majority of the embodied energy, this stored fuel could be retrieved and reused.[9] Furthermore, it could do so with no further uranium mining, refining, or transportation until the fuel had run its course, over scores of recyclings for centuries.[10]

Moreover, the EROI estimate of 100 could be greatly magnified—according to some estimates by an order of magnitude, thereby raising it to many hundreds or over 1,000. Indeed, in a subsequent paper a year later Weißbach and his coauthors describe a Genera-tion IV reactor with a potential EROI of around 2,000—a factor of 20 above LWRs and of 500–600 relative to wind and solar (Huke et al. 2014). The gain in EROI in their example comes mainly from a reduction of the EI required by newer reactor designs that incorporate far less material.

---

[8] Some helpful sources on the workings of nuclear reactors are Bodansky 2004, Hargraves 2012, Blees 2008, Tucker 2019, Cravens 2007. Thorium reactors are also breeders but not fast breeders, as the neutrons emitted from fission reactions need to be slowed by collisions with a moderator (see Glossary). These are similarly efficient at extracting and using virtually 100 percent of the energy in the fuel, rather than the trivial 1–2 percent extracted by an LWR on a single pass. Furthermore, they don't require periodic removal of fuel from the reactor to remove the fission products.

[9] Fuel recycling is unlikely to happen in the US unless the government takes over the process. President Carter issued an executive order in 1977 making permanent President Ford's temporary order banning recycling of once-through nuclear fuel. While President Reagan later reversed this order, recycling is too risky for a profit-making company. Several other countries, on the other hand, do engage in fuel recycling.

[10] Onsite storage makes transporting the U-235 unnecessary for an extended period, though depleted uranium (DU, mainly U-238) would require transportation. DU is the portion removed during the enrichment of uranium and is currently stored in New Mexico, near the US government's Waste Isolation Pilot Plant (WIPP). See the Glossary for these terms. In addition to DU, WIPP houses the aforementioned decommissioned nuclear warheads, mainly purchased from Russia, some homegrown. These are used as nuclear fuel (Chapter 7). No problem has ever arisen from transporting nuclear fuel, anywhere, though it is portrayed as a hypothetical hazard to promote fear. After all, natural uranium, almost entirely U-238, is all around us in the ground and oceans and constitutes the source of most of the natural background radiation in which all extant species have evolved.

This greatly diminished EI reduces to almost trivial magnitude the proportion of ER needed as input for the next cycle—at least one order of magnitude below 1 percent (1/2,000 versus 1/100). For fast breeder reactors, the land involved in mining/milling of uranium could remain untapped for decades if not generations, as discussed in Chapter 2. And when the retrieval and recycling of on-site-stored minimally-used uranium fuel, along with the abundant off-site-stored DU, are exhausted, the ground and oceans contain enough uranium to last the remaining lifetime of the sun's internal energy supply and hence the remaining lifetime of virtually the entire solar system, including the Earth.[11] This turns out to be on the order of 4.5 billion years, which coincidentally is roughly the age of the solar system already. And thorium is even more abundant in the earth's crust than uranium.[12] Thus, as explained in Chapter 2, nuclear energy is, in effect, no less "renewable"—more precisely, "indefinitely replaceable"—than wind/solar. Excluding it from the category of renewables functions (intentionally) to ostracize it.

While the purpose of an all-RE electrical system is defeated by its very low EROI, the very high EROI of fission energy does not capture all the advantages of nuclear. For example, hydro power can also enjoy a fairly high EROI and, in ideal circumstances, a high CF, but unlike hydro, nuclear is portable and has much higher power density.[13] The latter has ecological advantages and adds to the ease of replication in that nuclear faces far fewer constraints of land area and geographical placement—fewer constraints of placement even than renewables. Only fossil fuel plants compare in this regard.

Sustainability and/or Expansion of an All-RE System: The Trade-off between Net Energy and Energy for Device Replacement

While building the Roadmap's proposed RE electrical system is infeasible and socially undesirable even if supported by the matrix,

---

[11] The ubiquity of uranium in the ground is incidentally evidenced by the ubiquity of radon, the main source of natural background radiation received by virtually everyone on Earth. Radon is a gas that seeps up from the ground everywhere and is a product in the radioactive decay chain of uranium (discussed further in Chapter 8, with radioactive decay explained in the Glossary). Where there is radon, there is uranium

[12] While thorium is a decay product of uranium, its most abundant naturally occurring isotope (Th-232) has a half-life of 14 billion years, which is even longer than that of either U-238 (4.5 billion years) or U-235 (700,000 million years). That is, thorium is even more stable than uranium. Whatever the initial relative abundances may have been when the earth was first formed, the much slower decay rate of thorium accounts for its greater relative abundance today.

[13] Lakes Mead and Powell yield between 1 and 2 GW nameplate power through their dams but occupy roughly 250 square miles each.

after elimination of the matrix the now all-RE system would soon contract and collapse for lack of sufficient energy return to sustain or expand itself and at the same time to provide net energy for the society. This can be demonstrated by considering the energy needed for replacing obsolete conversion devices (turbines and PV panels) and whole farms. For these renewables, the energy needed for operation and maintenance (O&M) is minimal compared to the energy needed to construct and decommission the devices. This is also true for nuclear, since even though the main component of operation is fuel, we showed in Chapter 2 that the materials (and hence energy for extraction and processing) for nuclear fuel are far less than those for the plant itself.

While the input energy (EI) for O&M must be generated continually during the lifetime of the apparatus, that for replacement can either be generated continually or episodically. Either way, the apparatus has to generate the energy for its replacement before it becomes obsolete and requires that replacement. The importance of this difference is that for both wind/solar and nuclear the energy for replacement has to be available upfront before the replacement devices/farms or power plants are constructed, or the system could not sustain itself in perpetuity. Moreover, if expansion is needed, that much more energy for successive construction has to be generated prior to obsolescence.

While the energy for a farm's or plant's own O&M can be generated over its own lifetime and need not be generated by the farm's or plant's predecessor, this matters little, because the predecessor has to generate its own energy for O&M spread over its lifetime.

Much of the material in a nuclear plant (or fossil plant) would not need replacing as quickly as some of its internal parts. That is, the reactor (or combustion chamber) would require replacement before the building that houses them. In contrast, almost the entirety of moving parts of a wind turbine or the entire solar PV panel would require replacement (whether or not the material is recycled). As we will see below, because of their low EROI and their relatively short lifespans, the upfront energy needed to replace wind/solar farms is a great burden on their total energy return and lessens their net energy production, whereas for nuclear, with its high EROI and long lifespan, it is hardly a burden at all.

Among its many flaws, the Roadmap omits consideration of all these processes that would be called for after completion of the initial construction of the all-RE system and elimination of the matrix. Moreover, Jacobson plans that all energy-consuming applications will be electrified, thereby magnifying electricity demand by another factor of 5 or 6 (see Figure 2-1). This would include all heat-requiring industrial processes. Since fuel-based sources (fossils and nuclear) first generate heat that is converted in part to electricity, the heat portion could be used directly by such industrial processes instead of first being converted to electricity and back to heat. This double conversion would incur further losses to unusable forms of energy than direct heating. But because wind and solar PV (not CSP) produce electricity directly, this energy would have to be transmitted over long distances and converted to heat at the terminal consumption end, incurring such losses. The expansion of the electrical grid required by wind/solar would magnify the heat losses from long-distance transmission lines, requiring yet further expansion of the wind/solar farms.

The series of three bar graphs in Figure 5-4 shows what happens to a wind/solar system (once constructed using external energy from fossils and nuclear, but after their elimination) as progressively more power is diverted from EN and directed toward replacing turbines, PV panels, and batteries upon their obsolescence. Since power is energy per time, either production or consumption, and since the diagrams display power on the vertical axis and time on the horizontal

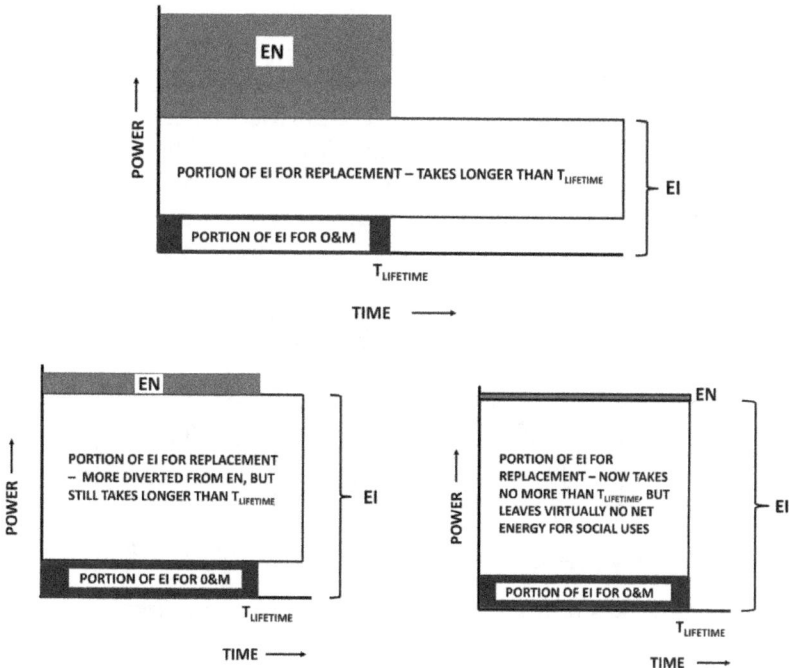

FIGURE 5-4

axis, energy produced or consumed over a particular time interval is represented by an area (power times time interval).

The energy needed to replace the obsolete conversion device plus storage apparatus is represented by the central areas, net energy for other social uses by the top areas, and energy for operation and maintenance by the bottom areas. Total EI is then represented by the sum of the central and bottom areas. The three diagrams are all drawn to represent the same power (total height) and the same device lifetime $T_{lifetime}$ (width of top and bottom), but they change from one to the next as the decision is made to divert progressively more power from EN toward the replacement process, to shorten it. $T_{lifetime}$ is an average over the lifespans of the system's individual conversion devices and batteries. Only by diverting more power from a device's EN can the time to generate replacement energy be shortened, and this must be done before the device ceases to function. Otherwise in each round of replacement the number of new devices and/or batteries would decline until the system collapses. Only in the third diagram does this occur, but it happens at the expense of virtually eliminating the net energy (EN), which makes the system energetically useless.

We use this extreme example to show how critically important an adequate EROI is, not just to provide the society with usable net energy, but even to be able to sustain, much less expand, the system. The EROI of wind/solar, including storage, is in the single digits and too low to constitute a practical electricity system.

In contrast, Figure 5-5 illustrates the case of high EROI and extended device longevity, both enjoyed by nuclear. Note that not only is the required power generation for O&M (height of bottom area) and for replacement and/or constructing a new plant

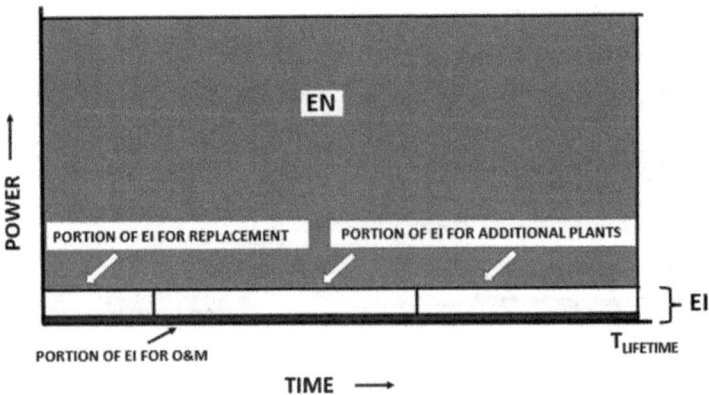

FIGURE 5-5

(height of central area) a much smaller portion of total power return (height of entire bar), and the net power generation a much greater portion (height of top area), but the device lifespan is also much longer (width of entire bar). Furthermore, for fossil fuel or nuclear plants, the material and land requirements are much smaller than for a wind or solar farm with the same output—they have much greater energy and power densities. So the energy needed for replacement and/or additional plants (central area) not only requires less power diverted from EN (less height), but if its power generation is allowed to proceed for the entire lifetime of a power plant, it can produce enough energy to replace itself and to build additional plants if needed. The energy required for one cycle of replacement (of whatever portions of the plant require it) is represented by the leftmost (narrower) portion of the central area (separated by the short vertical lines in the diagram), while the wider portions represent energy for completely new additional plants. So in this illustration, each plant would be capable of generating enough energy to replace itself and to build additional plants if required by expanding social need, even as the plant provides prodigious amounts of net energy to run the society.

If less rapid expansion were needed, more net energy would be available (the central area could decrease even further in height, and the short vertical lines would shift rightward, leaving fewer units for additional plants). If it were desired or necessary to increase the rate of expansion, the requisite energy input could be borrowed from the net energy (the height of the central area could increase, and the short vertical lines would shift leftward, making room for more units for additional plants). Little expense would be paid by net energy generation (top). This is a possibility not afforded by wind/solar without eliminating too much net energy, if not eliminating it completely. If the devices could achieve greater longevity, replacements would be required less frequently to sustain the system. Then even less power would need to be diverted toward replacements.

Thus, a system based on fossil fuels and/or nuclear—with their high EROI, extended plant longevity, and relatively low energy requirement for replacing a plant—can handily run a society and reproduce itself, either to sustain a steady state or to enable expansion, while a system based on wind/solar cannot both sustain itself (if it can even do that) and at the same time provide a decent and desirable amount of net energy. And it certainly cannot expand itself to keep up with a growing population, or permit progressive global industrialization, or provide for other expanding needs.

The low EROI of wind/solar sets the stage for an energy trap. Australian energy analyst Josh Floyd puts it this way (https://tinyurl.com/ns964ahp):

> If a "crash program" is instigated to roll out a large amount of capacity very quickly, then this will lead to a subsequent period around 25 to 30 years later (the typical expected operating life for the installations) where net energy available for purposes other than replacing electricity supply infrastructure drops precipitously. The depth of decline depends on the rate of initial expansion—the higher the expansion rate, the more pronounced the subsequent drop.

Floyd is here assuming that instead of being set aside continually, the energy needed for replacement is all diverted from net energy at the end of the device's lifetime. Thus, the only way for wind/solar to eliminate GHG emissions from the electricity system, if nuclear is to be eliminated as well, is to completely replace fossil fuels with themselves. But if they have only themselves to depend on, the system will collapse. This is the energy trap, the Catch 22. To put it succinctly: to eliminate fossil fuels, and by also eliminating nuclear, an all-RE system ends up eliminating itself.

In fact, end-of-life asset replacement is yet another way to cause EROI to drop below 1, or in other words to render renewables an energy sink instead of a source. Then, Floyd continues,

> the total net energy output from wind would be insufficient to supply the energy required to replace the turbines and storage systems as they reached their end of life. Supplementary energy would therefore be required from other sources, just to support the roll-over of assets.

But if generation of the energy by wind/solar farms to replace themselves were spread out over time, and might take longer than even the 20-to-30-year lifespan of an individual wind turbine or solar PV panel, rather than compressing this into a compact time frame at the end of 20-30 years, the effect alluded to by Floyd would be spread out over time and would act to keep EROI continually lowered throughout the time leading up to, and certainly after, the elimination of fossils and nuclear by 2050, when the "supplementary energy . . . from other sources" will have been eliminated.

┌───┐

Sidebar: An Example of Self-Negation due to Increasing Penetration

Consider the decision that a farm family has to make when their land is occupied by two crops, A and B. Crop A, by its chemical nature, takes a net amount of nutrients from the soil, leaving it depleted, while crop B returns more nutrients to the soil than it extracts. Crop B thereby enables the growth, year after year, of itself as well as of crop A, making A dependent on B.

If someone were to somehow convince the family that crop B is bad and that crop A is good, they might try to increase the land occupied by A, thus replacing B. That is, they might try to increase the penetration of crop A. But in doing so, the supply of nutrients donated to the soil by B would begin to decline, and by the time A were to approach complete penetration, and B were no longer sufficient to donate enough nutrients for itself and A, even crop A would cease to flourish and would soon die off altogether in the nutrient-depleted soil. Thus, increasing the penetration of A would be self-negating.

In contrast, however, increasing the penetration of crop B would incur no such self-negating effect, as crop B could be wholly self-sufficient. So is it the case with wind/solar (crop A) and nuclear (crop B). That is, while increasing the penetration of wind/solar is self-negating, as nuclear were to achieve greater and greater penetration, replacing both fossil fuels and wind and solar, and it no longer had to run on idle much of the time waiting to supplement the intermittent shortfalls of wind/solar, its CF would rise to close to 100 percent, and its EROI would rise along with it. It should be clear that nuclear enjoys enormous advantages in this regard, while wind/solar suffers equally enormous disadvantages.

└───┘

Levelized Cost of Energy (LCOE): Sorry, Wrong Number

Jacobson, Delucchi, and their colleagues explicitly dismiss consideration of EROI in favor of the levelized cost of energy (LCOE)—the monetary cost, per unit output of electrical energy and/or power, ostensibly of everything that is included in the life cycle of the various energy sources. This includes their conversion devices, their storage apparatus, and the added transmission lines. In their response to a critique of their work by Australian energy analyst Ted Trainer, Jacobson and Delucchi (2012) say:

> discussion of embodied energy is irrelevant, because with an indefinitely renewable energy resource with no external costs, the full lifetime cost as we have estimated is the relevant factor—there is no additional pertinence to embodied energy per se.

By the phrase "embodied energy" they mean the energy input (EI) required to obtain a particular total energy return (ER), which is the reciprocal of EROI (ER/EI), namely, EIOR (EI/ER). So Jacobson and Delucchi don't merely neglect consideration of EROI, they explicitly dismiss it, in favor of LCOE.

However, LCOE is not even an appropriate monetary metric for assessing the real costs of an all-RE system for several reasons.

- LCOE combines intrinsic with extrinsic features, thereby partly substituting price to the consumer for real input costs.[14]

- Despite its claim to include everything relevant to the life cycle of the conversion devices, LCOE omits a number of features, including cost of transmission and new grid additions, cost of backup and storage, cost to the environment, cost of extensive removal of land from other uses, and cost of decommissioning and recycling of certain parts of the devices.

- LCOE of wind/solar ignores the influence on lowering their price from inputs provided by the fossil/nuclear matrix.

- LCOE ignores higher government subsidies to wind/solar that artificially and differentially lower their price to investors.

- LCOE ignores arbitrary government rules and regulations that artificially and differentially boost the price of nuclear to consumers, including grid operator decisions to give priority to wind/solar input when it is available, thereby relegating some of the nuclear (or fossil) apparatus to intermittent backup status, which lowers their CF and hence their EROI.

Furthermore, in the quote above, Jacobson and Delucchi completely ignore the costs of the conversion devices, claiming that wind and sunlight provide an "indefinitely renewable energy resource with no external costs." As if nature also provides the installed turbines and PV panels along with the breezes and sunshine.

But perhaps the most fundamental reason that LCOE cannot serve as an unambiguous comparison of apples to apples (or dollars to dollars) is that there are several incommensurable categories of inputs—namely, material, land, time, labor, and energy. The qualitative differences among these inputs impose an arbitrariness in any attempt to measure them all according to the same common feature—in this case, money. How do we compare the relative

---

[14] Cost is a feature of input to a process, while price is a feature of output. The price set by the producer of a capital good, for example a windshield, becomes a cost for the purchaser and user of the good, for example a car manufacturer, who then produces a car that is sold for a price. Costs can be measured either in money or in the material manifestations that are associated with money—materials, land, labor, energy. Price is strictly measured in monetary terms.

monetary cost of a pound of copper to an acre of land to an hour of time to a person-hour of labor to a kWh of energy? Classical economists Adam Smith, David Ricardo, and Karl Marx recognized that only person-hours of labor could serve as such a common measure, but only for items that are in fact the products of human labor. A bare acre of land, an hour of time, and a kWh of energy are not, or not necessarily, products of human labor. So any assignment of monetary value must be partly arbitrary. Unlike EROI, LCOE is a storehouse of extrinsic influences and arbitrariness and, therefore, forfeits its pretensions as an unambiguous standard of comparison among various energy sources.

Of the extrinsic influences, the fossil and nuclear matrix is the most important, though proponents of LCOE simply take their availability for granted and ignore them. As long as the matrix enables the construction, maintenance, operation, and decommissioning of wind and solar farms, it serves to deduct that cost from the renewables' column. As long as the matrix fills in for the renewables' intermittency, it obviates the need for expensive storage apparatus. Meanwhile, the matrix suffers a reduction in its own efficiency, thereby elevating its own costs per TWh. Thus, the matrix lowers the apparent costs of renewables, even as its own cost is elevated. If the matrix could hypothetically be eliminated, the costs of wind/solar would soar. Indeed, if the accounting were done properly, and the artificial cost reduction were eliminated, the reckoning would more accurately reveal the inherent high costs of renewables. This is illustrated in Figure 5-6, in which we exaggerate the difference in the sizes of the bars marked "price" to clarify the point.

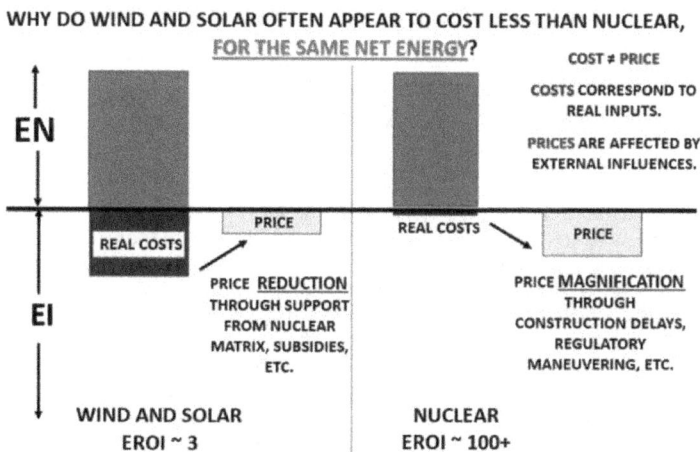

WHY DO WIND AND SOLAR OFTEN APPEAR TO COST LESS THAN NUCLEAR, FOR THE SAME NET ENERGY?

COST ≠ PRICE

COSTS CORRESPOND TO REAL INPUTS.

PRICES ARE AFFECTED BY EXTERNAL INFLUENCES.

EN

EI

PRICE

REAL COSTS

PRICE REDUCTION THROUGH SUPPORT FROM NUCLEAR MATRIX, SUBSIDIES, ETC.

REAL COSTS

PRICE

PRICE MAGNIFICATION THROUGH CONSTRUCTION DELAYS, REGULATORY MANEUVERING, ETC.

WIND AND SOLAR
EROI ~ 3

NUCLEAR
EROI ~ 100+

FIGURE 5-6

Note that while the price for wind/solar is often (but not always) less than for nuclear (for the same EN), the real costs for wind/solar are much greater than for nuclear. The difference between the relative costs to produce and the prices to the consumer, and the interventions that produce that distinction, are obscured by favoring LCOE over EROI.

Thus, as long as wind/solar farms remain at low penetration, their real costs are largely underwritten by, and charged to, energy supplied by the fossil and nuclear matrix. The nominal monetary cost of nuclear includes expenses that would disappear in the absence of wind/solar—and even more expenses would disappear were it not saddled with misregulation by the US Nuclear Regulatory Commission (NRC). The NRC, with no justification, aims to drive down the associated radiation exposure to the plant workers, as well as to the public, to a level that is below that of natural background radiation in most places in the world. In fact, the uranium-containing granite in New York City's Grand Central Station or the Capitol building in Washington, DC—both entirely unregulated—exposes those who work there to more radiation than is received by workers at nuclear power plants. Despite this comparison, the NRC bases their misregulation in part on the claim that radiation exposure inflicts cumulative harm over time, regardless of dose rate (Chapter 8 explains why this is not true). And they therefore justify their imposed requirements by professing to render the alleged harm "as low as reasonably achievable" (ALARA). At first glance ALARA might seem reasonable. But the judgment of what is "reasonably achievable" contains a measure of arbitrariness and has served as a curtain behind which the NRC, with the help of antinuclear environmental organizations, has helped to price nuclear out of the market in the US.[15]

The nuclear industry has generally capitulated and failed to stick up for itself, thereby exacerbating the problem (Devanney

---

[15] In addition to charging millions for licensing of power plants and commonly delaying approval of new reactor designs for more than a decade, the NRC is now considering a proposal to base their "risk analysis" on the worst conceivable plant-destroying accident, regardless of its vanishing improbability, assuming one to occur every year for the plant's entire lifetime (https://tinyurl.com/f73j3cjj). The unreality of the proposal, and its inherent arbitrariness (why not once a month, or week?), leads to a sky's-the-limit cost imposition on nuclear plants. This is not risk analysis; this is sabotage. If this were the accepted method of risk analysis for, say, air travel, only billionaires could afford to fly and airlines would go out of business. In flagrant contrast to the NRC, the Federal Aviation Agency (FAA) is infamous for its laxity in regulating aircraft manufacturing. Boeing planes have recently crashed more than once, and holes have appeared midflight in their fuselages, yet still the FAA relies mainly on Boeing to do their own inspections (https://tinyurl.com/ 5hchnwa6).

2023). This capitulation is partly due to the multiplicity of alternative energy sources on which utilities and equipment manufacturers depend for their profits. The nuclear-related industries, being thus diversified and cushioned against excessive corporate loss, are hardly allies in the attempt to expand nuclear energy, despite the antinuclear movement's painting them as the Goliath in the fray.

The price boosting, in turn, is causing the closure of nuclear power plants that can't make a profit in the capitalist market. And it is prolonging the life of fossil fuels. The NRC claims that it is simply trying to make nuclear power plants as safe as possible, ostensibly in order to avoid illnesses and deaths that would otherwise be magnified. But by leaving fossil fuels in charge, with their respiratory pollution and GHG emissions, the NRC is in fact increasing illnesses and deaths. Meanwhile the media and the rest of the government, along with the antinuclear environmental NGOs, shield the agency from the condemnation the commissioners have earned.[16]

Figure 5-7 is our modification of Figure 5-3, showing the effect of expanding renewables penetration on materials cost (the rising line), including the oversizing/overbuilding of renewables farms, the necessitated growth of storage apparatus, and the required additional transmission lines.

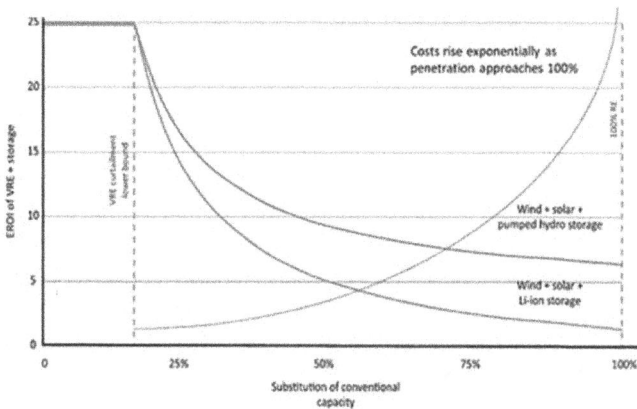

FIGURE 5-7

---

[16] There are countries where the price of nuclear energy is not artificially boosted by governmental regulatory policy. In fact, in Finland there was a recent decision by the operator of one nuclear power plant (Olkiluoto) to cut back on the production of electricity. Why? Because one month after its third and largest reactor began operation, the price of electricity fell to such a low level that the plant could no longer make a profit. The deliberate cutback was designed to boost the price back up to a profitable level (https://tinyurl.com/ 5y7jee8e).

If the material costs rise so greatly with greater penetration, par-ticularly as 100 percent penetration is approached, so must the over-all monetary cost of the system rise. Thus, Jacobson's (and others') claimed cheapening of the overall system at completion is an illusion rooted in the early gains in economies of scale for the manufacturing component of wind turbines and solar PV panels. But past the point of diminishing returns, dyseconomy of scale emerge.

As Floyd puts it (https://tinyurl.com/ns964ahp), many Green proponents deceptively apply the LCOE convention to a projected all-RE system that is "fundamentally different from those for which the convention has previously been established." He continues,

> The rules of the game change fundamentally if the fossil fuel sub-sidy is withdrawn. When different energy supply scenarios are compared on the basis of levelized cost, this is hidden from view. Inputs to LCOE calculations will by default have the financial sub-sidies for fossil fuels built in, but they have no way of accounting for the energy subsidy [from the matrix]. Nor do they include mecha-nisms by which this can be compensated, as the energy proportion from fossil fuels declines, when analyzing large-scale, long term transition to renewables.

Additional Problems with LCOE

Costs other than monetary may favor or disfavor a specific energy source. The rejection of fossil fuels is based on their effect on AGW. The indirect costs are measured, for example, in lost homes, livelihoods, and lives, impacts that can't be measured solely in monetary terms. Even with that proviso, the problem with LCOE lies less in the concept than in the application, since the potential soundness of the approach, done properly, serves to mask its fail-ures when done improperly—the failure to account for the totality of monetary costs and the failure to acknowledge the difference between intrinsic costs and those imposed from without.

An example of the manipulability of, and distortion by, LCOE is the use of 2009 as the baseline for the falling price of renewables. This was a year in which the LCOE peaked because of a rise in bor-rowing costs (the capital cost). Were 2002 to be used as the base-line, when the LCOE experienced an earlier low, the decrease by 2017 would have been virtually nullified (https://tinyurl.com/3y6x82ed). Only longer-term trends, rather than two-point com-parisons, can give a sense of the actual comparisons. That said, these distortions pale in comparison with the effect of eliminating the fossil/nuclear matrix altogether, which alone would render today's LCOE of wind/solar entirely irrelevant.

Fossil fuels have the greatest climate impact, and the changing real costs of AGW are simply incalculable. As we will see in Chapter 6, rather than the fossil industries being charged for this immense externality, they actually receive government subsidies.[17] It is in the arena of externalities that antinuclear individuals and organizations try to put the greatest onus on nuclear, but as we will show in Chapters 7 and 8, the various elements of these claims are mainly fabricated.

Unlike the rules imposed on wind turbines and solar panels, nuclear power plants are required by the NRC to provide and sequester upfront their costs of presumed eventual decommissioning eight to ten decades in the future, which further raises their upfront monetary price. While this is not unreasonable, renewables and fossil fuels are exempt from this requirement. If fossil fuel companies were required to be responsible for their own cleanup, they would not be able to remain profitable and would be forced to close down.

The unavoidable shuttering of the fossil industry, were they held financially responsible for their environmental impact, is precisely the reason they are not made responsible for their waste products. Not only would the massive profits yielded by fossils be eradicated, but we would all then be deprived of our main current source of energy (in the absence of nuclear expansion). Similarly, if manufacturers of wind turbines and solar PV panels were required to be responsible for their products' disposal upon obsolescence, this cost would be added to their temporarily declining price, which increase would have to be paid either by them or by wind/solar farm owners and would be passed on to consumers. As it is, only nuclear plant owners are forced to pay for their own waste disposal (from plant constituents rather than fuel). Thus, while relieved of having to pay a higher retail price

---

[17] An externality is a creation of the profit system, which is founded upon social production that is largely appropriated by private entities. The fragmentation of a closely-knit economy of interdependent parties into separate corporate units competing for profits produces two consequential results (among many)—in addition to private appropriation of the socially produced product, the consequent resource depletion and effluent waste stream are gladly handed over to the public, enforced by the state. In a cooperative economy that remains unfragmented, all consequences, both good and bad would remain internalities. Internalities can be dealt with to balance benefit and cost, but such a balance is obviated by the fracturing of the economy into corporate units, such that one class mainly benefits while the others mainly bear the costs. Beneficiaries will generally regard anything as worth the cost when paid by others. These considerations are discussed in more detail in Chapter 10.

for fossil-fueled or renewable-based electricity, the public is still forced to absorb the cost of waste disposal—both monetarily through taxes and through environmental degradation. This displacement from our visible monthly electric bills to an obscured portion of our taxes further fosters the illusion of relative cheapness of wind/solar, and indeed of fossil fuels. Meanwhile, 12 of the 104 US nuclear reactors in 2012 have been forced to terminate operation.[18]

In a recently published book, German energy economist Lars Schernikau and US environmental science professor William H. Smith (2023) seek a more complete metric which they call FCOE (full cost of electricity). As they put it:

> Levelized cost of electricity (LCOE) is a marginal cost measure and is inadequate to compare intermittent forms of energy generation with dispatchable ones when making decisions at a national or societal level. Using full cost of electricity (FCOE), which defines the full cost to society, wind and solar are not cheaper than conventional power and in fact, become more expensive as their penetration of the energy system increases.

How Long Does It Take to Build a Nuclear Power Plant?

While the real costs of renewables are obscured and underestimated, those of nuclear are popularly overestimated by focusing on the time it takes to construct US nuclear plants and claiming that such intervals are inherent. For example, Jacobson (2009) originally claimed that it takes 19 years to construct a nuclear plant, later reducing his estimate to 8 years.[19]

Whether 19 years or 8, such claims obscure the delays imposed by political rather than technical factors—by extrinsic rather than intrinsic features. These include expanded licensing delays and litigation-provoked prolongation of construction times, with concomitantly amplified borrowing costs. The litigation that results in drawn-out construction delays is often introduced by the very people and organizations who point to the long construction time as

---

[18] Though in the last two years, two new reactors have entered operation at the Vogtle plant in Georgia, making it the largest nuclear plant in the US, with four operational reactors totaling 4,536 MWe (megawatts of electricity) nameplate.

[19] The greatly exaggerated 19-year interval has been picked up and uncritically repeated by others. Among those clinging to the exaggeration is science historian Naomi Oreskes, in her column in the February 2022 issue of *Scientific American*. Ironically, her reference to this nearly two-decade duration is her only objection to nuclear.

an argument against nuclear, making it a deceptively self-fulfilling proposition. Furthermore, the licensing delays are compounded by added upfront (and continuing) expenses imposed by the NRC.[20] Their one-sided mandate is to focus on alleged danger, to the complete neglect of the benefits of nuclear energy and the safety it offers by replacing actually hazardous energy sources. And the five commissioners are given the sole authority by the Administration and Congress to license new plants and thereby determine the length of time it takes. It is not incidental that extrinsic influences that prolong the completion of nuclear power plants not only increase their monetary cost, but also lower their EROI (by amplifying the EI that it takes to produce the added safety equipment that is at best marginally effective).

Belying the 19-years-per-plant myth, both the US and France (the two largest producers of nuclear energy) took 19 years to build almost their entire fleets of nuclear power plants—112 reactors in the US and 56 reactors in France (Figure 5-8, A and B). There is no logical or practical necessity that nuclear plants be built in series

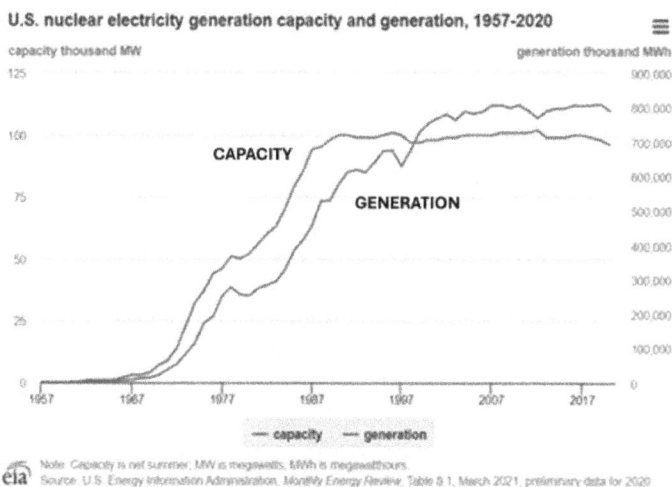

U.S. nuclear electricity generation capacity and generation, 1957-2020

Note: Capacity is net summer; MW is megawatts, MWh is megawatthours.
Source: U.S. Energy Information Administration, *Monthly Energy Review*, Table 8.1, March 2021, preliminary data for 2020

FIGURE 5-8 (A)

[20] As a federal agency with politically appointed members, the NRC generally responds to and reinforces nucleophobic pressures, even when they are not unanimous. The current five commissioners as of this writing, all appointed by a sitting president, include only one person, Annie Caputo, with a degree in nuclear engineering (bachelor of science). And she was only appointed in August 2022 to fill one of two empty seats. The other four possess degrees in divinity, forestry, environmental science, and political science, which reflects the political rather than scientific nature of the NRC (https://tinyurl.com/tebxpntc).

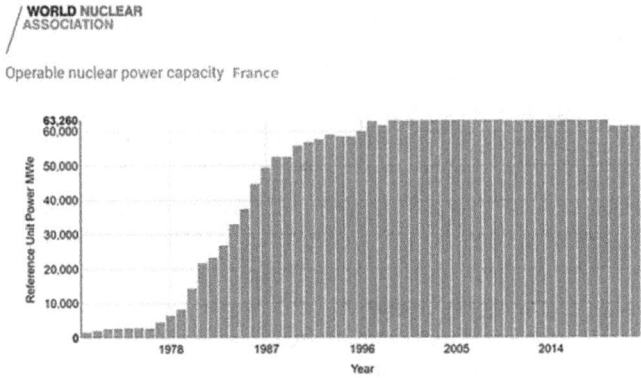

WORLD NUCLEAR
ASSOCIATION

Operable nuclear power capacity France

FIGURE 5-8 (B)

(each one waiting to begin until the previous one is finished) rather than in parallel (many being built simultaneously).

As an example of the speed with which nuclear plants can be built, Figure 5-9 shows the construction duration, from licensing to grid connection, for Japan's fleet of 54 commercial reactors (plus a few experimental reactors). The median duration was less than 4 years, and after the first quarter of the fleet was built, most reactors produce 1 GW of power or more. The graph is con-

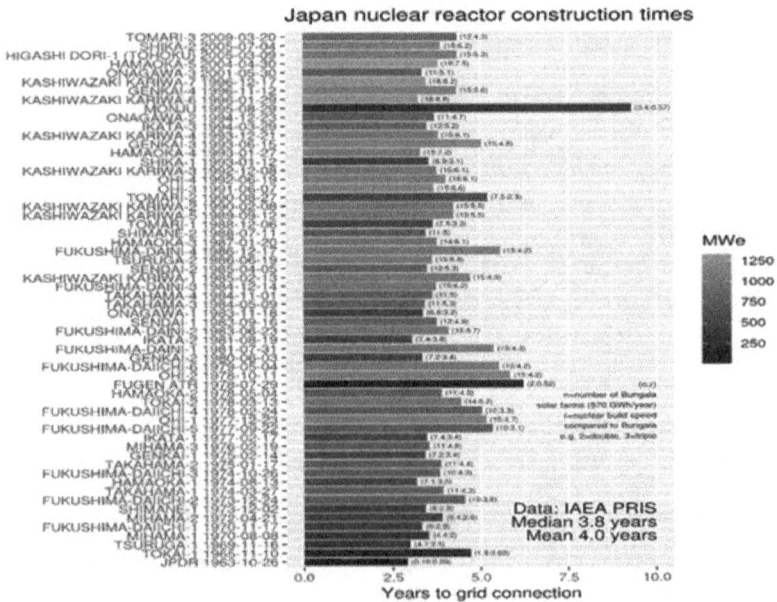

Japan nuclear reactor construction times

Data: IAEA PRIS
Median 3.8 years
Mean 4.0 years

Years to grid connection

FIGURE 5-9

structed from the International Atomic Energy Agency (IAEA) Power Reactor Information System (PRIS) database. This source shows that Japan is not alone in this construction speed, though many countries suffer obstacles that cause delays.

Furthermore, Figure 5-10 (Lovering et al. 2016) indicates the sudden sharp escalation in upfront costs and construction delays for nuclear plants in the USA in the wake of the 1979 Three Mile Island (TMI) accident.

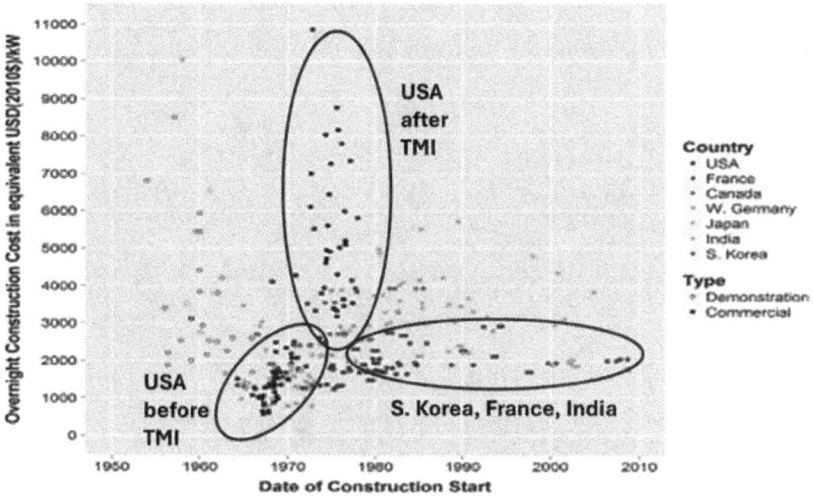

FIGURE 5-10

Note how the plants for which construction was begun between 1970 and 1979, but which were not completed before 1979, experienced massive increases in upfront costs compared to those finished before the events at TMI. And until 2010, no other nuclear plant construction was licensed in the US. Notice not only the sharp rise in costs in the US, but the much lower costs in other countries. This sudden jump reflected a combination of construction delays by antinuclear demonstrations and litigations, as well as contemporaneous imposition of unwarranted safety costs by the NRC. Such impositions not only favor wind/solar but tend to discourage venture capital from being put toward nuclear energy, particularly in light of the smaller government subsidies to nuclear.

Figure 5-11 from the same source exhibits the construction delays, which underlay a significant portion of the increased costs in the US. Following the fearmongering surrounding TMI, the duration of the construction process was prolonged anywhere from 5 to 10 years. The extra costs and construction time are based in the burdens imposed politically by both NGOs and the NRC. Yet they are often falsely presented as costs and construction durations that are intrinsic to nuclear energy. It is of note that Japan, among other countries, suffered no such sudden onset of construction delays (Figure 5-9, which covers reactors built between 1963 and 2009).

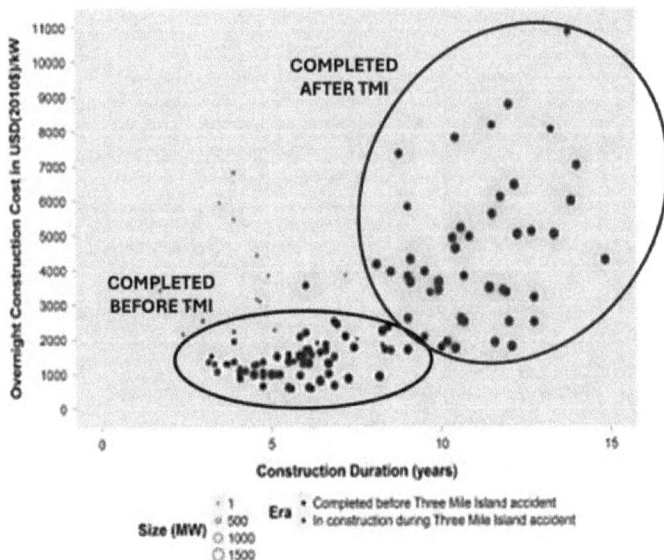

FIGURE 5-11

TMI provided an early opportunity for antinuclear organizations and individuals to promote unwarranted but highly effective and destructive fear. Yet TMI hurt no one, either plant workers or the public, at the time of the accident or in subsequent years, occasional unsupported claims by individuals to the contrary notwithstanding (Walker 2004; Cravens 2007). Indeed, a recently-released documentary movie features women with breast cancer who live near the TMI plant (*Radioactive: The Women of Three Mile Island*). The film attributes the women's breast cancers to the TMI accident and appeals for credibility by portraying the unfortunate cancer victims as the underdogs fighting a powerful corporation and the government (anti-establishmentarianism, Chapter 11). It lets the women's display of reasonable anger provide emotional reinforcement that masks the misinformation. But anecdotes, no matter their apparent plausibility, are not valid evidence, and at best are only suggestive of the need for further investigation. It turns out, however, that TMI's Dauphin County is among those with the lowest overall cancer rates in Pennsylvania, and even lower for breast cancer (https://tinyurl.com/bdvzps8d). (Figure 7-5 indicates just how minor was the release from TMI of radioactive material.) Moreover, the movie shamelessly exploits the fears that have haunted its subjects for decades.[21]

An additional point is that the cost of construction of a nuclear plant, apart from extrinsic delays and opposition, also depends on experience and a learning curve. During those intervals when countries build lots of them, the workers and designers benefit from this learning curve. But during those times when few to none are built, countries see this experience and skill fade and disappear. Then it takes time to re-establish it, during which the cost of a new plant again rises (https://tinyurl.com/4bksj5vk).

Figure 5-12 is a graph compiled by Australian mathematician and energy analyst Geoff Russell (2013). It compares the rates at which electrical energy per capita was added by a number of countries, some adding it through nuclear and some through wind/solar, particularly Germany. He chose various 11-year

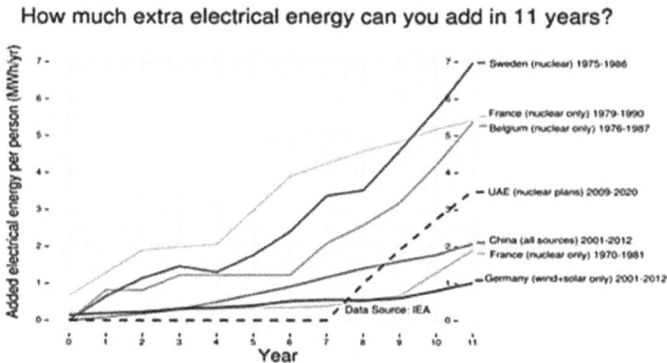

How much extra electrical energy can you add in 11 years?

FIGURE 5-12

---

[21] Against a background of ubiquitous disinformation and misinformation about the alleged carcinogenic nature of low-dose radiation (Chapter 8), several countries—including Japan (following the 2011 Fukushima Daiichi nuclear plant accident) and the US (following the testing of nuclear weapons in Nevada and the TMI accident)—have awarded compensatory payments to cancer victims, lacking a definitive consensus among scientists with regard to a causative role of these events. This has had a snowball effect in which expanding numbers of cancer victims have applied for such payments. While one cannot fail to sympathize with anyone who develops cancer (which includes the present authors), this has wrongly reinforced the attribution to low radiation exposures. Meanwhile, the actual causes remain obscure and escape remediation. See, for example, Daniel Miles 2008 and our discussion of downwinders in Chapter 7.

intervals, and those that seemed to best demonstrate the optimal speeds, so it is not cherry-picked to magnify one versus another. Though it's not up to date, it's quite instructive. Note particularly the rate at which Sweden, France, Belgium, and later the UAE, added electrical energy through nuclear, compared to the rate at which Germany tried to add it through wind/solar. Any claims that nuclear takes too long are taken out of context.

Other countries besides the US, most lacking similar hindrances to nuclear expansion, are continuing to build nuclear power plants. Figure 5-13 shows the number of reactors being built as of July 2024, totaling 59 (https://tinyurl.com/24taxb6y).

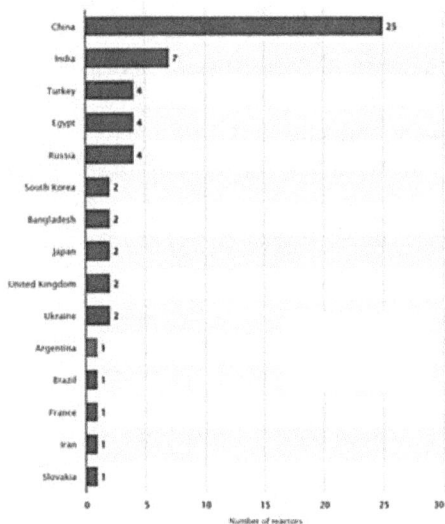

FIGURE 5-13

For wind/solar, the high upfront capital costs, apart from the prices charged by lenders, are derived primarily from the great requirements for material, land, labor, and most importantly, energy—that is, from inescapable technical rather than political or financial features. These are further magnified by the need to build excess nameplate capacities in an effort to obtain the desired actual average output, but without being able to provide reliable second-by-second electricity. The upfront costs, in the absence of the matrix, would be additionally magnified by the associated storage apparatus. In short, aside from their intermittency, the elevated upfront costs of renewables are manifested and magnified largely by the low energy and power density of wind/solar devices and farms, features at which nuclear excels. Able to handily provide ample energy for its own construction and operation as well as to run a society-wide electrical system, nuclear completely avoids an energy trap. So do fossils, but of them the world must rid itself.

6

Low EROI: Jobs and Wages

Renewables and *The Green Collar Economy*

Van Jones's book *The Green Collar Economy* (2008) is a clear expression of the Greens' assumption that an all-RE electricity system will produce plentiful high-paying jobs. Jones holds that a Green New Deal can solve the crises besieging the US, and, by extension, the world.

Green capitalism, he says, will halt AGW by replacing dirty and dangerous fossil and nuclear energy with clean renewables. Stagnation and stagflation, along with poverty, will vanish through the emergence of millions of well-paying green jobs (Jones uses a lower case "g"). He says that stagflation is caused primarily by high energy costs, which raise the cost and price of most other items, thereby discouraging economic activity and job production. As an incidental benefit, energy independence will wean the US economy off oil, thus avoiding the problem of peak reserves.[1] Private capital will be key, with government a partner to green entrepreneurship. Both the American dream and US global leadership will undergo renewal.

Jones counterposes green capitalism and its abundant jobs to the old capitalism, called the "grey economy" and, in his 2009 update, "hypercapitalism." In the first edition, he largely sidesteps the word "capitalism" in favor of dichotomizing "green/grey economies." As the 2008 financial crisis played out, he began to

---

[1] Beliefs in peak oil in the 2000s, shared by us at the time, proved premature.

articulate the c-word but made it clear that he supports capitalism per se and only opposes bad, or grey, capitalism, hypercapitalism, supercapitalism. Good, or green, capitalism is characterized by innovation and job-creating investment—entrepreneurial capitalism partnering with a just-do-it, go-getter government.

For Jones grey capitalism is connected to fossil fuels and big corporations in general, which represent a formidable barrier to change. But he views grey capitalism as doomed to failure, fundamentally anachronistic, spurring on a rapid transformation to the job-producing green economy.

As Jones points out in his first chapter (p. 10), we will need to "weatherize millions of buildings, install millions of solar panels, manufacture millions of wind-turbine parts, plant and care for millions of trees, build millions of plug-in hybrid vehicles and construct thousands of solar farms, wind farms and wave farms." That will require government-sponsored job programs to hire lots of currently marginalized workers, thereby helping to eliminate racialized poverty. Specifically, the prison-industrial complex will be dismantled by transforming two million prisoners (2.3 million in 2008), disproportionately people of color, into workers building the green collar economy—thereby blending the colors of all workers into one.

The new economy, Jones says, will be highly profitable even as it advances general solidarity. Government, green business, labor, and activists will all work together, with an honored place for the latter two—inspiring the rest of the world to follow suit.

The book's foreword, written by Robert F. Kennedy Jr., initiates the excitement. Since the grey carbon-driven economy is regarded as the principal drag on US capitalism, Kennedy suggests that the grey economy will be to the green economy what slavery was to capitalism. Slavery, says Kennedy, was "a ball and chain not only for the slaves, but also for the British economy, hobbling productivity and stifling growth," thereby impeding the creation of new jobs. Once slavery was eradicated, through a surfeit of moral courage, "[c]reativity and productivity surged." Entrepreneurial energies were unleashed, launching the industrial revolution and "the greatest wealth production in human history." With a new Green infrastructure in place now, US productivity and profits will surge, along with massive new job creation.[2] The infrastructure will run on

---

[2] Gains in productivity may or may not be accompanied by job proliferation. Productivity is a measure of the number of people served by the average output of a single worker. If a worker can only produce enough for herself, her productivity is 1. If another worker can produce enough to serve 100 people, his productivity is 100 and fewer workers are needed. Unless alternative products and services proliferate faster than productivity

renewable energy and a smart grid. This will tie together a decentralized energy system, backed by "battery and storage technologies" to compensate for intermittency.

Kennedy predicts rapid growth—based on the assumption that penetration of renewables will follow Moore's law and mimic the exponential growth of computer chip production and computing efficiency. He quotes energy entrepreneur Stephan Dolezalek who says, "The Internet boom caused information flow to increase exponentially, but the price per bit dropped to almost zero. The same thing, can happen with energy."

Smil (2010), for one, exposes the fallacy in Kennedy's analogy by noting that it "completely ignores the need [of renewables] for massive infrastructures to harness, process, transport and convert energies." Advances in computer chips are characterized by a decrease in materials, the "very opposite" of the situation prevailing in energy production, distribution, and consumption. Says Smil (p. 125):

> Coal and uranium mines, oil and gas fields, coal trains, pipelines, coal carrying vessels, oil and LNG tankers, coal treatment plants, refineries, LNG terminals, uranium processing (and reprocessing) facilities, thermal and hydro electricity generating plants, HV transmission lines and distribution lines and gasoline and diesel filling stations constitute the world's most extensive, and the most costly, web of infrastructures that now spans the globe. Its individual components number between thousands (large coal mines and large thermal power plants) and tens of thousands of facilities (there are about 50,000 oil fields) and its worldwide networks extend over millions of kilometers.[3]

---

grows, the number of jobs can decrease. Furthermore, the wages of the workers serving many fewer people can be as high as the wages of the higher-productivity workers only if the money comes from a source external to the production process, such as government subsidies. And while the production process itself would support higher wages for the more productive workers, they will almost never be forthcoming unless the remaining workers collectively fight their employers for that compensation.

[3] A further irony, pointed out more than a decade earlier by US journalist William Greider (1997), is that the decrease of materials in computer chips takes for granted an infrastructural matrix unacknowledged by Kennedy, nor even by Smil. Greider says (p. 484):

> The industrial paradox, however, is that in order to achieve these advancements in power multiplication and cost reduction, the companies must assemble increasing volumes of investment capital for the research and development and construction of the new factories. Thus for instance, a factory for 64k chips in 1980 cost about $100 million to build, the 256k factory in 1985 cost $200 million, the 1m factory in 1988 cost $300 million, the 4m factory in 1990 cost $400 million and the 16m factory in 1993

Our focus in this chapter is the Green claim that an all-RE electrical infrastructure would entail a massive proliferation of well-paying jobs. Certainly, Kennedy's choice of an analogy between higher renewables penetration and advances in computer chip design is both a flaw in his logic and a one-sided failure to see both the downward as well as upward pressures that higher penetration exerts on wages and jobs. As we will see, the jobs will grow, at least at first, but the wages will only follow if subsidized from outside the electricity system. In short, energy transitions have little in common with advances in chip power. For one thing the sheer scale of the former dwarfs that of the latter and therefore proceeds at a much slower pace. And our extreme dependency on uninterrupted energy exerts an even greater slowing effect.[4]

Acting as a further drag, were a rapid transition from fossils to wind and solar possible, all the capital tied up in, say, pipelines would be devalued and/or destroyed and replaced by a renewables infrastructure that would necessarily, because of its low energy and power density, be far more extensive. The dismantling of the fossil infrastructure would be hampered by asset inertia or stranded capital (Chapter 10).

It's not simply that hopeful Green innovators like Jones, Kennedy, and of course Jacobson gravely underestimate their projected transition times; it's that they seem oblivious to the dependency of low-EROI renewables on high-EROI fossils (or nuclear), and to the effect of eradicating the very foundation on which their plans rest. Nor do they seem to grasp that their low-EROI objective would condemn societies to net energy poverty and to consequent widespread socioeconomic poverty. Without a high-EROI economy, high wages can at best be won by a small part of the working class but not by the majority.

---

cost $700 million. A fully automated 64m factory in 1995 cost $1.2 billion and several new factories have been announced that will cost $2 billion. If the capital curve continues in this manner, companies may soon be spending $5 billion or more per factory to keep up with Moore's Law.

Neglecting production infrastructure for diminishing chip size is akin to neglecting energy conversion devices for turning completely free wind and sunlight into electricity.

[4] For example, the US alone has 3,000,000 miles of NG pipelines (https://tinyurl .com/3366fxp6) and 200,000 miles of oil pipelines (https://tinyurl.com/mrnz6rz4), along with more than 700,000 miles of electric transmission lines, most of which is local transmission lines (https://tinyurl.com/nhzyvvrp), though some sources suggest that in addition to the long-distance lines there are more than 5,000,000 miles of local distribution lines (https://tinyurl.com/jajrt5dx). While these estimates vary, the order of magnitude is, as usual, what counts here.

In Chapter 9 we critique the views of other Green thinkers who believe we can have low-EROI, low-productivity economies, and still have high wages and more leisure. In this chapter we focus on the assumptions of Jones and Kennedy about the impact of renewables on jobs and wages, as they are held by many renewables advocates.

Job Numbers, Wages, and Consumer Bills at Low RE Penetration

The numbers from 2016 of US energy production jobs per TWh are displayed in the following table, covering some 90 percent of the total.[5] The jobs per TWh have changed somewhat between 2016 and 2923, but the orders of magnitude remain the same.

NUMBER OF JOBS PER CONTRIBUTION TO US ELECTRICITY (2016)

ENERGY SOURCE	NUMBER OF JOBS	TWh OF ELECTRICITY	JOBS PER TWh	RELATIVE TO NUCLEAR
COAL	160,119	1,240 (30.4%)	129	1.36
NG	398,235	1,380 (33.8%)	289	3.04
WIND	101,738	204 (5%)	499	5.25
SOLAR	373,807	36 (0.88%)	10,384	109.3
NUCLEAR	76,771	806 (19.8%)	95	1

The numbers in the table describe the present situation in which the penetration of wind/solar is relatively low. Under these circumstances, the energy for the construction of wind/solar farms comes mainly from the fossil/nuclear matrix. The number of TWh produced by each energy source is an approximate measure of both their total energy output (ER) and the net energy they produce for the society (EN). That is, since ER comprises the sum of both EN and EI (ER = EN + EI), there is little to no difference

---

[5] The table is taken from the US Department of Energy's "US Energy and Employment Report" (www.tinyurl.com/52bkcbm0), which is referenced in a 2017 *New York Times* article (www.tinyurl.com/2h2xtp8n). The data in the third column comes from the DOE's Energy Information Agency (EIA) (www.tinyurl.com/18eupblo). Since wind's contribution is not provided apart from biomass and geothermal, our estimate for wind is an approximate average obtained by a web search of alternative sources that give slightly different estimates. The fourth column is the second column divided by the third, and the fifth is the fourth column relative to nuclear.

between the ER and the EN produced by wind/solar when the energy input (EI) to construct wind/solar farms and fill in for their intermittency is entirely provided by the high-EROI fossil/nuclear matrix and does not come out of the ER produced by the renewables. To put it another way, when the matrix is present to provide the EI for wind/solar, the ER produced by wind/solar is virtually all devoted to EN.

This changes dramatically as wind and solar approach and hypothetically arrive at 100 percent penetration. Under that circumstance, wind and solar would be forced to provide the energy to construct and replace their own conversion devices. And in the absence of the fossil/nuclear matrix they would further be forced to provide the energy to manufacture storage devices (batteries) to fill in for their intermittency. Thus, the portion of their ER devoted to EI rises from virtually nil to a substantial magnitude, thereby diminishing the portion of their ER devoted to EN. As a result, it requires more wind/solar farms, or larger ones, to produce enough ER to reestablish the EN demanded by the society. Furthermore, the additional need for energy to manufacture batteries increases the EI to an even greater extent. With that summary, let's examine the actual numbers—first at low penetration, as shown in the table, and then as penetration approaches 100 percent.

Under low penetration, for each TWh of total electrical energy produced per year (ER), the table shows that solar requires two orders of magnitude more workers than nuclear and for that matter almost that much more than fossil fuels. Wind requires many fewer workers than solar but still several times more than nuclear per TWh and more than for fossils. These figures apparently refer to actual electricity production rather than nameplate.

As we approach 100 percent penetration, and the difference between ER and EN becomes significant, the number of jobs per *total* TWh (ER) needs to be adjusted to indicate the number of jobs per *net* TWh (EN). From Weißbach et al. (Figure 5-1), the EROI for wind, at least in Germany, is 4 (given storage), meaning that EI is 1/4 of ER, leaving 3/4 of ER for EN. Thus, to estimate jobs per TWh of EN we have to increase the 499 in the table to **665** (4/3 x 499). And for solar, EROI is 2, meaning EI is 1/2 of ER, leaving the other 1/2 for EN. Thus, the 10,384 in the table has to be increased to **20,768** (2 x 10,384).

These adjustments, in turn, raise the ratio of jobs per TWh *of EN* for wind to that of nuclear from 5.25 to **7** (4/3 x 5.25) and for solar from 109.2 to **~220** (2 x 109.3). Since nuclear's EROI is roughly 100 for LWRs, the 1/100 required to generate more energy is trivial, so no adjustment is needed.

To put it the other way around, *under current conditions of low penetration* a nuclear worker produces over 5 times as many TWh/year—or, more manageably for individual workers, kWh/year—as a wind worker and over 100 times as many as a solar worker. From the second and third columns of the table above, we can calculate that each solar worker, on average, produces roughly 100,000 kWh per year (36×10^9 kWh/37×10^4 workers); each wind worker some 2,000,000 kWh per year (twenty times as much as solar); and each nuclear worker 10,500,000 kWh per year. The numbers under hypothetical conditions approaching 100 percent penetration are discussed in the following section.

Since the grid scrambles the various energy sources, one kWh is billed to the end-consumer the same as any other regardless of its source. In 2024, the average retail price of a kWh of electricity in the US was almost 17 cents—ranging from about 10 cents (Nebraska) to 44 cents (Hawaii). So, taking them in reverse order and using the average retail price, a nuclear worker generates about $1.8 million worth of electricity in a year, a wind worker about $340,000, and a solar worker about $17,000, on average. Much of the generated funds goes toward the energy input required to generate the energy output, with only the remaining portion available, at most, for annual wages. This, however, does not mean that all of the remaining portion is granted to workers as wages, just that it represents an upper limit on the portion that workers can attempt to obtain. And considering net energy return, rather than total, these numbers are even smaller for wind/solar.

So, as you can see (or surmise), even under current circumstances at low penetration a solar worker does not come close to generating a living wage for her/himself, nor does a wind worker. In contrast, only a nuclear worker does. That's why wind and particularly solar electricity require government subsidies if these jobs are to be "well paying." Without subsidies, the jobs would have to be performed by forced labor (by prisoners or slaves)—or workers would have to be coerced by extreme poverty.

Alternatively, if wages comparable to those for nuclear workers were to come from the solar or wind industry without subsidies, the consumer price per kWh of electricity from renewables would have to be close to two orders of magnitude higher for solar, and one order of magnitude higher for wind, than for nuclear. Then many, if not most, consumers would find renewable electricity unaffordable. Either the renewables production workers would suffer unlivably low wages or the consumers, who are mainly workers as well, would suffer unaffordably high prices. Neither of these outcomes would be desirable or tenable for the working class. So government subsidies are the only acceptable alternative, with those awarded to renewables dwarfing those to nuclear (see Figures 6-1 and 6-2). As consumers, then, we end up paying one way or another—through our electric bills or our tax bills—and the higher electrical prices reduce our real wages. Moreover, only with the continued existence of the matrix, before 100 percent penetration of wind/solar were to be reached, would there be enough net energy generated by the society to underwrite government subsidies.

In contrast to the US, Germany currently provides lower subsidies to renewable energy producers and instead subsidizes industry, mainly in the form of tax credits, to help it pay for the highest electricity prices in Europe. This Energiewende policy boosts the average price to 37 cents per kWh for consumers (https://tinyurl .com/ycyr5ukp).

According to Bryce (https://tinyurl.com/yf2zhemh), who includes the total electrical energy production by each source and refers only to the portion of subsidies represented by tax breaks:

> In 2018, the American solar industry got roughly 250 times as much in federal tax incentives as the nuclear sector, when compared by the amount of energy produced. Coming in a close second is the wind sector, which got about 160 times as much as nuclear.

Field installation of solar PV panels and wind turbines is achieved by direct manual labor using ancillary machinery. As a result, installation productivity is far less than that of manufacture in factories, where automation can take the place of large numbers of workers. So, being inversely related to productivity, the number of workers engaged in field installation will inevitably outweigh the number of workers in the manufacturing facilities, given that over time the number of panels and turbines manufactured will be the same as the number installed. It is partly the growing productivity in the manufacturing facilities that helps to account for the advertised progressive cheapening of wind and solar, but only for wind/solar farm owners who purchase the turbines and panels, not for electricity consumers. Cheapening also comes from features extrinsic to renewables, including support from the fossil and nuclear matrix, government subsidies, grid priority given to wind and solar by regional grid operators, and reduced borrowing costs.[6]

Job Numbers and Wages at High RE Penetration (Jacobson's Roadmap)

From the above EIA table of job numbers at low wind/solar penetration, and taking into account the required modifications when dealing with penetration approaching 100 percent, and therefore focusing on EN rather than ER—when it makes a difference, as discussed above—we can estimate what the actual corresponding job numbers (not just ratios) would be for wind/solar workers at 100 percent penetration, were completion of Jacobson's Roadmap possible in the US.

Jacobson projects that upon reaching 100 percent penetration wind and solar will each provide roughly half of US electricity. Since the US consumes about 4,000 TWh of electricity per year net (EN), almost entirely coming from commercial farms and power plants (with minimal coming from rooftop solar), 2,000 TWh is to come from wind and solar each. From the above calculations of jobs per TWh of EN at low penetration near 100 percent penetration 2,000 TWh from wind would require at least 1,330,000 workers (2,000 x 665) and 2,000 TWh from solar at least 41,500,000 workers (2,000 x 20,768), for a total of more than 42,830,000 workers in electricity production.

While today only ~18 percent of primary energy in the US is in the form of electricity (see Figure 2-1), the Roadmap aims to electrify almost all primary energy in the US, which would increase electrical energy consumption by a factor of ~5.6 (1/18 percent) —roughly equal to the factor 5.8 for the entire world (Chapter 4).

---

[6] It is the far greater number of field installers and maintenance workers that will therefore be the trend setters for wages in those occupations—again, tempered by government wage subsidies. But wages in low-productivity endeavors are severely limited compared to those in high-productivity processes—even if maximizing those wages is dependent on the level of organization and struggle by the workers themselves. Indeed, gains in productivity through automation raise the ceiling on wages, but they don't by themselves raise the wages, as many economists and the media often imply. It's a very rare company that will voluntarily raise wages to levels higher than circumstances demand—though when circumstances are altered by government subsidies, the downward pressures on wages due to profitability considerations can be partly relieved. However, if and when higher wages are won, those higher wages end up being spread among fewer workers, with the rest laid off and without wages altogether (except perhaps with unemployment insurance, another largely government subsidy). Moreover, the totality of wages in an automating factory ends up less than it was before the leap in productivity. The total reduction in wages is a central, if not the entire, purpose of automating in the first place.

At or near 100 percent penetration, storage would have to be greatly expanded, including manufacture of batteries that would require continual replacement, and additional oversizing/over-building of wind/solar farms not just to keep the average output equal to demand (the inverse of the CF) but also to keep batteries charged and available for any unpredictable wind/solar downtime. And since any system requires coverage for peak demand, this necessitates yet another layer of overbuilding/oversizing. (As mentioned above, the US presently has a capacity that is 2.5 times average demand.)

Thus, the electrification of almost all primary energy would raise that number to **~241** million workers in energy production (5.6 x 43 million). This exceeds by a wide margin the current number of employed workers in all US jobs, a little over 160 million, and represents close to three-quarters of the entire US population. Not only are there fewer people available in the US of working age, but it would leave no workers to perform the myriad of tasks required by a complex economy and which are the reason we produce energy in the first place.

This situation, however, could never even be approached, because long before wind/solar reached full penetration, other factors would emerge to halt the elimination of the fossil/nuclear matrix—such as the exhaustion of material resources and low EROI. Nevertheless, this exercise is useful in showing that even on the basis of required job numbers, such a project would fail.

But sheer numbers don't tell the entire story. Many kinds of work require training, so turnover would become another source of delay. The need for formerly unemployed workers, or those in jobs that pay too little for survival without government supplementation, to move their families in order to work in distant areas for largely temporary installation jobs causes both a delay in production and generally oppressive dislocations for such workers. Such dislocations have already occurred for oil fields that have exhausted their supply (see, for example, https://tinyurl.com/bdeezc2s). While wind and sunlight will not vanish, AGW is already producing a geographical shift in optimal locations, and such locations will sooner or later be exhausted, thus adding further to the dislocations.

So yes, growing penetration of renewables, on the way to an all-RE system, would, if it were possible to construct, reproduce, and expand such a system, create lots of jobs—creating, in fact, a severe labor shortage. But the other parts of the economy would change drastically as the energy poverty associated with renewables rose.

The severe labor intensity of a wind/solar-powered electrical system, plus the paucity and unreliability of energy produced, would fail to support many other jobs, thereby causing a rise in unemployment in other industries. This would be very disruptive of any economic system, not just capitalism.

Finally, added to the problem of sheer numbers of jobs, and added to the problem of supply chains for materials, is the problem of shortages of skilled labor. Furthermore, the expansion of the electrical grid to handle an all-RE system is vast. The US Department of Energy (DOE) estimates that it would require at least an additional 57 percent of long-distance transmission lines, on top of the 240,000 miles that exist in the US today. Bryce recounts the problems of supply chains for trained linemen, in addition to those for materials like copper and steel for wiring, transformers, and the like (http://tinyurl.com/34tncpuy). He also shows that the current rate of grid expansion would have to accelerate to a rate four to five times as great, even as the cost per mile is rapidly rising (https://tinyurl.com/38jah8h3).

Required State Subsidies for an All-RE Electricity System

Without external subsidies from government and from reliable fossil and nuclear backup, wind/solar farms would have been stillborn. Near day one they would have priced themselves so far out of the market that only a really untalented venture capitalist would have invested in them—until the mistake was recognized. As it is, this was the widely advertised experience of Texas oil tycoon and corporate raider T. Boone Pickens, whose investments in wind came crashing down more than a decade ago due to NG's price decline, causing him to abandon wind.

The value of actual US government subsidies assigned to renewables versus fossils and nuclear is indicated in a bar chart (Figure 6-1, borrowed from https://tinyurl.com/mvw9eepf.) If these values were to be normalized to the same energy output, the differences between nuclear and renewables would be greatly magnified.

The "other" category (O) represents technology-neutral energy subsidies, like transmission and conservation efforts, which aren't tied to specific sources. Note how the amount assigned to nuclear (N) was vastly lower than that assigned to renewables (R) and became successively more so throughout the decade. Even fossils (F) received more absolute subsidy than nuclear (more than twice as much energy produced by fossils), though since 2010 the subsidy was lowered for all energy sources.

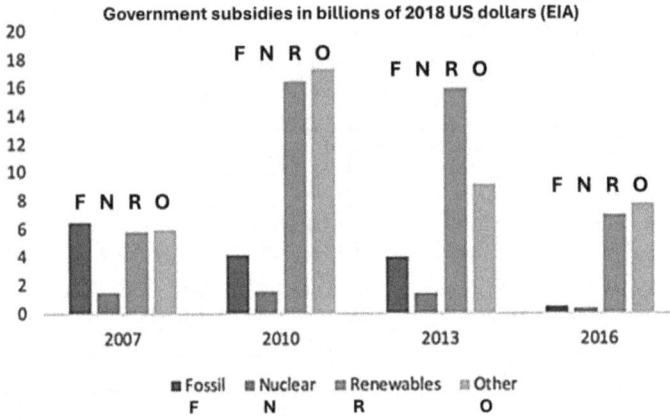

FIGURE 6-1

Government monetary subsidies can take the form of direct grants, tax credits, or cost-plus contracts, in which the builder of wind or solar capacity is rewarded with a certain fixed percentage above their costs. Among other costly policies for taxpayers, cost-plus contracts incentivize venture capitalists to spend as much as they can to build wind/solar farms, as that will maximize their profit.

The portion of subsidies in the form of tax relief in 2016 shows an even starker contrast (Figure 6-2, https://tinyurl.com/

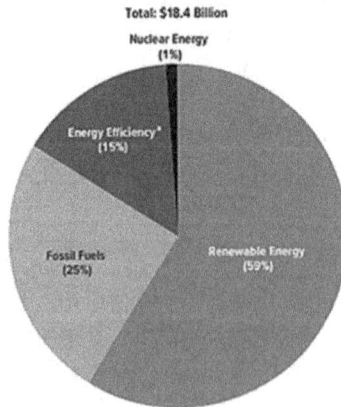

FIGURE 6-2

ycyev5s2). As the pie chart shows, the tax-preference component of the total subsidy to wind and solar is more than double that granted to fossil fuels, but it positively dwarfs that granted to nuclear in the ratio of almost 60 to 1. Given that in 2016 the total electrical output from nuclear was roughly twice that from wind and solar together, the ratio of wind/solar tax preference to that of nuclear is almost 120 to 1 per TWh produced (ER rather than EN).

Bryce (https://tinyurl.com/mufnzcrs) reports that in 2022, under the Inflation Reduction Act (IRA), the ratio of federal tax credits to solar outweighed those to nuclear (per quad of energy produced) by a factor of 302 to 1, and to wind by a factor of 70 to 1.

The Transition from Fossil Fuels to an All-RE Electrical System Is Self-Negating

So renewables, even at low penetration, are not a fountain of well-paying jobs—jobs yes, well-paying no. The Green fantasy can only be fully comprehended through comparisons of EROI (Chapter 5).

Moreover, a 100-percent wind-and-solar-based economy would guarantee energy poverty rather than abundance. The illusion that solar and wind can be scaled is premised on a simple extrapolation from low to 100-percent global penetration, ignoring the changes that necessarily accompany such a transition. The dream of plentiful well-paying Green jobs presupposes the continued existence of the matrix. The economies of scale that seem to cheapen wind turbines and solar panels come from the fossil-nuclear matrix. The green economy of Jones and Kennedy is parasitic on the grey economy and cannot replace it. In short, as their green economy expands its penetration it becomes self-negating.

While the rise of capitalism was partly fueled in England in the late 1700s and early 1800s by the mechanization of agriculture and legalized land enclosures at the hands of the state, the latter expelled large numbers of peasants (agricultural laborers) from the land and forced them into cities to work in relatively high-productivity factories. The opposite would be produced by the de-mechanization of energy production. Low-productivity installation of an energy infrastructure consisting of wind/solar conversion devices would pull large numbers of workers back into the rural areas. Ironically, while intended as a foundation for an industrial society, renewables would, in fact, undermine that objective. As we have said, it would return us to a kind of energy feudalism. But while agricultural feudalism lasted hundreds of years in the absence of a contemporaneous alternative, this would likely not last a few decades—if it could even come into being in the first place. Which, we argue, it could not.

Indeed, Bryce reports that the European Union is already seeing a fall-off in their industrial production and in their overall electricity consumption, both due to the rising price of electricity as cheaper Russian NG is cut off and renewables increase their penetration (https://tinyurl.com/bxjf9ckv).

The overriding point is that low EROI guarantees abundant low-productivity, low-paying, labor-intensive jobs that are, at the same time, insecure, tedious, and fatiguing. Low EROI, by depleting energy and shifting workers, also prevents the proliferation of a more interesting variety of tasks, and hence jobs, enjoyed by a complex society.

In summary, the decrease in EROI associated with the approach of an all-RE economy would be extraordinarily undesirable and therefore eventually vigorously resisted by virtually everyone, including members of both the exploiting and exploited classes. For the same energy invested in its own production, it would mean at least an order-of-magnitude less energy available not only for the extraction and manufacturing industries, but also for the private consumption of such things as cooking, heating, transportation, communication, entertainment, education, health care, and infrastructure repair and maintenance.

But resistance aside, the main reason that such a return to centuries past is even contemplated, let alone applauded, when a modern alternative is readily available—one with all the advantages and none of the disadvantages—is that this alternative is the victim of an utterly false and deliberately fostered defamatory label of "DANGER."

III

Nuclear Energy: Why We Should

7

Nuclear Energy: Nothing to Fear but Fear Itself

It ain't what you don't know that gets you into trouble. It's what you know for sure that just ain't so.

—Often attributed to MARK TWAIN

The Nature of Safety: Relative versus Absolute

First, let's get one thing straight. The most dangerous (and most prolific) fallout from a nuclear power plant accident is disinformation.

Second, let's acknowledge that nothing's absolutely safe, whether gas or electric stoves, fossil-powered vehicles and aircraft, self-driving cars, fireplaces, furnaces, electricity, food, water, medicines. Nor is cooking, eating, running, swimming, breathing, staying home, or a host of other normal activities. All involve risk, but we tolerate that risk because of the indisputable advantages. And as harmful side effects have occurred, we have learned collectively how to make each of these progressively safer.

Moreover, we rarely consider the risks of *avoiding* such everyday items and activities—of refusing to use stoves to prepare food, or vehicles to rescue us from undesirable or hazardous situations, or furnaces for warmth, or medicines that have side effects in some people. Avoiding masks and vaccines contributed to over 1.1 million US pandemic deaths. New technologies always produce unanticipated hazards, but such hazards, once comprehended, can be prevented or mitigated.

The Hindenburg was a blimp that caught fire in 1937 while docking. To make it lighter than air, it was filled with hydrogen, the lightest element of all. The accident killed 36 people, roughly a third of those aboard plus one victim on the ground. Realizing that hydrogen, known to be flammable, could be ignited by a spark, blimp designers turned to non-flammable helium, the next lightest element. No blimp fires have occurred since.

The Chernobyl plant can be thought of as the Hindenburg of nuclear reactors—with deaths directly due to the accident numbering roughly the same. Neither accident caused as many deaths as most airplane crashes or high-rise building collapses.

The Chernobyl plant was designed for a dual function—production of electricity and production of plutonium for nuclear bombs.[1] To enable timely periodic removal of fuel from the reactor for the latter purpose, the reactors lacked the outer building that surrounds all other commercial reactors. The reactors also used a graphite moderator (that facilitates the fission reactions) and a water coolant (that both absorbs heat from the core to keep the reactor from overheating and also, outside the core, turns to high pressure steam directed at the turbine that drives the electricity generator).

On the eve of the accident, operators were taking advantage of a routine maintenance reactor shutdown to test the auxiliary equipment that keeps the coolant circulating during such an event. To simulate a worst-case scenario, they disconnected the electricity to the pumps that normally send coolant through the reactor core, assuming that the auxiliary diesel pumps would come on soon enough to prevent the core from overheating. However, the diesel pumps were slower than anticipated to arrive at full power, and the diminished cooling led to a rise in temperature that caused a speedup of the fissioning and the formation of large steam pockets inside the core. This further reduced the cooling while leaving the moderating function of the graphite intact. The result was amplifying feedback, which was further worsened when the operators reinserted some of the control rods to slow the fissioning, but instead their graphite tips (intended for lubrication but also acting as moderators) caused it to accelerate. The rapid gain of temperature in the core increased the steam pressure, causing an explosion of coolant tubes and igniting a fire—a steam explosion, not a nuclear blowup. While firemen were trying to quench the fire, the

---

[1] However, most, if not all, plutonium for Soviet bombs was produced at three specially designated plants in the South Urals and east of there in southern Siberia (https://tinyurl.com/yc78by9t). Much of it was produced before 1977, when the first Chernobyl reactor started up (https://tinyurl.com/yw7ryxbz). While electricity was the Chernobyl plant's primary, if not only, product, its dual-function design was intended as backup if needed.

graphite finally ignited. Graphite fires are notoriously difficult to put out, and quenching it took 10 days. During this prolonged effort firemen in helicopters, dumping sand and other quenchers, were exposed to extreme amounts of radiation emitted through the destroyed roof of the reactor. The combination of graphite moderator and vaporizing water coolant mimicked the role of the Hindenburg's hydrogen. We leave the ensuing events to Chapter 8.

Lessons were learned from both the Hindenburg and Chernobyl accidents. No other reactor designs outside the Soviet Union permit amplifying feedback, and the accident led to modification of all Soviet reactors to eliminate that feature. All other reactor designs include a moderator/coolant combination that exerts countering, rather than amplifying, feedback if the reactor core starts to overheat. This is achieved either by combining the moderating and cooling functions so that diminution of the cooling function is always accompanied by diminution of the moderating function, and/or through a design that imposes a second automatic cooling mechanism, such as a heat-caused expansion that decreases the density of the fuel below the critical density required for the fissioning to proceed. A temperature gain is thereby automatically countered if the coolant happens to form steam pockets or leaks, or if the core starts to overheat for any other reason.

Just as no further Hindenburgs have been produced, so have no further Chernobyl-type reactors been built. And just as no fire can ever again be ignited by the gas in a blimp, no fire can ever again be ignited by the gas process in any other reactor. The only way a fire can be ignited at a reactor site, or an explosion take place, is through a chemical combination of oxygen with some element or compound, but not by the fission process. In any case, the physics of the design, rather than people's actions, prevents a nuclear explosion in a power plant. It is no more possible than a steel bar's catching fire from a match or your bicycle's standing up by itself and rolling up a hill.

Thus, the chemical explosion at the Fukushima Daiichi nuclear plant in 2011 was due to the exposure of fuel rods to steam as the cooling water boiled away. The water boiled away because the auxiliary diesel cooling pumps for units 1 through 5, negligently located in the plant basements, were submerged by the tsunami. Only the pump for unit 6 was located higher and escaped the flooding; moreover, it was able to keep cooling the fuel in both units 5 and 6. The contact of fuel rods and steam produced free hydrogen, both combustible and potentially explosive.

In contrast to Chernobyl, all six reactors at the Fukushima Daiichi plant were enclosed not only in containment domes but also in outer buildings. Hydrogen gas inside units 1, 3, and 4 exploded when it met sparks (as in the Hindenburg), blowing holes in their outer buildings. The properly designed containment domes remained intact, and because of the countering-feedback reactor design, the cores of 1 and 3 (as well as that of unit 2) automatically stopped fissioning. Units 4, 5, and 6 were already shut down for maintenance. But the highly radioactive fission products (see Glossary and explained below) continued to generate heat as they decayed, and the drowned diesel pumps could no longer circulate the water coolant to absorb that heat. Had all the pumps been located as high as that of unit 6, there would have been no accident at the Fukushima Daiichi plant, despite the tsunami.

Sidebar: Chain Reaction, Fission, and Fusion

This phrase "chain reaction" often elicits fright, but all it means is "the domino effect," in which an initiating event gives rise to a second event of similar type that in turn gives rise to still another, and so on. A fission reactor involves a set of parallel chain reactions that, once initiated, continue at a steady rate unless deliberately sped up or slowed down by the operators.

Conventional usage of the phrase fails to distinguish between each event's causing only one further event and each event's causing multiple further events. (More precisely, a set of chain reactions need only encompass one reaction causing one further reaction *on average*. Some may cause more than one while others cause none, as long as the process remains in a steady state.) To keep these separate, we term the latter a branching reaction rather than a chain reaction. A chain is one-dimensional, while branching can only be depicted as at least a two-dimensional image (like a family tree).

A chain reaction lends itself to human control, while branching may be controllable under certain circumstances but may also be able to escape human control. After initiation, both chain and branching reactions are self-sustaining, until fuel is exhausted or rendered inaccessible (through dispersion), or the conditions of their progress are otherwise terminated. A branching reaction in a bomb requires only a split second.

A fireplace blaze is a mix of the two types of reaction. Once enough heat is provided by a match to a properly arranged stack of wood pieces, one or more events are enabled that consist of the combining of a molecule in the wood with an oxygen molecule from the air. That combining event (in this case, a chemical process) releases at least as much heat as the match and enables still further combining events, in a process called combustion. While initiation of the process requires the input of heat to the system (the wood stack), its continuation provides the heat needed for the subsequent events, and they will continue as long as wood and oxygen are available. The initiating event that comes from outside the system (the match) is called in climate science a "forcing," while the term "feedback" is applied to the subsequent set of self-sustaining events that are internal to the system. Feedbacks can be either amplifying and accelerate the process or countering and decelerate it.

Fire can be controlled in a set of parallel one-dimensional series of events, or chain reactions, as in a stove or fireplace once it stabilizes. Or a fire can be uncontrolled in a proliferating multi-dimensional series of events, a set of branching reactions, as in a house or forest conflagration, until it either stabilizes or begins to run out of combustible material and dies down or is extinguished by human effort that interferes with the supply of oxygen.

The parallel chain reactions in a nuclear fission reactor keep going once the process is initiated, as long as the fissionable nuclei are properly arranged geometrically and as long as non-fissionable nuclei don't interfere. Just as you cannot get a fire going in a fireplace without the proper arrangement of wood pieces—kindling below logs—neither can nuclear fission chain reactions occur without the proper arrangement of fissionable nuclei in the fuel. The packaging of the fuel requires a far more intricate series of actions than putting logs in a fireplace, and includes such things as adequate quantities, suitable geometric arrangement, and enough enrichment in the fissionable, or fissile, portion of the fuel (see Glossary for "enrichment" and "fissile").

In a fire, two particles chemically combine into one (for instance, a fuel molecule with an oxygen molecule), while in a fission reaction one nucleus divides into two approximately equal-size nuclei (fission products). Thus, nuclear fuel is self-contained, while combustion fuel requires an external agent, oxygen (or another oxidizing agent). The chemical combining process involves the electron clouds surrounding the immensely smaller and denser atomic nuclei, and leaves the nuclei unchanged, while the nuclear processes are independent of the surrounding electron clouds (though the clouds are thereby disrupted). Nuclear processes are energetically much stronger than chemical processes, by approximately six orders of magnitude (millions), as alluded to in Chapter 2.[2]

Because combustion fuel generally requires exposure to air (in particular, oxygen), the products disperse as fine smoke particles and gases. In contrast, fission requires no exposure to air and the products remain inside the reactor until deliberately removed. Unless contained through special expensive effort, combustion creates unconfined waste products, that are harmful to health and disruptive to the environment, while fission products are confined. Fission products can be, and are, properly handled to obviate any harm to either health or the environment. Enclosed inside the metal cladding of fuel rods, or within liquid salts,

---

[2] In the latter part of the seventeenth century, an early attempt to explain the process of combustion held that along with the smoke there is a substance dubbed phlogiston that is released from the wood or other fuel and drifts away. It wasn't until roughly a hundred years later that French chemist Antoine Lavoisier weighed the combustible substance before and after it burned and found that the ashes outweighed the original substance. He naturally concluded (and it has been upheld ever since) that combustion involves combination not separation. Thus, the existence of oxygen was recognized, and the concept of phlogiston dissolved into thin air, so to speak.

fission products are far more easily contained, manipulated, and stored than the fine particulate and gaseous waste associated with fire.

Now, nuclear fusion (as opposed to fission), which occurs in the center of the sun, is a combining process like combustion, in which two or more light nuclei combine with each other to form a heavier nucleus, albeit one with less mass than the sum of the constituents. This loss of mass results in the gain of prodigious energy, which is released during the fusion event, part of which constitutes the solar energy that arrives on Earth. The other part is in the form of kinetic energy of the fused nuclei (heat) that remains in the sun's core.[3]

Since chain reactions can be controlled, they can be used to generate electricity. Branching reactions are uncontrollable once they start, which makes them useful for destructive purposes. Whether fusion will ever be achieved in a feasible and sustainable chain reaction, and therefore become a source of energy on Earth, usable for anything other than destruction, remains to be seen. See the note at the end of this chapter concerning the recent fusion breakthrough at Lawrence Livermore National Laboratory's (LLNL) National Ignition Facility (NIF).

While combining versus separating is a key difference between fossil burning and nuclear fission, a nuclear power plant is at the same time similar to a coal or NG plant in that both are used merely to generate heat—heat that is then used to create steam to drive the turbine of an electrical generator, as indicated in Figure 7-1. Note that the two diagrams only differ in their lefthand portions. Thus, the final common pathway of both types of electrical generating plants is the conversion of steam to electricity, and both types of fuels do nothing other than generate heat. But fission, being self-contained, does not release GHGs.

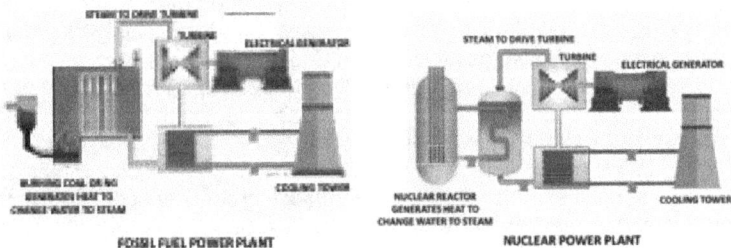

FIGURE 7-1

---

[3] While this gain of energy may seem to violate the principle of energy conservation (the first law of thermodynamics), the expanded view of energy and mass considers each to be a different form of the same entity. Thus, together mass and energy are conserved in any transaction, even if one is partly or wholly converted to the other. This is the meaning of Einstein's famous equation $E = mc^2$ (where E is energy, m mass, and c the speed of light in a vacuum), which tells us, when some of one disappears, how much of the other newly appears.

Suffice it to say, nuclear explosions, as opposed to chemical or pressure explosions, cannot possibly happen by chance—meaning without human intention—and none ever has on Earth, not even at Chernobyl or Fukushima. Half a century ago, more than a dozen instances of spontaneous continual and nonexplosive nuclear fission reactions were identified in the Oklo underground uranium deposits in Gabon, Africa. The uranium has since been extracted. The deposits demonstrated that such natural reactors and their fission products do not migrate or negatively impact their surroundings. No other examples of spontaneous earth-bound fission have been discovered. Yet.

The Comparative Hazards and Benefits of Various Technologies

It's useful to realize that the initial invention of automobiles in the late 1800s led to fears of extreme danger. While they could move little more than 10 miles per hour, they were thought to be unacceptably hazardous. Yet today, there is no organized movement to ban cars—or any other mode of transportation—despite the fact that almost 40,000 people are killed by auto accidents every year in the US alone, with some 1.35 million killed annually throughout the world. This translates to some 3,700 people killed each day, or one person killed by an auto accident somewhere in the world every 23 seconds (https://tinyurl .com/bdhne2b9). If anything should be banned based on danger alone, it's road vehicles. But in appraising technologies, we don't generally consider danger alone. Common sense and need tell us to compare risk with benefit.

Around the turn of the twentieth century, when miles of wire were first installed to distribute electricity, either underground (Edison) or overhead (Westinghouse), the latter were regarded as particularly hazardous. Indeed, electricians who installed such wiring suffered electrocutions in far greater numbers than occur today, though electrocutions still occur on rare occasions, and not just among electrical workers.[4] Yet, because of its recognized benefits, there remains no organized questioning of the safety of distributed electricity.

Commercial airplane crashes rapidly declined. Beginning in 1959, approximately one crash per 25,000 take-offs declined in

---

[4] It's estimated that electricity causes some 1,000 deaths in the US every year, of which somewhere between a quarter and a half are among electrical workers (https://tinyurl .com/4te4yvn3).

five years to one-tenth that rate, and it has since further declined to less than one per 4,000,000 take-offs over the last six decades. The improvement by a factor of roughly 160 in air-travel safety—per take-off, and even better per passenger mile (for one reason, because planes are bigger now)—exemplifies once again the fact that we learn from mistakes. And this enables us to progressively improve the risk/benefit ratio.

The direct death rates due to the manufacture and burning of fossil fuels kill roughly six times as many of the world's people each year as road accidents (8 million compared to 1.35 million).[5] And there has been little in the way of diminution of these death rates, yet neither respiratory pollution nor fire danger is the main reason that environmental organizations call for their elimination. Moreover, as we will see below, the death rate from fossil fuels is tens of thousands of times greater than from nuclear on a per-TWh basis.

While planes crash, ships collide, trains derail, cars roll over, even stationary buildings collapse, the question is not whether a particular technology is hazard free, but rather whether a dangerous outcome is sufficiently unlikely or, if one occurs, is sufficiently mild and the advantages of its use sufficiently great to warrant the employment of such a source. In the cases just cited, among many others, most people recognize that they're extremely useful and even necessary for modern life. This recognition is beginning to grow for nuclear energy as well, while the antinuclear movement focuses solely on dangers, and ones that they generally exaggerate or fabricate. In fact, if safety were their main concern, they should be *demanding* nuclear, not opposing it.

The inculcation of mass fear requires little time or effort, and its sponsors seem to feel no responsibility for the consequences of their actions. On the other hand, erasing fear requires prolonged time and significant effort. Social media today can turn collective fear into a branching reaction. But scary is not the same as dangerous, as Joshua Goldstein puts it in the documentary movie *Nuclear Now* that he and filmmaker Oliver Stone recently produced. More dangerous can be less scary and more scary less dangerous. Scariness is subject to extrinsic influences that often distort reality, while danger is intrinsic (echoing the relationship between LCOE and EROI).

---

[5] A 2021 study, one among many, by Harvard and several UK universities estimated that 8 million deaths occurred in 2018 from fossil fuel pollution, representing 20 percent of all deaths that year (https://tinyurl.com/225ysu5s).

Deaths and Illnesses Due to Nuclear Energy, Fossils, and Renewables

Awaiting an energy source that carries zero risk of death or injury would leave us without energy altogether, and lack of energy would incur an even higher death rate. The late French philosopher Paul Virilio put it this way, "To invent the sailing ship or the steamer is to invent the shipwreck. To invent the train is to invent the rail accident of derailment. To invent the family automobile is to produce the pile-up on the highway." Yet we continue to invent, to adapt, to improve, and to look forward to ever newer inventions that make life easier and healthier.

Fear can distract us from asking, How many ships make it safely to shore? How many trains are on time? How many car trips see only wear, tear, and refuelings? These successes often escape our attention but form a dense matrix in which the few mishaps are embedded. Escaping attention is also the uneventful production of clean electricity by some 440 nuclear reactors around the world, which form a dense matrix of success and safety. Additionally what escapes attention is the fact that only one commercial nuclear plant accident has directly resulted in death in almost 70 years. So, why does nuclear energy scare so many, especially left-leaning people (discussed further in Chapters 9 and 11)?

How do we make choices in our individual lives about transportation, foods, medical procedures, drugs, energy sources? Often without our awareness, we apply a risk/benefit estimate, based implicitly on our understandings of both. We also sometimes do so consciously. For some of us, because the time advantage does not seem to warrant the risk of flying, fear may lead us to drive, though deaths and injuries per passenger mile are far greater for driving. For those free from excessive fear, the benefit of flying on long-distance trips far outweighs the risk, so we fly. Either way, we make a risk/benefit estimate.

But double standards cloud the picture. Fossil-related accidents are given a pass; wind-turbine-related accidents are due to carelessness; solar-panel-related accidents are unfortunate; hydroelectric dam ruptures are the luck of the draw; explosions of improperly stored fertilizer happen elsewhere. But when nuclear accidents are seen as uniquely ghastly, harming people halfway around the world, then rational decision is out the window. The scope of a nuclear accident, we are told, dwarfs all others. While we have already covered the relative advantages, we need to ask, What are the relative hazards of nuclear, fossils, and renewables, as well as of stored fertilizer and ruptured dams?

First, illnesses due to the routine use of nuclear energy are virtually non-existent, and the rare cases of radiation sickness have been from weapons tests or experiments, not from commercial reactors (with the exception of one instance). With few exceptions, even those cases have been temporary with generally complete recovery and normal longevity. For a thorough history of nuclear accidents, see the book *Atomic Accidents* by US research scientist James Mahaffey (2014). The Chernobyl accident has been the only one associated with a (primarily) commercial reactor that directly caused any deaths in the entire 82-year history of nuclear energy (since Fermi's first sustained fission reaction).

This is to be contrasted with the ongoing illnesses and deaths from coal, oil, and NG, including black lung and mine collapses among coal miners, the hundreds of thousands of bystander respiratory illnesses and deaths annually from motor vehicle exhausts, and the all-too-common explosions due to NG leaks.[6] Such illnesses and injuries rarely see recovery, and usually shorten lifespans. Moreover, even the sole wartime employment of nuclear weapons by the US at Hiroshima and Nagasaki caused deaths from the blasts and firestorms, but almost none from the associated radiation and fallout (discussed more fully in Chapter 8).

Despite prolific irresponsible pronouncements by its foes, nuclear energy has been the safest source of energy for electricity, and by orders of magnitude—in terms of illnesses, injuries, and deaths. It is stunningly safer than any fossil fuel, significantly safer than hydro, and even safer than wind and solar. To give an order-of-magnitude idea of the comparisons, here's one record compiled in 2010 by Switzerland's Paul Scherrer Institut (PSI) for the 32 years up to 2000.

The table is divided into the highly industrialized countries that are part of the Organisation for Economic Co-operation and Development (OECD) and those that are not. Differences can be found in other compilations, depending on definitions. This set of figures refers only to immediate or shortly delayed fatalities due to traumatic accidents. Most importantly, the figures include only fatalities that have already happened, and not to predictions of

---

[6] In the last 20 years, according to the US Department of Transportation's Pipeline and Hazardous Materials Safety Administration (PHMSA), there have been a recorded 680 NG explosions in the US alone, one every 10–11 days, that together have killed 260 people, more than one a month (https://tinyurl.com/46t6v2b8).

Accidents by Energy Source for Electricity 1969–2000

	OECD		Non-OECD	
	Total fatalities	Fatalities/ TWy	Total fatalities	Fatalities/ TWy
Coal	2,259	157	18,000	597
Oil	3,713	132	16,500	897
NG	1,043	85	1,000	111
Hydro	14	3	30,000	10,285
Nuclear	0	0	31	48

The third and fifth columns express the numbers of deaths in terms of the energy generated by the various sources, in terms of terawatt-years.
(Data from Paul Scherrer Institut, in OECD 2010.)

allegedly related future deaths.[7] Nor do they include the millions of slow deaths, actual deaths, due to fossil fuel pollution.

As far as the table's figures for nuclear are concerned, nothing has changed since then, not even due to the 2011 events at Fukushima, where some 20,000 deaths were due to the tsunami and over 2,000 deaths to nucleophobic, government-sponsored forced relocations, but not a single one to radiation from the damaged coastal power plant (https://tinyurl.com/yc6dedkm and Chapter 8).

Here's another set of figures compiled from various sources in 2008 by technology analyst Brian Wang. Differences with PSI's figures reflect the following: Wang includes wind and rooftop solar; he includes deaths from pollution and other off-site impacts in addition to accidents; he draws numbers from different time intervals; and finally, his numbers measure death rate per terawatt-hour (TWh) rather than per terawatt-year (TWy), and these represent total energy production not just electricity. Neither the PSI

---

[7] Actual counts of past events are completely different from guesses about future events, even if the guesses are based on a validated formula. Chapter 8 discusses the many predictions of delayed radiation-caused deaths and shows that they are based not on actual empirical evidence, but rather on a formula (linear no-threshold, or LNT) that not only has no empirical evidence behind it—a feature even admitted by most of its proponents—but that actually has abundant empirical evidence against it.

nor Wang's statistics include deaths from extreme weather events, tornadoes, floods, storm surges, or other consequences of fossil-caused AGW. Note that despite all these differences in approach, the comparative orders of magnitude remain the same. Despite the differences, nuclear still comes out the safest by a significant margin (https://tinyurl.com/mr4yxct6):

Comparing Deaths per TWh for All Energy Sources

Energy Source	Death Rate (deaths per TWh)
Coal – world average	161 (26 percent of world energy, 50 percent of electricity)
Coal – China	278
Coal – USA	15
Oil	36 (36 percent of world energy)
Natural Gas	4 (21 percent of world energy)
Biofuel/Biomass	12
Peat	12
Solar (rooftop)	0.44 (less than 0.1 percent of world energy)
Wind	0.15 (less than 1 percent of world energy)
Hydro	0.10 (European death rate, 2.2 percent of world energy)
Hydro - world	1.4 (about 2500 TWh/yr)
Nuclear	~~0.04~~ 0.0013 (5.9 percent of world energy)

Wang's original figure for nuclear, 0.04 deaths per TWh, included not just actual deaths, but also the prediction by the UN's World Health Organization (WHO) of future deaths from Chernobyl, that is, future deaths attributable to radiation exposure at the time of the accident. More than three decades later these have not come to pass, and they were based on the false claim of linearity (LNT, mentioned in the preceding footnote and discussed thoroughly in Chapter 8). The UN Scientific Committee on the Effects of Atomic Radiation (UNSCEAR) in 2008 corrected their conflation of past-actual and future-projected deaths, down-revising the figure to 0.0013 deaths per TWh. But even with the exaggerated figure, nuclear was still the lowest, and with the correction is two orders of magnitude lower than its nearest rival, wind power.

Displaying the key figures graphically:

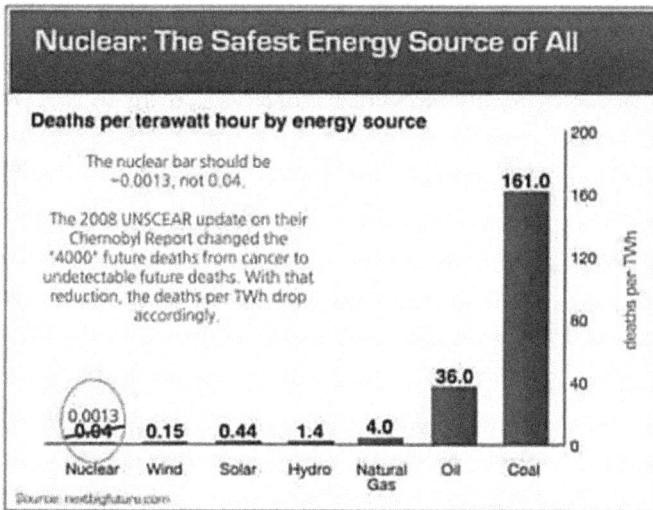

Nuclear: The Safest Energy Source of All

Deaths per terawatt hour by energy source

The nuclear bar should be
~0.0013, not 0.04.

The 2008 UNSCEAR update on their
Chernobyl Report changed the
'4000' future deaths from cancer to
undetectable future deaths. With that
reduction, the deaths per TWh drop
accordingly.

0.0013 0.04	0.15	0.44	1.4	4.0	36.0	161.0
Nuclear	Wind	Solar	Hydro	Natural Gas	Oil	Coal

Source: nextbigfuture.com

FIGURE 7-2

While wind and solar are relatively safe compared with fossil fuels, as Wang indicates, gravity and altitude can be a potentially fatal combination. Recall the photo from the first section of Chapter 4 (reproduced here on the left). The righthand photo shows two workers embracing before one jumped to his death and the other was incinerated trying to access the ladder inside the pole.

The injuries and deaths associated with installation of rooftop solar PV panels have also been primarily due to falls.

With increasing penetration of renewables, not only would the absolute number of associated injuries and deaths rise, but the death and injury rate per TWh would multiply rapidly as the fossil and matrix backup were to diminish and the need for more storage were to grow. Battery fires (Chapter 4) could become an additional source of renewables-associated accidents, both injurious and fatal, barring significant advances in battery technology.

In sum, if an average coal plant causing 161 deaths per TWh were to be replaced by an average nuclear plant causing 0.0013 deaths per TWh, almost 161 lives would be saved during the time it took to generate that much electricity. For plants generating 1 GW of electricity, that would require about 6 weeks, or about 42 days or about 1,000 hours. Think of it, almost one life saved every 6 hours or so, through the replacement of a single coal plant with nuclear. If all the coal plants in the world, producing roughly 2,000 GW in toto, were replaced by nuclear, one life would be saved every 10 seconds or so, or roughly three million lives per year. Even if nuclear were as hazardous as antinuclear organizations claim, the death rate would be orders of magnitude lower than with coal.

In his 1990 book *The Nuclear Energy Option* the late US physicist Bernard Cohen (about whom more in Chapter 8) provides a relevant comparison of different hazards. Referring to inhaled toxins he says,

> One often hears that in large-scale use of plutonium we will be creating unprecedented quantities of poisonous material. Since plutonium is dangerous principally if inhaled, it should be compared with other materials which are dangerous to inhale. If all of our electricity were derived from breeder reactors, we would produce enough plutonium each year to kill a half trillion people. But as has been noted previously . . ., every year we now produce enough chlorine gas to kill 400 trillion people, enough phosgene to kill 18 trillion, and enough ammonia and hydrogen cyanide to kill 6 trillion with each. It should be noted that these materials are gases that disperse naturally into the air if released, whereas plutonium is a solid that is quite difficult to disperse even intentionally.

The point is not to create more fear of chlorine, phosgene, ammonia, or hydrogen cyanide, but rather to defuse the fearmongering around plutonium. And to expose the double standards of nuclear fearmongers.

The Conflation of Nuclear Energy with Nuclear Weapons

A major fear-inducing distortion arises from the conflation of nuclear generation of electricity with nuclear weapons. Shortly following the US Air Force's atomic (nuclear fission) bombing of Hiroshima and Nagasaki in August 1945, US geneticist Hermann Muller, in his 1946 Nobel Prize acceptance speech, implicitly conflated the two. The prize was awarded for his work two decades earlier exposing fruit flies to very high-dose ionizing radiation in his laboratory and examining the resulting chromosomal changes. He mischaracterized these almost macroscopic changes as atomic-level mutations, a process that take place on a much finer microscopic level that he was not able to examine (Calabrese 2022; 2023). In his otherwise laudable desire to end the atmospheric testing of nuclear weapons, combined with his more self-serving aspirations as a scientist, Muller resorted to a deliberate untruth (Calabrese 2022; 2023).

In particular, Muller claimed that the evidence that he and his colleagues had gathered led to the "inescapable" conclusion that regardless of how low the exposure, radiation causes permanent damage in the form of mutations in humans and other organisms. Yet his experiments never employed low exposures. Nor had he examined DNA on the molecular level where mutations occur. Moreover, he knew that more recent experiments by his colleagues suggested the possible falsity of his conclusion in the low-dose range. In short, the conclusion that the available evidence confirmed permanent mutations due to low-dose exposures was far from "inescapable," and indeed has been resoundingly escaped not only in subsequent years but to a considerable degree even in years prior to Muller's pronouncement (see for example, Sacks, Meyerson, and Siegel 2016; Calabrese 2022, 2023; and Chapter 8 in this volume).

Jacobson plays on the same conflation of electricity and bombs as Muller. In section 4d of his paper in *Energy and Environmental Science* (2009), titled "Effects of Nuclear Energy on Nuclear War and Terrorism Damage," Jacobson says (we quote this paragraph in its entirety, borrowed from https://tinyurl.com/9a9zmcbn):

> The explosion of fifty 15 kT [kiloton] nuclear devices (a total of 1.5 MT [megatons—*sic*, should read 0.75 MT], or 0.1 percent of the yields proposed for a full-scale nuclear war) during a limited nuclear

exchange in megacities could burn 63–313 Tg [teragrams] of fuel, adding 1–5 Tg of soot to the atmosphere, much of it to the stratosphere, and killing 2.6–16.7 million people. The soot emissions would cause significant short- and medium-term regional cooling. Despite short-term cooling, the CO_2 emissions would cause long-term warming, as they do with biomass burning. The CO_2 emissions from such a conflict are estimated here from the fuel burn rate and the carbon content of fuels. Materials have the following carbon contents: plastics, 38–92 percent; tires and other rubbers, 59–91 percent; synthetic fibers, 63–86 percent; woody biomass, 41–45 percent; charcoal, 71 percent; asphalt, 80 percent; steel, 0.05–2 percent. We approximate roughly the carbon content of all combustible material in a city as 40–60 percent. Applying these percentages to the fuel burn gives CO_2 emissions during an exchange as 92–690 Tg CO_2. The annual electricity production due to nuclear energy in 2005 was 2768 TWh yr-1 [per yr]. If one nuclear exchange as described above occurs over the next 30 yr, the net carbon emissions due to nuclear weapons proliferation caused by the expansion of nuclear energy worldwide would be 1.1–4.1 g CO_2 kWh-1 [per kWh—*sic*, should read 1.1–8.3], where the energy generation assumed is the annual 2005 generation for nuclear power multiplied by the number of yr [years] being considered. This emission rate depends on the probability of a nuclear exchange over a given period and the strengths of nuclear devices used. Here, we bound the probability of the event occurring over 30 yr as between 0 and 1 to give the range of possible emissions for one such event as 0 to 4.1 g CO_2 kWh-1. This emission rate is placed in context in Table 3.

Aside from the two arithmetic errors bracketed above, note the precision with which speculation parades as fact (recall the final section of Chapter 3 on significant figures). And note the odd juxtaposition of wide ranges of numerical values with excessive numbers of significant figures—"2.6–16.7 million people" dead. The wide range represents imprecision, while the number of significant figures suggests precision.[8]

Jacobson disregards the last three-quarter century's history of political and economic standoff. As Clack et al. (2017) note, "in

---

[8] While Jacobson employs fictional levels of precision in his all-RE advocacy, when his goal is to engender fear of nuclear energy he resorts to extreme imprecision. So while his LOADMATCH computer program, for example, leads him to make claims about electricity demand 30 years into the future with a precision of three to four significant figures, here he claims that nuclear energy will cause (!) a nuclear war to happen in the same 30 years, with a probability somewhere between 0 and 1. You can't get more imprecise than that.

the almost 60 y[ears] of civilian nuclear power (two of the assumed war cycles), there have been no nuclear exchanges. The existence of nuclear weapons does not depend on civil power production from uranium."

And Jacobson's claim that "we bound the probability of the event occurring over 30 yr as between 0 and 1" is strange, since all probabilities are so bounded—even the one that a hippopotamus could lay an egg. It's like saying, "We meet every week on a day somewhere between Sunday and the following Saturday."

Jacobson's audacious rhetoric implies (without quite saying) that a nuclear exchange like the one described, *caused* by civilian nuclear power, will happen *every* 30 years, with something between a certainty of its occurrence and a certainty of its nonoccurrence. This is like the provocative question to the witness that the judge orders be withdrawn and stricken from the record and that the jury is ordered to disregard, while the questioning attorney is admonished. However, the judges here—Jacobson's peer reviewers and editors—shirk even that duty.

While Jacobson focuses a spotlight on the specter of nuclear war to further promote the fear of nuclear energy, this kind of argument also helps empower nations locked in interimperialist conflict to invoke the threat of nuclear weapons to strike fear in, and impose their will on, their rivals, as Russia is currently doing in its invasion of Ukraine. Thus, Jacobson engages in an approach that inadvertently serves to increase rather than reduce the risk of war.

Additionally, with his attribution of carbon emissions to nuclear energy (since fossil fuels, he claims, must provide baseload for the 19 years of nuclear construction delays), Jacobson (2009) employs a double standard. Insofar as fossil fuels are required both to construct wind and solar farms and to fill in for their intermittency throughout their lifespan, he frees renewables from responsibility for the far more extensive GHGs emitted by the supporting fossils, laying that charge instead on the matrix. But if nuclear were to provide the energy for its own construction (or even for that of renewables as long as they persist), with no fossils needed, GHG emissions would be abolished—at least from electricity production. And could be abolished from other energy applications when they are electrified.

Jacobson also recapitulates President Carter's justification in 1977 for prohibiting the recycling of nuclear fuel and President Clinton's 1994 inveigling of Congress to withdraw funds from Argonne National Laboratory to discontinue developing the second experimental (fast) breeder reactor (EBR II) outside of Idaho

Falls (see Glossary). Both presidents, whether or not they were personally subject to it, played on popular fear, claiming that their actions around electricity generation would impede the proliferation of nuclear weapons. They instead impeded deployment of a technology, pioneered in the US by scientists from many countries, that could slow and eventually eliminate GHG emissions. They ended up retarding the use of a fuel whose energy density is so high that it *lowers* the motivation for resource wars. However, despite the ill-informed, or dishonest, political opportunism of the US chief executives, several countries are developing commercial fast breeder reactors—China, Russia, India, South Korea, and Japan. France once had a commercial fast breeder reactor, called the Superphenix, but has discontinued its use. A few US companies, along with those in other countries, are further developing fast breeders in the form of small modular reactors (SMRs), for introduction to commercial use (https://tinyurl.com/ 4n75hp9n). But until US policy changes, all the US models will necessarily end up being manufactured elsewhere or exported.

The claim that nuclear power leads to nuclear weapons proliferation is analogous to the claim that the use of fireplaces and stoves leads to the use of napalm for the firebombing of innocent populations. Development and application of any type of weaponry is the result of deliberate political, not technical, decisions. Nor does the prohibition of a technological application for social use sequester the scientific knowledge underlying that technology or prevent it from being put to antisocial use.

Finally, in Jacobson's Table 3 (2009), he assigns at most a 2–6 percent contribution from the possible wartime explosion of nuclear weapons to the overall imputed carbon emissions associated with nuclear energy. So one can only guess at his motive for raising it in such speculative detail. But whatever the motive, his speculative reference to "2.6–16.7 million people" killed by the burned-fuel-turned-to-soot serves as a poster child of nuclear fearmongering. It is perhaps worth noting that the fallout from the largest nuclear weapon ever tested atmospherically—the 50 MT AN602 (later called the Tsar Bomba), exploded in 1961 in an unpopulated area by the USSR as part of the Cold War—had zero known victims, yet the force of this single bomb exceeded the total of Jacobson's hypothetical fifty 15 kT bombs by a factor of 67. Granted it involved a restricted geographic area, but had there been any victims, they would surely have been enumerated, if not exaggerated, in Western propaganda.

Jacobson commits the logical fallacy that conflates necessity with sufficiency. Knowledge of nuclear technology is unarguably necessary to produce, albeit not to purchase or be given, nuclear weapons. But knowledge is not a sufficient condition for either production or acquisition of nuclear weapons. And while possession is a necessary condition for their wartime use, it is not sufficient for their deployment. While not himself possessing a nuclear bomb, Jacobson nevertheless brandishes the weapon of nuclear fear.

Historically, possession of nuclear reactors for electricity has never led to, and is almost never followed by, the acquisition and/or development of nuclear weapons. Of the 33 nations which have either nuclear weapons or nuclear reactors, or both, the rulers of India and Pakistan are the only two who developed nuclear weapons after, but not because, they already had nuclear electricity. Their weapons emerged from their political rivalry, not from their possession of nuclear reactors.

Referring to the table below, the other 31 nations either manufactured weapons first, followed by reactors for electricity (five: the US, UK, France, China, and Russia), or have weapons without reactors for electricity (two: Israel and North Korea), or have reactors for electricity without weapons (twenty-four: Argentina, Armenia, Belgium, Brazil, Bulgaria, Canada, Czech Republic, Finland, Germany, Hungary, Iran, Japan, Mexico, Netherlands, Romania, Slovakia, Slovenia, South Africa, South Korea, Spain, Sweden, Switzerland, Taiwan, and Ukraine).[9] Thus, the seemingly

	Nuclear Energy	No Nuclear Energy	
Bomb	US RUSSIA FRANCE UK CHINA INDIA PAKISTAN	NORTH KOREA ISRAEL	TOTAL 9 Bomb
	7 TOTAL	2 TOTAL	
No Bomb	JAPAN SWEDEN GERMANY SWITZERLAND ...	. . . . .	TOTAL 186 No Bomb
	24 TOTAL	162 TOTAL	

TOTAL 31 Nuclear Energy   TOTAL 164 No Nuclear Energy

---

[9] Germany has shuttered its nuclear plants, at least for the time being, though two-thirds of the German population favor reopening them (https://tinyurl.com/mr2ccens).

reasonable belief that nuclear energy automatically, or even proba- bly, leads to nuclear weapons is without foundation.

In fact, the reverse has happened. Not only have no new nations acquired nuclear weapons since around 2006—almost two decades ago, when North Korea first tested a nuclear weapon—but since the end of the Cold War in 1991, the dismantling of significant por- tions of the Russian and US nuclear arsenals has provided uranium and plutonium for fueling nuclear reactors that generate electricity. As much as half of the nuclear-fueled electricity in the US has come from dismantled warheads, mostly purchased from Russia, with a small portion from dismantled US bombs (https://tinyurl.com/ mryau7mj). Since nuclear is the source of almost 20 percent of US electricity, dismantled bombs had provided roughly 10 percent of all US electrical output, at least through 2013.

The claim that a technological development forces or at least encourages, rather than simply permits, a political application is an example of technological determinism, a form of the *post hoc ergo propter hoc* fallacy (following, therefore because of). Political devel- opments derive from political goals; if a technological development has salutary applications, it should be pursued and the political mis- use fought and prevented. If technology necessarily fostered mis- use, we would be banning land vehicles, airplanes, gasoline, road construction explosives, hunting rifles, knives, and so on, all of which have been used to kill millions of soldiers and civilians.[10]

Moreover, the very existence of chemicals and pathogenic organisms, like viruses and bacteria, permits their weaponization. World War I saw the use of chemicals for gas warfare, but they were indiscriminately self-destructive. Blowback led to the interna- tional prohibition of such weapons, honored until modern times when violations occurred during the Iran-Iraq war of the 1980s and assaults on defenseless populations in Syria.

Similarly, after centuries of use of biological agents (see, for example, https://tinyurl.com/3726rpmu), their most recent use was by the Japanese military in China during World War II, with none others proven to have occurred since. Nevertheless, while this encouraged laboratory research into biological weapons in sev- eral countries, an international treaty in 1975 has been effective in banning their further use, for the same reason—uncontrollable mutual self-destructiveness.

---

[10] We return to the technological determinist fallacy in Chapter 9, where we examine the claim that certain technologies, like nuclear power and biotechnology, are inherently undemocratic.

Nuclear Power Plant Accidents and Disinformation Fallout

Some may say, okay, nuclear energy should not be conflated with nuclear weapons, but what about Chernobyl! What about nuclear waste! What about nuclear terrorism!

Nuclear opponents pretend that these part-questions/part-exclamations/part-denunciations fill nuclear defenders with enough shame and guilt to silence us. Such exclamations are intended as both knockdown arguments and synecdoches for the argument to terminate nuclear energy. No other source of energy, certainly not fossil fuels, faces this strange synecdochical structure. Its effectiveness feeds on the way exaggerated fear turns occasional accidents—whose dangers to the public, as well as to nuclear workers, are at best immensely exaggerated and at worst completely fabricated—into inevitable apocalypse, if not now, then tomorrow, if not tomorrow, then tomorrow night, if not . . .

Two key antinuclear propagandists were the late US nuclear physicist Ernest Sternglass and the likewise deceased US biologist John Gofman, whose proclamations are still offered as truth. Rather than nuclear plants, their focus was Cold War nuclear weapons proliferation, but the rhetoric moves seamlessly between them.

Sternglass worked at the University of Pittsburgh and was a colleague of Bernard Cohen, a pronuclear advocate (discussed in Chapter 8). According to Spencer Weart (2012, pp. 184–86), Sternglass became "famous with his [September 1969] *Esquire* article, titled 'The Death of All Children.'" He argued there, and repeatedly in his many media opportunities, that infant mortality had failed to decline in the American South because, said Sternglass, "the South lay in the path of fallout from the Nevada bomb tests." Forget that the South was nowhere near the path of the Nevada fallout and that, said Weart, it "received a far more serious dose of poverty than of radiation," as well as racial discrimination, both below Sternglass's opportunistic radar.

Unembarrassed and extrapolating to the world, Sternglass concluded that "one out of every three infants who had died in the 1960s had been killed by bomb test fallout." But why stop there when you have a responsive audience, or at least gullible media? He crescendoed, "even if ABMs destroyed every missile with nuclear explosions high in the atmosphere, the ABMs' own fallout would doom every baby born for decades in both America and Russia."

"[T]horoughly discredited" by other scientists, Sternglass switched to infant mortality statistics near nuclear power plants—though the exposure within fifty miles of such a plant approximates that from your home's americium-containing smoke detectors.[11]

Weart continues, "by the time one set of stats was refuted, the professor had a new set that he found even more persuasive. Not one other scientist was ever convinced by his figures, but there was always some newspaper reporter who would publish Sternglass's latest horrific calculations." Unburdened by evidence, Sternglass claimed that the military sponsored all nuclear research. His conflation of military with civilian threw up obstacles to nuclear research that spurred others to action. Among those spurred was John Gofman, formerly at Berkeley and Livermore (now Lawrence Livermore National Laboratory, LLNL).

Appropriately repelled by Hungarian-American physicist Edward Teller's enthusiasm for hydrogen bombs, Gofman said of Livermore workers, "They can only think how to obliterate, control, and use each other." Gofman asked his collaborator, biophysicist Arthur Tamplin, to review Sternglass's data. Though Tamplin knew Sternglass's argument "was nonsense," he too enlisted uncertainty and doubt as tools. Tamplin suggested, says Weart, that "stray radiation might well have killed some thousands of babies even if that could not be proved."[12]

When the US Atomic Energy Commission (AEC), overseer of peacetime uses of nuclear energy from 1946 to 1974, tried to stop the publication of such admitted and unfounded speculation about weapons testing fallout, Gofman and Tamplin turned attention toward nuclear reactors for electricity. Of course, in today's conspiratorial atmosphere, the AEC's attempts to interdict such fabrications might be offered as "evidence" of their validity.

Nuclear fear rhetoric has hardly changed. In the aftermath of the Fukushima Daiichi accident, the scientific consensus among many independent sources is that there have been no detectable deaths or illnesses attributable to radiation, nor will there be. However, the forced population relocations by the Japanese government, by their own admission, directly caused some 2,300

---

[11] Americium-241 (with one more proton than plutonium) is radioactive, an alpha particle (helium nucleus) emitter. The alpha particles in a smoke detector continually ionize air molecules, and the positive and negative ions are attracted to opposite plates, forming a current. When the ions are neutralized by smoke particles, the current is interrupted, causing the alarm to sound.

[12] See also Calabrese and Giordano 2024 and Calabrese 2023b.

immediate deaths, and the dislocations have negatively impacted countless lives. The non-lethality of the radiation has not discouraged warnings of up to one million cancers from Fukushima and widespread Pacific dead zones from radioactive (tritium-containing) water released into the ocean in the aftermath of the accident.

So, let's talk about Japan's tritiated water.

The Tritium-Containing Water at Fukushima Daiichi

Now that the Tokyo Electric Power Co (TEPCO) has begun to release their store of cooling water into the Pacific from the Fukushima Daiichi plant, the media offer conflicting stories of danger versus safety. Here we explain the process, and by using numbers and relevant comparisons we show how extremely safe the release is, for sea life as well as for humans.

First, all other radioactive substances have been filtered from the tritium-containing water. Tritium, an isotope of hydrogen containing one proton and two neutrons, is the smallest radioactive nucleus and the weakest emitter of all—a sort of nerf ball ejector. Each decay of a tritium nucleus emits an electron (beta particle) that can only travel a quarter of an inch in air and cannot penetrate skin. So one would have to drink the water for it to have any effect.

But how much imbibed tritiated water crosses the threshold to increase cancer risk? More than 11 gallons (42 liters), much more, and it would have to be drunk within a matter of hours (see box below for this calculation). The figure of 11 gallons is derived as a safe margin of radioactive exposure from the advice of the National Council on Radiation Protection and Measurements (NCRP, an advisory organization chartered by the US Congress in 1964). However, the NCRP is overcautious for regulatory purposes, as evidenced, for example, by radiation therapy's failure to cause permanent damage or any clinically significant increase in cancer risk in normal tissue that is incidentally exposed to the high-dose radiation. Also, acute radiation syndrome (ARS) from whole-body exposure doesn't begin to occur till about 30,000 mrem/day (https://tinyurl.com/2zk3dx2f), a factor of more than 7,300 times the NCRP's 4.1 mrem/day.

Even 11 gallons per day is far more than a person can take in. Consider the often recommended (but more than you need) daily intake of eight 8-ounce glasses of water (half a gallon). Even at that

Sidebar: Calculation for Those Interested

The NCRP recommends an upper limit of 4.1 mrem/day of whole-body exposure.[13] In units of energy deposited in each kilogram of tissue, this translates to 25.6 x 10^4 GeV/kg per day (GeV = giga, or billion, electron volts). In an average 70 kg person (154 pounds), this is a daily safe limit of **1.8 x 10^7 GeV/day**.

How much tritiated water would one have to drink to take in 1.8 x 10^7 GeV/day? The Fukushima Daiichi water contains 7.6 x 10^{14} Bq of tritium in 8.6 x 10^8 liters of water, or 8.8 x 10^5 Bq/liter of tritium.[14] When a tritium nucleus decays, it emits a beta particle with energy 5.7 keV (kilo electron volts, a very small amount of energy). So each liter of the water emits 8.8 x 10^5 decays/sec times 5.7 x 10^3 eV/decay = **5.0 GeV/sec**. Since there are 86,400 seconds in a day, this is equivalent to 5.0 GeV/sec-liter x 86,400 sec/day = **4.3 x 10^5 GeV/day per liter of water**.

And since NCRP's safe limit allows taking in 1.8 x 10^7 GeV/day, this amounts to 42 liters (1.8 x 10^7/4.3 x 10^5 = 42), or **11 gallons/day**.

excessive rate, drinking 11 gallons would take you 3 weeks. To take in 11 gallons (42 liters) within a day you would have to drink about 2.6 liters per each of 16 waking hours. However, drinking as much as 2 liters per hour would kill you from water toxicity, which arises from excessive dilution of your electrolytes, including sodium (called hyponatremia), that causes abnormal heart rhythms. During 16 waking hours, 2 liters per hour would add up to 32 liters total, which is a little more than 8 gallons a day. At this rate of intake, the water alone (never mind the tritium) would be toxic, likely fatal. So if somehow taking in 11 gallons per day causes tritium radiotoxicity and only 8 gallons per day causes water toxicity, the water in the storage tanks at Fukushima Daiichi is more toxic than the tritium it contains.

Now, what happens when the tritiated water is released into the ocean? Same answer: any sea creature that drank the water would have to drink many gallons at a time (without releasing water), adjusted by their body weight relative to that of a human being. It would have to drink even more, much more, because the tritium-containing water becomes instantly diluted as it hits the ocean, and

---

[13] A mrem, or 1 millirem, is a small unit of initial exposure damage that omits consideration of the body's repair or removal of that damage within hours (Chapter 8), and rem is an acronym for "roentgen equivalent in man."

[14] A Bq, or becquerel, is a unit of radioactivity that equals one emission from a radioactive isotope per second. It does not contain any information whatsoever about the energy released by that event, only the frequency of events.

during its passage through a kilometer-long underwater pipe it is deliberately being mixed with fresh ocean water, diluting the tritium further by a factor of over 500 before it hits the ocean. This intra-pipe dilution is done to produce a safety factor that is physically unnecessary and done with the intent to reassure the public, given widespread radiophobia.[15]

Nor can tritium concentrate in sea life or accumulate up the food chain, since it mainly exists in the form of water (HTO instead of HHO, i.e., H_2O), and as any sea creature drinks and excretes, water is continually exchanged between it and its surroundings, while none accumulates in its tissues. The level of tritium soon reaches an equilibrium concentration, which is the same concentration for the larger predators, on up the food chain. For a chemical to bioaccumulate up the food chain it would have to lodge permanently in particular organs. So this is not a problem for sea life either, nor for anyone who consumes seafood from the area.

In fact, many nations release far more tritium into the oceans than TEPCO is doing, because it is perfectly safe to do so (Figure 7-3, https://tinyurl.com/2f4u94kt).[16]

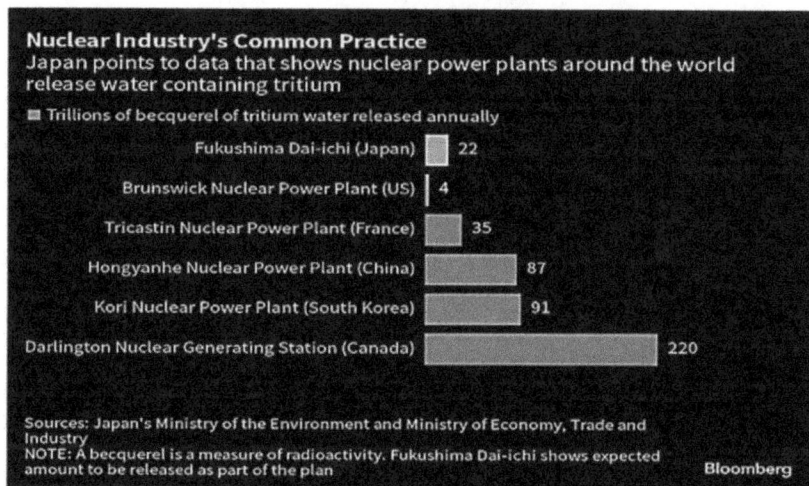

Nuclear Industry's Common Practice
Japan points to data that shows nuclear power plants around the world release water containing tritium

■ Trillions of becquerel of tritium water released annually

Fukushima Dai-ichi (Japan)	22
Brunswick Nuclear Power Plant (US)	4
Tricastin Nuclear Power Plant (France)	35
Hongyanhe Nuclear Power Plant (China)	87
Kori Nuclear Power Plant (South Korea)	91
Darlington Nuclear Generating Station (Canada)	220

Sources: Japan's Ministry of the Environment and Ministry of Economy, Trade and Industry
NOTE: A becquerel is a measure of radioactivity. Fukushima Dai-ichi shows expected amount to be released as part of the plan   Bloomberg

FIGURE 7-3

---

[15] However, such types of reassurance can backfire, as they suggest a hazard that is in fact nonexistent.

[16] This chart exposes the hypocrisy of China's government, which is raising objections to the Fukushima release and refusing to import seafood from Japan, though a single Chinese nuclear plant releases four times as much tritium into the Pacific per year as all of Japan is planning to release. While less directly hypocritical, the South Korean government is watching mass protests against Japan's release by their citizens, while also releasing more

Moreover, the Pacific already contains vast amounts of natural radioactivity. As described in a recent article in *Science* (https://tinyurl.com/yrp4pmnb) and shown in Figure 7-4 borrowed from that article, the Pacific contains natural radioactivity in the form primarily of K-40, Rb-87, U-238, U-235, and C-14. These add up to ~8,100,000 PBq (petabecquerels, or 10^{15} Bq). Of that total, 3,000 PBq is already naturally occurring tritium, and the entire store of tritiated water at Fukushima Daiichi would add less than another 1 PBq (actually 0.76 PBq, to be released 0.022 PBq per year for decades). In other words, the entire store of tritiated water at Fukushima Daiichi contains less than one-three-thousandth of the naturally occurring tritium and one-eight- millionth of the total radioactivity in the Pacific (recalling that radioactivity is not the same as energy release, and the energy of tritium decay is trivial compared to that of the other radionuclides in the ocean).

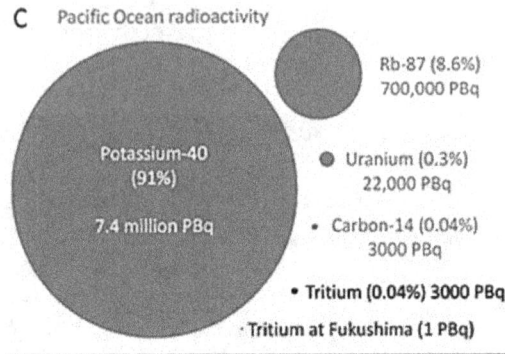

C   Pacific Ocean radioactivity

Rb-87 (8.6%)
700,000 PBq

Potassium-40
(91%)

Uranium (0.3%)
22,000 PBq

7.4 million PBq

Carbon-14 (0.04%)
3000 PBq

Tritium (0.04%) 3000 PBq

Tritium at Fukushima (1 PBq)

FIGURE 7-4

To establish how much energy each of these isotopes emits per second during its radioactive decay, we have to multiply by the energy per emission or decay event for that specific isotope. K-40 emits more than 1,300 keV per decay event, roughly 230 times as great as a tritium decay (5.7 keV). C-14 and U-238 emit, respectively, more than 27 and 700 times as much per decay as tritium. So to appreciate the relative *energies* released by the radioactive decay of the these isotopes, the circles representing potassium, carbon, and uranium would have to be enlarged by those respective factors, leaving tritium even smaller by comparison.

---

tritium from a single nuclear plant, and apparently withholding this information from the demonstrators.

Incidentally, the scheduled annual release of 22 TBq from Fukushima Daiichi amounts to one-sixteenth of a gram of tritium (63 mg). Into the Pacific Ocean. Per year.

And if this is not enough to defuse the fear, as we said above, the NCRP's recommended safe daily limit is greatly underestimated, and according to the experience of radiation cancer therapy, even 7,300 times that limit, or 80,300 gallons, would be safe. So taking in one-eighth of an Olympic swimming pool in one day, with the concentration of tritium contained in the Fukushima water, would still not raise one's risk of cancer, but then neither would jumping over a 50-story building.

Moreover, the Pacific Ocean is a far greater hazard than the water to be released. After all, if you were to drink a lot less than 11 gallons of this very salty water you would die from salt toxicity, or if you swam out too far and got caught in a rip tide you would drown. Yet there is no mass terror surrounding the Pacific by itself, just varying degrees of caution. If anything is safe, for both humans and sea life, it is the tritiated water at Fukushima Daiichi.

By this time you may be screaming, "Alright already!" But it's important to know that the human body contains radioactive potassium and carbon (K-40 and C-14) obtained from our food. In a person weighing 70 kg (154 pounds), their K-40 and C-14 would be continually releasing and depositing energy in the body at the same rate as that contained in a little over a liter of the tritiated water at Fukushima.

The concentrations of K-40 and C-14 in sea life are roughly the same as in humans, and we eat seafood without a qualm. Of course, most people are unaware of the radioactive elements in our food and our bodies. But, by the same token, this gap in knowledge leaves us subject to the fearmongering surrounding the water at Fukushima.

Other Creative Fearmongering

Prominent among the emulators of Sternglass's and Gofman's shame-less distortion and cherry-picking are US science writer Joseph Mangano and the late US oncologist Dr. Janette Sherman. To argue that Fukushima's radiation was killing US babies, they compared the weekly infant deaths before and after the Fukushima Daiichi accident (March 11, 2011) in 8 West Coast cities. They chose the 4 weeks before and the 10 weeks after. Why 4 weeks versus 10? Well, by chance the number of deaths was significantly lower in those 4 weeks than in the earlier weeks. Comparing the 10 weeks before to the 10 weeks after would have revealed an actual decrease (by chance). So their selection of the 4-week interval suggests deliberate cherry-picking, in order to deceive and ter-rorize. And by choosing weeks immediately following the acci-dent, they didn't even allow time for the radioactivity to cross the Pacific. For other fabrications by these two authors see https:// tinyurl.com/bdf6yuv5, and the linked https://tinyurl.com/ nhe6fs99.

Some scientists who have debunked the fearmongering have received death threats. Science—far from infallible but with no better basis for determining policy—gasps for air in such a pol-luted atmosphere. While fear purveyors routinely claim that a formerly radioactivity-free Pacific is being contaminated by human-made radioactivity, the entire earth (land, sea, and air) is bathed in radiation that is inescapable and comes from naturally occurring underground and oceanic radioactivity and from the continual influx of cosmic rays from outer space. While our minds may be unaware of natural background radiation, our bodies have evolved to respond to it with biological defenses, and our immune systems seem to depend on it for their very development (Chapter 8).

As US physician Robert Gale and US author Eric Lax note in their book (2013), the 20 to 40 trillion Bq, mainly from Cs-137 (cesium), that had leaked into the Pacific from Fukushima Daiichi over the two years since the 2011 tsunami make up 0.0000003 percent of the ocean's natural radioactivity and add only 0.003 percent of the trivial amount of Cs-137 already there. Cs-137 contributes so little natural radioactivity because its half-life is 30 years, compared to K-40's 1.3 billion years, and the K-40 is continually replenished by rivers (along with the small amount of other isotopes).

The following diagram (Figure 7-5) illustrates the comparative contributions to the radioactivity in the world's oceans (https://

tinyurl.com/mrya9eej).[17] Note that the range for Fukushima's total contribution of radioactivity is significantly greater than Gale and Lax's figure for cesium two years earlier, yet it is still trivial compared to the natural radioactivity from K-40 and U-238. And the contribution to be expected from the tritium (0.022 PBq/year) is even smaller. It indicates that natural radioactivity renders virtually undetectable all that added by human activity, including past nuclear weapons testing. Certainly human-added radioactivity is insufficient to cause harm to either sea life or human beings compared to the natural radioactivity in which ocean creatures evolved and thrived.[18]

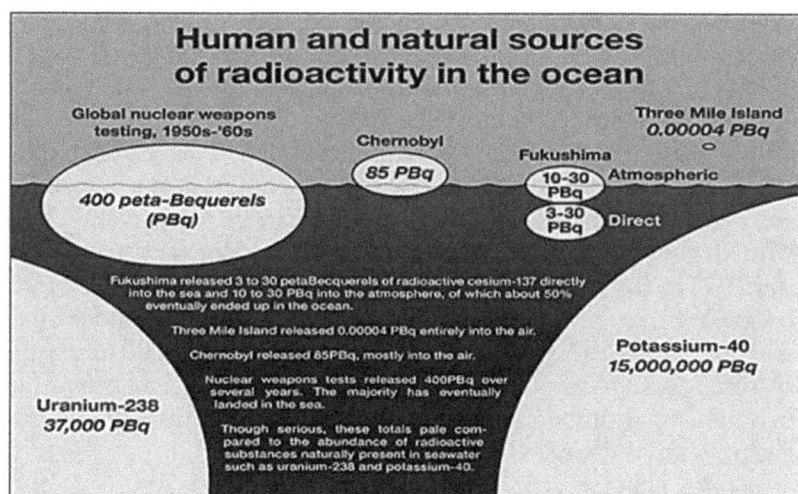

Human and natural sources of radioactivity in the ocean

Global nuclear weapons testing, 1950s-'60s

400 peta-Becquerels (PBq)

Chernobyl **85 PBq**

Fukushima **10-30 PBq** Atmospheric **3-30 PBq** Direct

Three Mile Island **0.00004 PBq**

Fukushima released 3 to 30 petaBecquerels of radioactive cesium-137 directly into the sea and 10 to 30 PBq into the atmosphere, of which about 50% eventually ended up in the ocean.

Three Mile Island released 0.00004 PBq entirely into the air.

Chernobyl released 85PBq, mostly into the air.

Nuclear weapons tests released 400PBq over several years. The majority has eventually landed in the sea.

Uranium-238 37,000 PBq

Though serious, these totals pale compared to the abundance of radioactive substances naturally present in seawater such as uranium-238 and potassium-40.

Potassium-40 15,000,000 PBq

FIGURE 7-5

In addition to radioactive half-life, biological half-life plays an even more important role for many radioactive isotopes, including cesium. Biological half-life is the timescale during which a substance remains in the body before being excreted or exhaled. For cesium the biological half-life is approximately 70 days, much shorter than 30 years.

In nearly every radiophobic discourse, mega-numbers and hyper-adjectives are utilized out of context. And attributions of guilt or innocence are employed in one-sided ways, whether it

---

[17] We found this illustration in 2015, though it is undated. It was published by the Woods Hole Oceanographic Institute in Falmouth, Massachusetts.

[18] Note also the minuscule release from TMI, mentioned in Chapter 5.

focuses on Fukushima-leaked radioactivity while ignoring the far greater natural radioactivity or focuses on imagined danger from nuclear waste while ignoring the far greater amounts of waste and toxicity associated with wind/solar.

A striking example of radiophobic discourse comes from US environmental journalist Robert Hunziker (http://tinyurl.com/ 3tw4kjhp), whose prolific inventions are happily published by the leftist online magazine *CounterPunch*. The present authors submitted a rebuttal of this 2015 article, which the magazine's editor, Jeffrey St. Clair, refused to publish. He even refused to answer our emails asking whether he had received it. Our rebuttal was, however, subsequently published in two pronuclear blogs (http:// tinyurl.com/yktret5n and http://tinyurl.com/ 3ehw95tj).

James Conca, in an article published in 2022 in the *Nuclear Newswire* of the American Nuclear Society (ANS, http://tinyurl .com/5n76uan8), contrasts the reactions to the fallout from above-ground US nuclear weapons testing in Nevada that took place in the 1950s upwind from St. George, Utah—in the southwest corner of the state with the lowest cancer rates in the US— with that of the Japanese government's response to the effect of the 2011 quake and tsunami on Fukushima Daiichi. In St. George, people were simply advised to remain inside until the cloud passed, while in Japan, the panicked reaction caused the immediate deaths of some 2,300 people (http://tinyurl.com/y9xdh954), with an even greater number subsequently among the resulting refugees (http://tinyurl.com/4k2wdrtv).

Conca, citing an article by US radiobiology experts Antone Brooks and Bruce Church, both of whom grew up in St. George, says,

> The maximum dose rate in St. George was 3.5 mSv/h (May 19, 1953), while in Fukushima it was about 1–10 mSv/h at the main gate on March 11, 2011—and four days later, on March 15, 2011, about 25 miles downwind, it was 0.045 mSv/h.[19]

In other words, at first Fukushima exhibited dose rates that bracketed the maximum dose rate in St. George, and it quickly fell off with both time and distance to levels approximating the highest natural background dose rate in the world. Brooks and Church deliberately excluded internally deposited fallout, which would

---

[19] The term "mSv/h" means millisieverts per hour, a measure of radiation dose rate. For comparison, the average natural radiation level in the highest known natural background area in the world—Ramsar, Iran—is about 0.03 mSv/h.

have raised the exposures in St. George to levels far exceeding those in Fukushima. The internal deposition would have included Sr-90, Cs-137, I-131, Pu-239, plus other less abundant isotopes. And at that, there was no increase in the very low cancer incidence in St. George. Ignoring this absence and ignoring all the other carcinogenic agents in their environment, those who happened to develop cancer attributed the cause to this fallout. Without this attribution they would have been ineligible for government compensation (discussed by Miles, 2008).

A recent upwelling of nuclear fear came from the dramatic TV miniseries "Chernobyl" in 2019. Despite its horror-movie theme, writer-producer Craig Mazin says he is pronuclear and sees nuclear energy as necessary to combat climate change (http://tinyurl .com/yw25w26w). He says he hoped the audience would not take his dramatic fiction seriously. Given the nature and intensity of nucleophobia, his expressed hope was at best extraordinarily naïve. Indeed, antinuclear writers amplified the fictional plot as though it were investigative reporting.

Critiques of the miniseries have been plentiful, so we will focus on its major scare claim: if the corium (melted mix of fuel with other substances in the reactor core) from the (non-nuclear) explosion were to fall into the pressure suppression pool underneath the reactor, there would be a thermonuclear (fusion) explosion of 2–4 MT that could destroy everything in a 30 km radius, and the radiation would likely be fatal to the inhabitants of Kiev and Minsk (100 km and 400 km away, respectively). The event, the story continues, could impact Poland, the Czech Republic, Hungary, and most of East Germany by disrupting food and water supplies, and could render Belarus uninhabitable for 100 years. This is like saying that a bullet accidentally dropped onto the floor could ricochet, fatally piercing the hearts of 29 people while setting fire to their homes, flattening all the local supermarkets, and draining the city's reservoir.

This fictional miniseries echoes Jacobson's straight-faced prediction of 2.6–16.7 million dead from fifty fission bombs with a total force of 0.75 MT, less than the 2–4 MT fusion bomb described in the miniseries. While Mazin says he hopes his audience will realize he is offering up fiction, Jacobson apparently hopes his audience will regard his account as fact.

One character in the miniseries describes the potential for Europe to be rendered uninhabitable for 500,000 years. This is the stuff of Frankenstein-cubed, nor need we fear Dracula's return, though these gothic characters likely caused no comparable conflation of fantasy with reality.

The Chernobyl steam explosion put the emergency workers at risk, but neither a fusion nor a fission explosion can occur in a nuclear reactor. As physicist David Bodansky, author of a popular textbook on nuclear energy (2004), explains, referring only to fission explosions (p. 373):

> In the light of possible misapprehensions, it is worth noting that a bomb-like nuclear explosion cannot occur in a nuclear reactor. In a bomb, a critical mass [20] of almost pure fissile material (U-235 or Pu-239) is brought together violently and compressed by the force of a chemical explosion, and the chain [really branching] reaction develops fully within one-millionth of a second—quickly enough for much of the fuel to fission before the mass is disassembled.

The "almost pure fissile material," by definition, requires enrichment to above 90 percent in a uranium bomb and above 93 percent in a plutonium bomb. In other words, the impurities have to be reduced to below 10 percent and 7 percent, respectively—impurities consisting, among other things, of other, non-fissile isotopes of uranium and plutonium. To give some idea of the effort involved in the enrichment process, mined uranium consists of only 0.7 percent U-235 plus 99.3 percent U-238. Plutonium does not occur in nature (anymore) and is only produced from U-238 in a nuclear reactor, which also produces Pu-240 almost as quickly as it produces Pu-239. In this context Pu-240 is an impurity which, in excess of 7 percent, interferes with the intended explosion.

So in a uranium bomb (Hiroshima), U-235 has to be enriched from 0.7 percent to over 90 percent, which is not achieved by doing anything to the U-235, but rather by separating and removing the vast majority of the U-238 while minimizing any U-235 removed with it. In a plutonium bomb (Nagasaki), the newly created Pu-239 has to be removed from the reactor quickly after formation to keep it above 93 percent of the isotopic mixture with Pu-240. And to transport a bomb to its target, the fissile material either must be separated into two subcritical masses that are blasted together (Hiroshima, uranium) or it must initially be of subcritical density and implosively compressed (Nagasaki, plutonium).

The conditions required for detonation of a fission bomb are no closer to being met in a reactor than in an ice cream parlor.

---

[20] The minimum mass, given adequate density, of fissile material needed to enable fissioning.

Conditions for a thermonuclear (or hydrogen) bomb are far more restrictive, as they aim at an explosion three orders of magnitude more powerful. U-235 in Chernobyl's RBMK reactor was enriched from its naturally occurring 0.7 percent to about 2 percent; there was no detonator, no precise geometric placement of the chunks of uranium, nowhere near the required weapons-grade (bomb) enrichment, and, once the reactor was shut down, there was no longer firing of a neutron "spark plug." The corium was no longer fissioning but rather was undergoing radioactive decay—not splitting into two roughly equal atomic halves in a set of parallel chain reactions but rather ejecting small subnuclear particles plus electromagnetic radiation, each ejection an independent event.

As at Fukushima Daiichi, it was the radioactive decay that generated the heat once the reactor was shut down and the fissioning had ceased. Radioactive decay initially generated heat at roughly 6–7 percent the rate of the fissioning process, decreasing over hours.

A thermonuclear device, in contrast, requires incorporation of deuterium (H-2) and tritium (H-3) into the specially designed geometry, and a fission explosive device is needed to provide sufficient energy to initiate the fusion of the hydrogen isotopes, much as a match is required to light the wood in a fireplace. Moreover, the damage estimates for a real thermonuclear device are themselves tremendously exaggerated (https://tinyurl.com/nwfrx5ex).

Most importantly, all other nuclear reactors in the world today have their components physically arranged so that, even without human intervention, there will be countering (rather than amplifying) feedback if the fuel, for any reason, starts to overheat.[21] (And the eleven RBMKs have since been modified to render a similar accident virtually, though perhaps not entirely, impossible.) A design incorporating countering feedback prevents a runaway fission reaction, not dependent on wishful thinking or meticulous management, but on the physics of the design. Overheating of the fission process in all other commercial reactors is no more possible

---

[21] Consider the familiar operation of a thermostat. Inside a mechanical thermostat an electric switch opens and turns off the furnace or heat pump if the room gets hotter than the desired set point. The switch opens because of the heat-caused expansion of a metal component, just as the fuel rods in some reactors are designed to expand and push each other away, thereby reducing the critical density necessary for the chain reactions. When the temperature of the room again falls below the set point and the metal component cools, the switch recloses, turning the furnace or heat pump back on. Similarly, the fuel rods approach each other once again and the chain reactions resume. In a digital thermostat, the on-off switch is controlled by an electronic, rather than mechanical, reaction to the room's temperature, but the countering feedback is the same.

than a sudden cessation of the Earth's rotation on its axis—another subject

An additional upthrust of nuclear fear a few years ago came courtesy of long-time antinuclear campaigner Harvey Wasserman, who was interviewed by Cenk Uygur, former commentator on MSNBC (http://tinyurl.com/4dpk5znd). Wasserman's target was the Diablo Canyon nuclear plant on the southern California coast that, he claims, could have an accident that would kill 10 million people in the "blink of an eye." Numerous commenters took Uygur to task for hosting this fountain of mendacity.

Yet another example of fearmongering is found in a New Yorker article by Daniel Ford, former executive director (1972 to 1979) of the Union of Concerned Scientists (UCS) and an economist rather than a scientist of the type implied by the name of the organization (http://tinyurl.com/2j644xur). The occasion for Ford's article was the current war in Ukraine, in which nuclear power plants are in the range of shelling from adjacent battlefield engagements. Ford emphasizes the probability of destruction of a nuclear plant, merely implying horrible effects. Ignoring the history of actual events, he relies on innuendo, with phrases like "potentially catastrophic," "the technology's biggest safety issues," "inherent dangers," "risk of a bad accident," "the damage that might result," "dispersal of radioactive fallout over a large area," and "major accident." He speculates on a worst-case scenario that he attributes to Brookhaven National Laboratory in 1965, "a devastating meltdown could cause forty-five thousand deaths, with radioactive contamination creating a potential 'area of disaster the size of the State of Pennsylvania.'"

If we told you that a chemical plant fire "could" cause 200 million deaths and destroy an area the size of Africa, who wouldn't believe us mainly because we claimed that the chemical company's denial was a lie? And why is it that the scarier the statement the more believable it seems to be? Surely truth does not require such a criterion; so why should believability? Is it that no one would risk making such outlandish statements if they were not true? The recent untethered proliferation of manifest falsehoods in US political life should render that judgment quaintly obsolete.

The relationship between the consequences of an untoward event and the probability of its occurrence is important. We have learned to accept technologies with potentially significant consequences if the probability of an accident is small and others if the probability of occurrence is significant but the consequences are not. Consider the fact that the largest airplanes of all, the Boeing 747 and the Airbus 380, in the event of a crash, would almost certainly result in more deaths than that of any other aircraft. However, these two transportation giants have the best safety record of all per mile traveled. That is, the small frequency of crashes far outweighs in importance the possible consequences, so they are well accepted by the public. On the other hand, computer crashes are not so improbable, but their consequence, particularly for human safety, is not significant. Antinuclear propaganda exaggerates both the probability of an accident and the possible consequences.

With nuclear energy, first, the probability of an accident, while not trivial, is far smaller, per TWh of electricity produced, than an accident involving a fossil fuel plant, mine, or refinery or a renewables farm. Second, and even more important, the likely consequence of a nuclear accident is far smaller than for fossil-associated accidents, or even those associated with renewables farms when considered in the aggregate. The combined direct consequence of the three most prominent nuclear power plant accidents (in the world's approximately 20,000 land-based reactor-years, with another 15,000 or so seagoing reactor-years) is less than that of a single 747 or 380 crash, and all other nuclear accidents combined hardly even made the news because of the lack of serious direct consequences.

That a significant number of deaths have resulted indirectly from the nuclear accidents is attributable to nucleophobia rather than to nuclear energy, or to radiophobia rather than to radiation. So, in summary, while the probability of a nuclear power plant accident—though smaller than that associated with its competing energy sources—is not trivial, its consequences generally are.

Ford goes on to say,

> Perhaps most alarmingly, the accident at Fukushima nearly caused a fire in a spent-fuel storage pool that was outside the strongbox. According to recent analyses by [Princeton physicist, Frank] von Hippel and others, if that had happened, the release of radioactive material could have multiplied a hundredfold, potentially requiring the relocation of as much as a quarter of the Japanese population, depending on which way the winds were blowing.

Note the uninhibited use of hypothetical terms such as "nearly," "if," "could have," "potentially," "as much as." These permeate the rhetoric of this apocalyptic fantasy ("a quarter of the Japanese population").[22] Another rhetorical element key to the production of nuclear fear is the vagueness of characterization, promoting the uncertainty and doubt that are intended to accompany fear (FUD). Ford gives no hint of the conditions under which a "fire" could occur in a spent-fuel pool (SFP, though the fuel is far from spent, discussed below), no sense of the conditions under which radioactive material would be emitted and escape containment, and no sense of what this radioactive material would do once "released." Vague description, no analysis, and the appeal to a voice of authority.

There is no way to defuse unwarranted fear other than by examining in some detail the claimed disastrous possibilities, so bear with us.[23]

First, the reference to a possible fire may have been suggested by an actual fire on the third level of unit 4 that extinguished itself "naturally" within a few hours (Nuclear Energy Agency 2015, p. 61). As we know, a good deal of radioactive material did escape, courtesy of the hydrogen explosions following the partial meltdowns of three of the six units. We also know, according to UNSCEAR among other official organizations, that there have been no detectable deaths that could be attributed to the radiation release. Note that no specific SFP is named in Ford's account. If we're forced to fill in the picture, we would guess this refers to the SFP in unit 4, which had the highest heat content at the time of the accident and where the water in the pool reached the lowest level about one month after the onset of the accident, 1.5 meters above the stored fuel assemblies. At that point, the pool was refilled to capacity. Noted in the NEA report (p. 87):

> SFP accidents that result in radioactive release are highly unlikely. The SFP accident progression from initiation, to alarm, to dose consequences is slower in comparison to that of most reactor core accidents, progressing for days before any significant consequences occur. Due to the slow progression, operator intervention

---

[22] Leading antinuclear guru Helen Caldicott went Ford one better when she prophesied shortly following the events at Fukushima that Japan might become uninhabitable. Never mind that Hiroshima and Nagasaki are once again thriving cities and were rebuilt in the years following their atomic destruction—a hint, incidentally, of the falsity in the claim of long term uninhabitability around Chernobyl.

[23] As a general point, to gain perspective when some writer imagines worst-case scenarios for nuclear energy, try doing the same for the competing energy sources and try imagining the loss of all those things we take for granted in modern life if we lacked sufficient energy.

is highly likely to arrest the accident in its early stages, before onset
of fuel degradation and radioactive release.

The slow accident progression of SFPs is a design feature, an element of defense in depth. And in this case, this feature probably
did prevent partial uncovery of the fuel.

Detection of radioactivity release from SFP 4 was hindered
because it was overshadowed by the release due to the partial meltdown and hydrogen explosions in units 1, 2, and 3. Capitalizing
on the confusion, antinuclear critics like US nuclear engineer Arnie
Gundersen[24] claim that a criticality incident must have occurred in
SFP 4 to cause the hydrogen explosion, though the one has nothing to do with the other. In other words, the once-through fuel,
he maintains, must have reacquired a critical mass/density, initiating neutrons must have been injected, and fission must have
recurred in the stored fuel assemblies. But from the NEA report,
among Japan's "criticality controls," there was, and still is, borated
aluminum in the racks (p. 169), which prevented recriticality.

Boration (impregnation with the element boron) absorbs stray
neutrons that might otherwise initiate fission events, the same
function performed by borated control rods (and borated coolant)
in a reactor to regulate the number of parallel chain reactions and
hence the rate of heating and the amount of electrical power produced. The pool water was also borated. Furthermore, the hydrogen was vented from unit 3 through pipes joining units 3 and 4,
and the reactor in unit 4 had been down for maintenance.

Moreover, as Gundersen knows, the reason that once-through
fuel has been stored in the SFP in the first place is precisely because
the buildup of fission products in the fuel impedes the chain reactions, and maximum power can no longer be extracted from the
fuel. In other words, the fuel has difficulty maintaining criticality
even in the operating reactor. Additionally, the water level never
fell below 1.5 meters above the stored fuel rods, so the stored fuel
could not have achieved sufficient temperature to melt the zirconium cladding and release hydrogen.

Gundersen conflates a zirconium fire or at least zirconium oxidation, either of which releases volatile hydrogen, with criticality
(resumption of fissioning). The zirconium events are possible if the
fuel rods are too densely packed in the racks, enough fresh fuel is

---

[24] Gundersen is a nuclear engineer with several decades of experience. This gives him the
imprimatur of expertise, which he has chosen to use in service of fear-inducing distortion.

present, and the water level drops to expose the rods, but the boration prevents recriticality. While SFP 4 did have abundant fresh fuel that was densely packed, the water prevented a zirconium fire. Even if water levels drop too far, loose packing (called "open racking") can allow the surrounding air to cool the rods sufficiently to prevent bursting of the cladding with release of the radioactive fuel.[25] And in the case of hydrogen release by reaction of the zirconium cladding with steam, there are devices called hydrogen combiners that catalyze the formation of water and act as a further layer of defense in depth.

As to Ford's reference to von Hippel's hypothetical projection of the relocation of a quarter of Japan's population, which would have been 32 million people in 2011, the *actual* 10-day reactor fire at Chernobyl, involving both zirconium and graphite, saw the relocation of about 1 percent that number of people (over 300,000) with horrendous results for the victims. But they experienced no radiation-related deaths or even radiation sickness. The only radiation-related, but preventable, deaths at Chernobyl were suffered by the fire fighters and other responders (total deaths less than 60), plus at most a similar number in the surrounding population due to the attributed, but preventable, thyroid cancers (Chapter 8).

Yet another of Ford's stratagems involves his mathematical extrapolation from the time intervals between TMI, Chernobyl, and Fukushima Daiichi—the three most widely advertised nuclear power plant accidents, each with a different cause. He uses this extrapolation to yield a guess as to the future frequency of such events, echoing Jacobson's prediction of a future nuclear war. TMI was followed in 7 years by Chernobyl, and Chernobyl was followed in 25 years by Fukushima, an interval 3½ times longer. Ford tries to obscure the significance of this longer interval by noting that the latter event involved three reactors while the first two events only involved a single reactor each. As though the number of reactors neutralizes the fact that the expanding interval between accidents reflects a learning curve that makes accidents less likely.

This "concerned scientist" misses the point that if the learning curve is ignored, the expanding number of the world's reactors should result in a contracting interval, not an expanding one. In other words, the more reactors the greater the chance of an accident, so more accidents should occur per unit time. Ford concludes, "the world should expect one full or partial meltdown

---

[25] See https://tinyurl.com/y87cfzy5, https://tinyurl.com/3n5nz8ea, https://tinyurl.com/mtfdy9ev, and https://tinyurl.com/r74rx493.

every six to seven years," and by implication for the foreseeable future. By that reasoning, we could have once confidently predicted that the world should expect one plane crash every 100 takeoffs, whereas the figure has diminished to one per 4 million. Outside of Ford's imaginary world, learning curves make new technologies ever safer.

Indeed, the fact that TEPCO had placed the auxiliary diesel cooling pump for unit 6, the last to be built, higher than the others reflected a learning curve, and it paid off in preventing either a core meltdown or uncovery of the fuel rods in the SFPs of units 5 and 6. Had this lesson been learned earlier, as we said above, the entire accident at Fukushima Daiichi would never have occurred.

Finally, Ford is prompted to ask whether Biden's revival of nuclear energy is "desirable, given that technologies exist for harnessing the sun and wind which don't raise daunting safety issues in the first place." While he is likely correct about the absence of daunting safety issues, despite notable exceptions for the workers harmed by turbine fires and PV panel installers falling off roofs, he takes for granted the feasibility and sustainability of an all-RE electrical system.

Before we leave this inventory of antinuclear fabricators, we should at least mention Pakistani-American nuclear engineer Arjun Makhijani, who, like Gundersen, occupies a rare category. He has testified before a Congressional committee to warn that commercial nuclear reactors manufacture plutonium that is usable in bombs. Here is a critique of his testimony showing how, if he was sworn in, he was guilty of perjury, or if not sworn in, mere prevarication: https://tinyurl.com/yxvh5xhr. And for a critique of his recent book ostensibly on the severe hazards of tritium, see https://tinyurl.com/9xzjphyw.[26]

The Fear of Nuclear "Waste": The Dybbuk of Nuclear Energy

There are several misunderstandings concerning nuclear waste (or spent fuel). First, what is generally considered waste is fuel that has

---

[26] The antinuclear stance may be likened to a large group's need to cross an ocean but refusal to employ an immense cruise ship moored at the dock. Instead, they prefer to take a number of rowboats, encouraged by a sign on the cruise ship reminding people of the insalubrious trip on the Titanic and the salutary effects of upper-body exercise. The sign has been hung by the oar manufacturers. Offering further encouragement, the gurus of rowboat technology explain the superiority of this mode of transportation, which incidentally affords the crossers more time to enjoy the sea breezes and sunshine, even if their cell phones will soon run out of charge.

been through one cycle in an LWR but still contains almost 99 percent of the original energy. That is indeed a waste, but a voluntary and premature one, and the fuel is certainly not spent. Capturing and converting the remaining energy to usable form requires periodic removal of the fuel to separate it from the built-up fission products and reinserting the cleansed fuel into a fast breeder reactor. In fast breeders, the major portion of the fuel—the fertile material (for instance U-238)—is transmuted over time to fissile material (such as Pu-239) and the energy extracted. But as we have said, recycling is currently banned in the US, though some other countries are building fast breeders and recycling.

Second, this wasted fuel is handled safely, as we explain below. There are no recorded injuries, illnesses, or deaths from this once-through fuel from commercial reactors anywhere in the world, even though direct exposure to it could be lethal within weeks after such an event. This is both similar to and different from vats of molten material in steel plants, stores of ammonium fertilizer, chemical plant materials and refuse, NG pipelines, crude oil in refineries, and so on. However, each of the latter has in fact caused deaths, some numbering in the thousands (for example, at the Union Carbide pesticide plant in Bhopal, India, in 1984).

Third, the highly radioactive fission products decay away to near natural background intensity in a few centuries and need only be stored and shielded for that length of time. The chemical toxicity of the various components is no worse than that of other heavy metals.

Let's examine these points in more detail. To begin with, the fission process inside a reactor consists of breaking up heavy nuclei (uranium, plutonium, or still heavier elements) into fission products (cesium, strontium, iodine, and so forth) that are approximately half the size of the parent nuclei; these roughly half-size nuclei are known as daughters. The daughters are far more radioactive than their parents—they decay faster as they are far more unstable. Their greater instability is a result of an imbalanced ratio of neutrons to protons that at first approximate the ratios in their parents. Neutrons serve as a kind of glue to overcome the mutual repulsion among the electrically charged protons. Parent nuclei, having roughly twice as many protons and therefore more mutual repulsion to be overcome, require a higher ratio than do their daughters—roughly three neutrons to two protons. So when a parent breaks into two roughly half-size daughters, the latter try to reestablish relative stability by shedding those unneeded neutrons. The readjustment involves the emission of photons (gamma rays) and energetic beta particles (electrons); the latter event transforms neutrons into protons, thus adjusting the ratio toward greater stability for these smaller nuclei. Since each element is defined by the number of protons, this process also changes one element to another.[27]

Since the gradual buildup of fission products, which are not themselves fissile, eventually impedes the chain reactions by intercepting neutrons, the fission products have to be removed from time to time, usually after a few years. This requires removing, in sequence, the fuel rods that contain the partially used fuel from the reactor, and either replacing them with new rods or recycling them through one of several possible processes to remove the fission products and then reinserting the cleansed rods back into the reactor (though it has to be a fast breeder reactor to sustain the production of electricity). However, today in the US when the fuel becomes laden with fission products, the rod is permanently removed from the reactor and stored. As we have said, it is first stored onsite for a few years in 40-foot-deep pools of circulating water to let the highly radioactive fission products decay for a while and cool down (see photo; the fuel rods are within the borated lattice at the bottom of the pool).

---

[27] Heavier nuclei that are radioactive also can emit alpha particles (helium nuclei). Alpha particles travel only about one inch in air and cannot penetrate the skin. Beta particles are electrons or positrons, and if energetic enough they may travel a few feet in air, but unless they are extremely energetic they cannot penetrate the skin either. In other words, externally emitted alpha or beta particles cannot do damage to our bodies, but taken internally through inhalation, ingestion, injection, or absorption, they can. Photons, called gamma rays when emitted by a nucleus, or x-rays when they come from the electron clouds around nuclei, can penetrate the skin and the entire body if energetic enough. They are the only products of radioactivity that, coming from the external environment, can damage internal cells. But, as we explain in more detail in Chapter 8, as long as the dose or dose rate of gamma rays is not too great, in other words does not exceed a particular threshold, the body is generally capable of repairing that damage or, that failing, of destroying and removing that cell. This all happens within a matter of hours, leaving the organism unharmed, or even in an improved condition. As a result of these capabilities, safety from fission products can be achieved by keeping enough distance from the source and/or shielding the gamma radiation with water, concrete, glass, or a metal.

Figure 7-6 (borrowed from Devanney 2023) shows the approximate proportions of components in a once-through fuel rod (labeled "Spent fuel"). But you can see why we say that the fuel is hardly spent at all. Repeated recycling into a fast breeder reactor can transform virtually all the U-238 into Pu-239 plus a small amount of other transuranics (such as americium and curium, illustrated in Figure 7-7 below), which can continue fissioning and releasing energy.

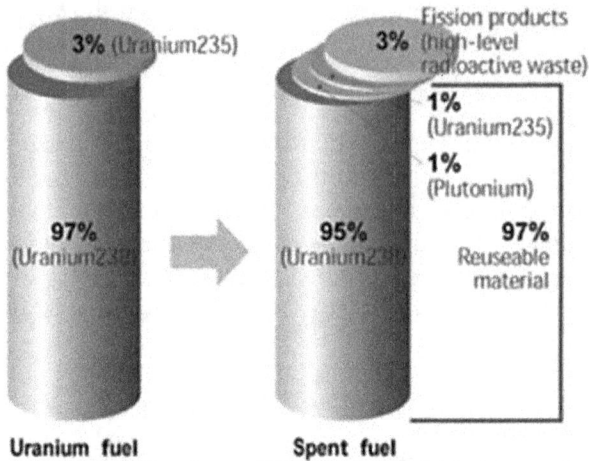

FIGURE 7-6

This once-through fuel needs time to cool because of the radioactive decay process, which upon initial removal, as noted above, generates heat at a rate about 6–7 percent that of the fission process. But the aggregate radioactivity diminishes over time, and the rate of heating therefore declines. After a few years, when the aggregate heating rate sufficiently declines, the fuel is removed from the pools and stored more permanently in steel and concrete casks, from which no detectable radiation can escape (see photo).

The once-through fuel could also be encased in glass and dropped onto the deep ocean floor where its decay products would be prevented from affecting anything around it (glass lasts for millennia and would long outlast the vast majority of the radioactivity in the fission products). The once-through fuel could also be stored in a salt mine or in a volcanic rock cave, such as that beneath Yucca Mountain in Nevada.[28]

Any of these methods is virtually completely safe, but a terrible waste. This is why we reject the terms "spent fuel" or "nuclear waste" in favor of "once-through fuel." Throwing it away is analogous to repeatedly driving 4–5 miles and then siphoning the unused gasoline out of your fuel tank, storing it in gas cans, and refilling the tank with fresh gasoline.

The main point is that the more radioactive a source, the faster it decays to insignificance. So, the far more radioactive fission products, which are mixed with the fuel removed from the reactor, disappear in a much shorter time than the fuel itself—a couple of hundred years versus tens of thousands to billions of years. But by the same token, the long-lived isotopes that constitute the initial fuel, having very little radioactivity (per mass of material), are not dangerous, since they don't overwhelm our body's ability to repair itself (heal the injury). This alone puts the lie to the claim that nuclear once-through fuel will have to be safely stored for thousands of years or more, threatening hundreds of future generations. After all, uranium and thorium are all around us in the ground and oceans and constitute the source of much of our natural background radiation (NBR). Responsible for over half of ground-based NBR is radon, a daughter product, or more suggestively a great granddaughter of uranium and granddaughter of thorium (Chapter 8).

Figure 7-7 illustrates the relative rates at which a couple of common fission products, strontium and cesium (US spelling), are transformed to more stable (nonradioactive) forms compared to the rate of decay of two heavy nuclei, americium and curium, both so-called transuranics (elements heavier than uranium) that are also bred alongside the fission process. The areas represent-

---

[28] This effort was abandoned by President Obama in 2010 at the behest of Nevada Democratic Senate Majority Leader Harry Reid, who also prevailed on Obama to appoint Gregory Jaczko as chair of the NRC. Jaczko's main qualification from Reid's point of view was his staunch opposition to nuclear energy. The citizens of Nevada, incidentally, were in favor of the federal funds that their hosting of Yucca Mountain would have brought to the state, but Reid was unmoved by the sentiments of those he claimed to represent. It was Jaczko who persuaded the Japanese government to forcibly relocate the tens of thousands of residents around Fukushima Daiichi, with all the negative consequences that action entailed. Removed from the NRC at the urgings of the other four commissioners, who regarded him as a tyrant, Jaczko continues to propagandize against nuclear energy.

ing the various elements are proportional to the amount of the element remaining after the designated number of years. The figures below each column represent the rate of heat production in terms of watts (W), which is directly proportional to the aggregate radioactivity of the several isotopes. The relative quantities of each element and the specific rates of heating at each point in time are what would be found from 50 liters of the once-through fuel as it is removed from the reactor, though only the rates of decay are important here. As you can see, the rate of heating and radioactivity drops off dramatically in only a few centuries, not in the "hundreds of thousands of years" claimed in antinuclear fearmongering.

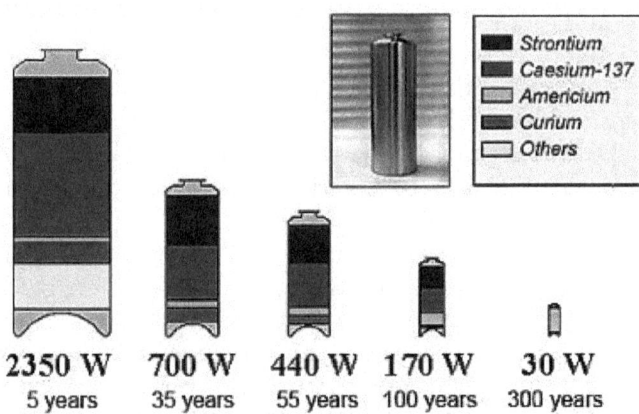

http://tinyurl.com/4thp3w54

FIGURE 7-7

Fossil fuel waste largely consists of GHGs and particulate matter, both released freely into the atmosphere. These are considered externalities, while the alleged hazards of nuclear "waste" are, by regulatory fiat, considered to be a part of the process of nuclear generation of electricity. The hazards of fossil fuel waste, apart from the ongoing impact on respiratory health and fresh water supplies, are indirect and emerge only slowly in the form of AGW with its attendant "natural" disasters. The deaths from these effects outweigh those due to nuclear energy to an incalculable degree, yet in an upside-down account, the waste from fossil fuels is rarely considered to be as hazardous as that from nuclear.

The late nuclear submarine pioneer Ted Rockwell (Admiral Rickover's righthand man) once said:

The US has many real problems. Nuclear waste is not one of them. The real waste problem is the money being spent on silly ideas like million-year isolation vaults. A simple, fenced-off area to store the glass logs would do nicely, with perhaps an OSHA sign reading: do not eat the glass.[29]

Nucleophobia and Controllability

One of the roots of nucleophobia is the belief that nuclear energy is uncontrollable, a belief that presumably distinguishes nuclear from wind/solar. However, as we explained earlier in this chapter, because reactors, unlike a bomb, are designed to permit only chain reactions and prohibit branching reactions, the nuclear process in a reactor is controllable. And even in a bomb, initiation of the branching reaction is deliberate and controllable. In fact, nuclear bombs that have accidentally fallen out of planes have failed to explode (http://tinyurl.com/ycys7xh9).

It's ironic that the most feared form of energy, nuclear, is the safest of all for electricity, even as it could be extremely destructive as a weapon.[30] Indeed, it may be justifiably said that the three biggest nuclear power plant accidents in the world—Three Mile Island, Chernobyl, and Fukushima—indicate not how dangerous nuclear energy may be, but rather how safe it is.

It follows that those who oppose the building of nuclear power plants to replace coal and NG plants unwittingly share responsibility for the deaths of millions of innocent people every year from

---

[29] From a letter Rockwell wrote to *The Washington Post* in 1996 that they did not publish. It was published, however, by US nuclear engineer Rod Adams in his blog *Atomic Insights* (https://tinyurl.com/4nsdkdpr).

[30] However, even as a weapon, nuclear fission (atomic) bombs, while tremendously destructive of structures, have been no more deadly than the chemical firebombings (using napalm, jellied gasoline, a form of fossil fuel) of some 65 Japanese cities. The firebombings culminated in the destruction of Tokyo with the resulting deaths of some 100,000 persons, in March 1945, five months prior to the atomic bombings in August. While estimated deaths vary according to source, the number of deaths in Tokyo by fossil fuel bombs, either roughly matched or slightly exceeded those in Hiroshima and Nagasaki by nuclear bombs. Granted Tokyo was a much larger city than either of the other two, and it took many bombs to accomplish in Tokyo what required only one bomb each in Hiroshima and Nagasaki. But the point is that chemical explosions, when so designed and used intentionally, can be at least as deadly. It is the political intent that matters and not the particular technology. In fact, as recounted by Iris Chang in her 1997 book *The Rape of Nanking*, the Japanese military murdered some 250,000 Chinese civilians in Nanking over six weeks, beginning in December 1937 as World War II was about to take off, and they did so with low-tech bayonets, swords, and bullets.

fossil pollution alone, and many more from the unconstrained effects of AGW. And that the rapid building of nuclear power plants would save those millions of lives.

* * * * *

The Recent Fusion Breakthrough at the Lawrence Livermore National Laboratory (LLNL)

LLNL's December 5, 2022, breakthrough at its National Ignition Facility (NIF), in producing energy from nuclear fusion on Earth, stirred media focus and prompted a claim by the Biden administration that we are nearing acquisition of a new energy source that can displace nuclear fission energy.

This is a misstatement of the nature of the achievement and requires correction, particularly since some antinuclear propagandists claim that in contrast to fission, fusion will be safe, as it does not, they say, increase radioactivity (but see below).

NIF scientists and engineers may have seen a scientific breakthrough—with even greater advances since (https://tinyurl.com/mu2kzty6)—but their mandate concerns the design and effects of nuclear weapons, not the powering of an electrical system. A breakthrough for one purpose is not necessarily a breakthrough for a different purpose. In particular, the breakthrough at NIF lacks two key requirements for a source of usable electrical energy—first, the EROI must exceed 1 by a significant margin, and second, the fusion reactions must be sustainable over extended periods of time, not just be a single split-second event.

In order to achieve EROI > 1, either each energy-releasing fusion reaction must yield more energy than its igniting source requires as input for each ignition, or the reactions must be self-sustaining once ignited, in which case the energy yield of the series of self-sustaining reactions will sooner or later exceed the energy consumed by the igniting source, including the energy to construct the entire apparatus. That is, self-sustaining means that each fusion event has to be the igniting event for subsequent fusion events without any further need for the original ignition source, as explained above in the box on chain reactions.

The mass media and the chief Administration spokesperson, Secretary of Energy Jennifer Granholm, asserted after the December 2022 success that the fusion did release more energy than the igniting laser required, but in fact it only released more energy than the laser released. In other words, their claim focused on the energy put into the fusing nuclei by the laser, but it ignored the electrical energy required to fire the laser.

In round figures, the laser released 2 MJ (megajoules) of energy—a little over $1/2$ kWh—and the fusion reaction released 3 MJ ($5/6$ kWh), an excess that caused the NIF scientists to celebrate a breakthrough and signifying that, in net, none of the energy released by the laser was lost (or more accurately, if any laser energy was lost, it was more than made up for by the fusion event). This result signified an immediate EROI of 1.5 (3 MJ/2 MJ), greater than 1 but still too small to produce energy to run a society. But the electrical energy required to turn on the laser was 300 MJ (83 kWh), with a resulting net EROI of 0.01 (3/300), in other words, an EROI << 1 rather than > 1. Media attention has focused solely on the gain from 2 to 3, neglecting the drop from 300 to 2. As long as this process requires energy in excess of that released in the fusion reaction, it constitutes an energy sink rather than a source, and a very deep sink at that.

In short, the nuclear fusion event at NIF, rather than being a method of energy production that could be converted to electricity, was an end-use application for some purpose concerning nuclear weapons that apparently remains classified and not released to the public. The response achieved by NIF scientists and engineers is not necessarily without a benefit that is as yet unrevealed, but it cannot serve as a source of usable energy.

As to the possibility of finding a method by which earth-bound fusion might someday be able to sustain itself, and thereby eventually produce more energy than it takes to initiate the process, it's much too soon to tell. Many alternative approaches are under experimentation around the world (https://tinyurl. com/y69xx827).

That said, it is worth noting that not even the sun's fusion energy sustains itself. That is, the solar energy we receive here on Earth is not the result of a set of chain reactions in the center of the sun. Rather, the fusion reactions are sustained by repeatedly tapping the heat energy created and stored when space dust condensed under its own gravitational attraction to form the sun and the rest of the solar system 4.5 billion years ago. Thus, with very rare exceptions, the fusion reactions are not ignited by each other, as they are spatially too rarefied within the sun's core. The heat energy is analogous to the lasers at NIF and the gravitational condensation analogous to the electricity required to activate the lasers. If hypothetically the temperature of the sun's core were to diminish sufficiently the fusion reactions would cease.

With each fusion event, the energy released comes from the conversion of a small portion of (hydrogen) mass to energy, part of which pays back the borrowed heat and part takes the form of electromagnetic energy (photons). A small portion of the photons, or more accurately their descendants, eventually arrive on Earth as sunlight, though the vast majority go off in other directions.

The sun's internal heat energy is only a part of the decrement in gravitational potential energy during the sun's formation, much of it having been long since radiated off into outer space during the process of condensation. This is similar to the way that the vast majority of the 300 MJ that switched on the laser was lost to heating the surroundings (roughly 298 MJ), with only 2 MJ released by the laser to ignite the fusion reaction.

As we described earlier in this chapter, think first of a set of dominos that are arranged in a single file so that once the front domino is given enough energy from an external push to fall backwards, it provides the energy (push) that fells the second domino, the second the third, and so on. That would earn the label "chain reaction." Think next of a set of dominos that are arranged side by side, such that an external push on the first domino ends with no result other than the falling of that first domino (and the heating of the domino and the surface on which it lands). A similar external push on each domino is needed to make each fall. This is not a "chain" reaction and is not self-sustaining over time. The side-by-side positioning of the dominos is analogous to the process taking place in the sun's core.

No sustained fusion chain reaction anywhere in the universe has, as far as the present authors know, been discovered, or created. Yet. Nor has there been any indication that such is possible, though the advice to "never say never" is still to be respected. A fusion chain reaction is certainly not prohibited by energy considerations, since more energy is released by each fusion event than is required to overcome the electrostatic repulsion of hydrogen nuclei (positively charged protons) and thereby induce a further fusion event. But the appropriate architecture of the apparatus (analogous to the single-file arrangement of the dominos) has yet to be found. Such an architecture, for example, would have to somehow retain enough of the heat energy released by the fusion events so that further fusion events would have an ongoing fund of heat to tap. That could constitute a chain reaction, or more accurately, a set of parallel chain reactions. The fact that each fusion reaction would release a neutron is of no use for sustaining fusion, only for sustaining fission. The only path to sustaining fusion reactions is to raise the kinetic energy of the hydrogen isotopes to adequate levels for them to collide and fuse, overcoming their mutual repulsion.

Furthermore, finding the appropriate architecture would constitute a different kind of breakthrough from the recent one at NIF, this time for the possible production of sustainable fusion energy on Earth for the purpose of energizing an electricity system. The recent NIF breakthrough is simply not relevant to this goal, nor, as we said, was the generation of electricity the purpose of their research and experimentation.

Additionally, while some of those who oppose nuclear energy, meaning fission energy, mistakenly appeal to the NIF fusion breakthrough as a safer way to create nuclear energy, since there are no radioactive fission products, they are not even correct that fusion is necessarily safer. The pathway from hydrogen to helium on Earth is different than that inside the sun. On Earth the preferred path (for example, in a thermonuclear bomb) forces a deuterium nucleus to fuse with a tritium nucleus. Each fusion event results in the creation of a helium nucleus and a free neutron, both with high kinetic energy. The free neutrons can be absorbed by atoms in the surrounding apparatus, turning it into radioactive material. The radioactivity will be intense and therefore relatively short-lived, like that of fission products. This material should, however, be no more difficult to contain and shield than the confined radioactive products in a fission reactor. But either way, the process is not the clean event portrayed by those eager to do away with, or indefinitely postpone, fission energy. Whether an alternative pathway can be found that does not involve the release of neutrons is a question for the future.

Thus, just as wind/solar is incapable of serving as the sole energy sources for an electricity system, so is fusion energy on Earth—so far—although for reasons of architecture rather than EROI. At least wind/solar is not an energy sink like the NIF fusion achievement. But as the consideration of EROI demonstrates, the energy investment in wind/solar to obtain the energy output renders them wholly inadequate to serve social needs, or even to sustain themselves, without nuclear energy (or fossil fuels). Sustainable fusion on Earth, at this point in time and possibly forever, is even further removed than wind/solar from serving as a source of electricity. It is still the case that nuclear fission is the only feasible energy source today that by itself can both power societies and halt AGW. Recognition of this truth is all the more urgent because we don't have the luxury of delay, as AGW continues to create ever- increasing havoc with our environment and our lives.

8

Radiophobia and the Linear No-Threshold Myth

Is All Ionizing Radiation Harmful?

Let's begin with an apocryphal story. Suppose we discover a chemical that turns out to be poisonous. Next we discover its presence in many vegetables and meats. As a result we name it pantothenic acid (from the Greek, meaning found all over) and announce that it should be avoided at all costs. This would necessitate avoiding many foods. But since this would be difficult, people might try to ignore such advice, though it might cause some fear until time has passed and they turn their attention elsewhere.

Then suppose we find that people on diets lacking pantothenic acid—for example, all-rice diets—suffer illness and early death. Can a poison really have an amount insufficient for health? Well, it turns out that nothing is poisonous unless consumed in excessive amounts and that many things are critically necessary in moderate amounts. So we are forced to seek the lower limit of the necessary range and the upper limit of the safe range, for virtually everything.

Pantothenic acid, incidentally, is known as vitamin B5; its deficiency causes neurodegeneration and dementia.

Ionizing radiation[1] is also found to have both a safe and necessary range of exposure, just as do all vitamins, oxygen, water, sunlight, and innumerable other agents. Too little, we die young. Too much, we die young. Just right and we maintain or enhance our health status and prolong our lives. This phenomenon is well known, but it's been suggested without evidence that radiation

---

[1] Throughout, "ionizing" is to be understood unless otherwise specified.

uniquely lacks these three ranges of exposure. The midrange is popularly known as the Goldilocks zone, and its existence is denied by the claim that exposure to radiation is linear in its harmful effects from zero dose to very high doses, with no relief at any dose. This is the demonstrably false linear no-threshold concept, or LNT, of the dominant school in radiobiology.

Sidebar: LNT Denies What Makes You You

A key characteristic of living matter is homeostasis (homeo = same, stasis = stays)—the tendency of live organisms to react to perturbations by trying to restore themselves to their former condition.

Analogs of homeostasis have been incorporated into mechanical and electrical inventions, such as the thermostat (described in Chapter 7), a device that restores a room's temperature to its desired range if it wanders outside of it.

Similarly, any external or internal process that perturbs a living organism's condition is met by countering feedback (an internal process) that attempts to restore the organism's former condition. Sickness or injury initiates a healing process. Overheated we sweat or pant to shed heat. Overcooled we shiver to generate muscular heat. Dehydration causes thirst, though we have to voluntarily respond to quench thirst. Depletion of our main energy source causes hunger to replenish it, again requiring voluntary action. Deplete our oxygen and we automatically breathe faster. Without homeostasis we would not live long, so the only species that have made it through evolution's natural selection are those capable of homeostatic responses.

The reason that you remain yourself throughout life is that every change is met by a tendency to retain your major characteristics. LNT proponents focus on the perturbation, either ignoring or underrating the homeostatic processes – the ones that make them them.

The public's awareness of radiation is difficult to separate from the dropping of two atomic bombs in World War II. Excessive radiation, as with virtually all other agents, can assuredly cause illness or death. But the vast majority of people in Hiroshima and Nagasaki who were close enough to the center of the blasts to receive a fatal, or even sickening, dose of radiation were instead killed quickly by the blast and fires. And the vast majority of those far enough away to survive the blast and fires were also far enough away to receive harmless, if not ultimately beneficial, doses of radiation (see, for example, Sutou 2020).

The misidentification of the lethal agent helped some scientists to convince others that all radiation was poisonous, to be avoided at all costs (Calabrese 2022; 2023). But we cannot escape it. Radiation comes from the ground and sky; it's in our food (K-40 and C-14), making all of us radioactive. We may be able to keep our distance from others, but not from ourselves. Only those biological species that have evolved protective responses to radiation's undeniable damage have escaped extinction. While we are all continuously exposed to natural background radiation—greatly exceeding that from normally operating nuclear plants—we take this sea of radiation for granted, like the sea of air surrounding us, or we are unaware of it. This leaves us vulnerable to manipulation.[2]

The intensity of radiation and/or quantity of radioactive material is often greatly exaggerated. While the area around Chernobyl is now called a wasteland, military photographer Bob Hall, some six weeks after the August 1945 atomic bombing, took pictures of Hiroshima, many of which hang in the National Museum of Nuclear Science and History in Albuquerque, NM. His film wasn't even fogged, despite film's being more sensitive than humans (and incapable of healing itself). Nor did Hall suffer any health detriment (https://tinyurl.com/mumhuzkb). Hiroshima and Nagasaki were rebuilt and occupied within four years, while repopulation of the areas around Chernobyl and Fukushima Daiichi is still largely prohibited by the authorities, 38 and 13 years later, respectively.[3]

---

[2] Shortly after this sentence was written, a news item appeared in October 2022 about a suburban St. Louis elementary school whose grounds contain "unacceptable levels" of radioactive waste. Neighboring Coldwater Creek flows into the Missouri River, upstream from which World War II waste materials had been dumped some 80 years ago. It seems that "unacceptable" was the paid judgment of a single engineer, author of past antinuclear articles. Quantitative estimates are absent from his report and from the associated news articles. Adjectives without numbers and numbers without context or comparison predominate. To judge what level is "acceptable" or "unacceptable" requires knowledge of dose rate and the threshold for harm. This engineer was deliberately chosen from among those who deny the existence of a threshold—chosen by lawyers pursuing a class action against the school. Otherwise, he would have been useless to the plaintiffs. Two subsequent independent investigations, one by the Army Corps of Engineers, found the levels of contamination in the school to be below the threshold for harm. Of course, conspiracy theorists would take this as proof of the opposite, and some parents of the school's students are hesitant to trust this judgment of harmlessness, yet neither is skeptical of the hired engineer's paid claim of a hazard. This fear and doubt results from decades of denial by some scientists that there are thresholds below which there is no harm and possibly benefit, as we discuss below.

[3] Several hundred residents of Pripyat, two miles from the Chernobyl plant, insisted on returning to their homes, and around one hundred still live there. There are anecdotal reports that they are outliving those of their peers who accepted the relocation or didn't think they had a choice. The stresses on health of forced abandonment of their homes is greater than that of the residual radiation, which is well within the range of natural background radiation on Earth. More on this below.

The idea that all exposures to radiation are harmful was invented in the early twentieth century. It was later made into dogma by Hermann Muller (mentioned in Chapter 7), who affirmed this conclusion during his 1946 Nobel Prize acceptance speech. It was knowingly purveyed in the mid-1950s by geneticists carefully chosen and funded by the Rockefeller Foundation (RF). Letters and internal memos exchanged among the involved parties, discovered over the last few years by U Mass Amherst toxicologist Edward Calabrese and colleagues, reveal deliberate dishonesty, rationalized as service to an assumed higher cause than truth (i.e., ending the testing of nuclear weapons).[4]

The current editors of the US journal *Science*, where the RF genetics panel's conclusions were first published in 1956 (by a third party and without the panel's review), have been urged to retract the article and disassociate themselves from its conclusions. Despite direct requests by Calabrese and others, they have refused to do so. They sidestep the evidence of fraud (Calabrese 2022; 2023), justifying their position with the fact that the authors are no longer alive and able to defend themselves.[5] In so doing, the editors place the reputations of a handful of deceased and dishonest geneticists above scientific truth.[6]

The science of radiobiology is no longer in its early stages, yet partly as a result of the editors' inaction, the public and most radiation-related medical personnel and scientists remain in thrall to the LNT fiction.[7] Drawing its authority from this myth, there is now a worldwide regulatory and advisory industry on which tens of thousands of jobs and careers depend.

---

[4] This history is thoroughly covered by Calabrese in numerous scientific journal papers and in an interview, recently recorded and sponsored by the Health Physics Society, in 22 segments, each about 20 minutes long (https://tinyurl.com/8z6u757r).

[5] Personal communication from Calabrese.

[6] Scientific frauds are rarely revealed to the public, but they are not uncommon. See Broad and Wade (1982), who document numerous frauds detected only decades later. Scientists are reluctant to level such accusations at each other, and most are forced to pay attention to their own work if they are to receive research grants. This allows little time or interest in exposing fraud. Fraud is most often first suspected when experimental or observational results cannot be replicated, though this is more often the result of honest error—a completely different category. But honest error is corrected when discovered. Fraud more often is defended and reinforced. As mentioned in Chapter 1, a recent report in *The Washington Post* refers to more than 10,000 fraudulent papers published in 2023 that, on discovery, were retracted, resulting in the shutting down of a number of scientific journals. *The Wall Street Journal* had a similar report a month earlier on May 14 (https://tinyurl .com/4znxzcrz).

[7] Prominent among exceptions are radiation oncologists, who use precisely targeted, high-dose radiation to kill cancerous tissue. They unavoidably expose normal tissue near the cancer that virtually always recovers from its exposure without producing a second cancer. As a result, they are perhaps the medical specialists least susceptible to the illusion of harm

Natural background radiation varies with location. Places in the Rocky Mountains like Denver experience anywhere from three to ten times the US's average dose rate, while coastal places like New Orleans have something like one-third the average, for a ratio of high to low near ten to one. The Rockies contain abundant uranium and its radioactive decay products, including thorium, radium, and radon, and the higher altitudes provide less atmospheric shielding from cosmic rays.

David, Wolfson, and Fraifeld (2021) found that

> life expectancy, the most integrative index of population health, was approximately 2.5 years longer in people living in areas with a relatively high vs. low background radiation . . . This radiation-induced lifespan extension could to a great extent be associated with the decrease in cancer mortality rate observed for several common cancers (lung, pancreas and colon cancers for both genders, and brain and bladder cancers for males) . . . Exposure to a high background radiation displays clear beneficial health effects in humans.

With regard to the dose and dose-rate ranges surrounding the Goldilocks zone—too much and too little—all three ranges exist for virtually all agents. Specifically, too much oxygen causes blindness and death; too little causes asphyxiation. Drinking too much water can dilute electrolytes and cause cardiac arrhythmias and possible death; too little causes dehydration and risks death. Further, while the respiratory tract requires a certain amount of water for its proper functioning, inhaling too much water results in drowning. Exposure to too much sunlight can cause skin cancer, too little can produce vitamin D deficiency (unless the vitamin is provided in food), leading to malformation of developing bone in children, known as rickets.

Many vitamins were discovered in populations whose diets or environment were deficient in that vitamin, and who, as a result, suffered from a specific disease—scurvy from a deficiency of vitamin C, rickets from vitamin D, beri beri from vitamin B1 (thiamine), pellagra from vitamin B3 (niacin), and so on. Yet every vitamin also has its overdose range, with likewise harmful effects (for an overview see Calabrese, Nascarella, et al. 2024).

Not only does this characterize everyday entities like oxygen, water, sunlight, and vitamins, but it is also true of psychological

---

from low-dose and low-dose-rate radiation. Most radiologists and nuclear medicine physicians simply point to the greater immediate benefit of radiological imaging, despite their belief in the myth (personal and professional experience, Sacks).

states like fear and physical agents like light. Too little fear and a person can fall off a cliff or step too near a rattlesnake. Too much and a person can become paralyzed, afraid to move. Between these is a level that promotes caution and wellbeing. Too little light prevents visibility, too much produces blinding glare, and so on.

Those who claim that radiation is an exception to this triphasic pattern should bear the burden of proof. Yet the burden has been shifted onto its defenders by the political power of the regulatory and advisory industries. Radiation scientists who recognize the health-promoting effect of radiation at moderately low doses and dose rates are faced with the illegitimately authoritative dictum that even low-dose or low-dose-rate exposures result in net harm. In the face of growing evidence of the contrary, some LNT defenders grant that the risk of harm may not be quite linear with increasing dose or dose rate, but they still argue that as either increases so does the risk of harm. That is, they deny the existence of a range in which harm diminishes with increasing dose or dose rate.[8]

LNT implies that all radiogenic damage is permanent, precisely because of the hypothesized absence or limited effectiveness of any repair or removal mechanisms. If true, permanent damage would accumulate throughout one's lifetime, ultimately causing cancer. This misimpression has sometimes led physicians to withhold medically needed x-ray studies from children or patients who have had multiple previous exposures. Their reluctance has sometimes resulted in misdiagnosis and actual harm.

---

[8] Nobelist Hermann Muller made two errors. First, as discussed in the previous chapter, he thought that because there were chromosomal changes in fruit flies following irradiation, there must also be mutational changes at the atomic level. Second, he assumed that because high doses of radiation produced these presumed mutations it must be true that even low doses produce the same effect. Moreover the effect on chromosomes appeared to be linear in the high-dose range, so he assumed the effect was linear down to zero dose. That is, without evidence he extrapolated the relationship outside the range of his observations. However, even if he had been correct about high-dose mutations, all relationships have their range of applicability, outside which different relationships obtain. So we need to find the boundaries of the applicable range rather than assuming they obtain throughout the range from zero to very high, in other words assuming that there are no such boundaries.

A joke about John and Mary's first flight illuminates the error of overextrapolation. An hour after takeoff the plane undergoes a violent shaking and the pilot announces that they have just lost one of the four engines, but nothing to worry about, they'll just be about 20 minutes late landing. A while later another violent shaking and the pilot announces that they just lost the second engine, so they'll be about an hour late getting in. Still another interval and the third shake brings the announcement that they'll be two hours late. John says, "Good grief, Mary, do you realize that if we lose the last engine we could be up here all night."

While linearity (meaning proportionality) may characterize initial damage—twice the dose causes twice the initial damage—experimental and observational studies have repeatedly found that cellular repair, or removal of unrepaired cells, requires on the order of 24 hours. Moreover, the organism, through an overprotective response or a response of extended duration, ends up, for prolonged periods, in a healthier state than before the exposure and with lower risk of future cancer. This is partly due to the now better prepared immune system if it encounters future invaders or injuries. In addition to radiation, this response is also elicited by other physical or chemical damage.

The duration of the effect has not been well studied, but for some sources of damage it lasts for months to years, perhaps many years, much like a vaccine that may last a lifetime for some diseases. The repair/removal response is also sufficient to reduce the ever-present reservoir of endogenously (metabolically) damaged cells awaiting correction, which is the main reason that the organism is at least temporarily healthier (see, for example, Löbrich, Rief, and Kühne 2005; Sponsler and Cameron 2005; B. Cohen 1995; Chen, Luan, et al. 2007).[9]

Oxygen is needed to provide cells with energy to perform their normal tasks, but a side effect causes incidental toxic damage to various cellular structures in both plants and animals. Conversion of energy into usable forms, through oxygen metabolism (chemical processing), takes place within the hundreds of mitochondria (tiny energy conversion factories) that are found in all types of cells in the body except red blood cells.[10] As oxygen is metabolized, numerous aggressively reactive compounds are formed, called reactive oxygen species (ROS). ROS can escape from mitochondria and—while acting as necessary signaling chemicals between cellular components—cause damage to various cellular structures, particularly the cell's DNA. Since DNA is the sole molecular structure that appears in just one copy per cell, and since it plays a central role in the cell's normal

---

[9] If endogenous damage is continually being repaired or removed, one might well ask why there is a steady-state reservoir of damage in the first place? It is analogous to the residual unemployment rate of 4 percent that is considered to represent full employment. The reason the reservoir (or residual unemployment rate) persists is that there is a lag between damage and repair/removal, just as between a lay-off and the finding of a new job. So a snapshot in time will always find some damage (unemployment) that is soon to be repaired/removed.

[10] Red cells, before entering the bloodstream, shed their nuclei and other internal structures to make room for wall-to-wall hemoglobin molecules that carry inhaled oxygen from lungs to tissues and carry the waste product CO_2 from tissues to lungs for exhalation.

functioning, damage to DNA is the most consequential. During one's lifetime, oxygen inflicts the most extensive and continual damage to our cells, making it perhaps the most toxic element in the world for most living organisms, even as it is a necessary nutrient—a matter of dose and ability to repair or remove damaged cells.

Radiation either can produce its own ROS by collision with water molecules (the most prevalent molecules in the body) or it can inflict damage through direct collisions with DNA and other cellular structures (Borrego-Soto, Ortiz-López, and Rojas-Martinez 2015). Extant species have evolved multiple layers of protection from the ubiquitous natural background radiation coming from the atmosphere, ground, and oceans, and indeed from countless agents in our environment—protective processes extending from the cellular nucleus to the entire cell to the surrounding tissue to the whole organism.

Damage to DNA molecules can result in breakage of one or both strands of their double helix—simply termed single-strand breaks (SSB) or double-strand breaks (DSB). DSBs are, unsurprisingly, less likely to be repaired properly than SSBs, which are properly repaired most of the time, using the other intact strand as a guide. But DSBs are tens of thousands of times less common than SSBs. When not repaired, cells containing unrepaired DNA may undergo a process that results in cell suicide, known as apoptosis. Apoptosis can be produced either by internal cellular processes or with the aid of neighboring cells, called the bystander effect. The latter involves an exchange of signaling molecules between the damaged cell and an undamaged neighboring cell. There are both beneficial and detrimental bystander effects. Following apoptosis by either pathway, the cellular remains are removed by the immune system. This removal is part of a cascade of events that sacrifice certain cells while protecting the organism from possible pathological outcomes, including cancer. Even if apoptosis fails to occur, the now abnormal cells are generally removed by the immune system anyway, either before they begin to multiply in unconstrained fashion (the signifying feature of cancer) or during such a development. Or sometimes the immune system merely holds the tendency toward malignant development at bay. At still other times the immune system is tricked or overwhelmed by an incipient tumor, and a clinically detectable cancer develops. Were it not for this occasional failure, no one would ever develop cancer.

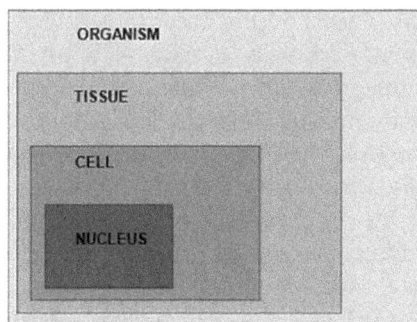

The nested panoply of defenses undermines the claim of elevated risk of cancer from low radiation exposures. Because repair processes are backed up by removal processes, the occasional failure to repair DNA need not herald the permanence of a "scarred" cell.

For the first two billion years of Earth's 4.5-billion-year existence, oxygen was largely absent from the atmosphere. This changed when cyanobacteria emerged and began to spew out Earth-girdling volumes of the gas. The evolution of multicellular life forms was enabled by the new atmosphere of oxygen combined with radiation, each of which functions as both necessity and scourge, depending on dose. The only species that survived the natural selection filter were those capable of either detoxifying oxygen and radiation or repairing/removing their damage. And detoxify, repair, or remove we do—continuously, and without our having to lift a finger. As a result of these self-activated internal rescue operations, most of us remain blissfully unaware of the continuous efforts our bodies exert to save us from early death.

Endogenous DNA damage is extensive, causing something like a million destructive events in every cell every day. While natural background radiation varies from one geographic location to another, it accounts for only hundreds or thousands of destructive events per day. Thus, the endogenous processes outnumber damage from external radiation by factors of thousands to tens of thousands (Feinendegen, Pollycove, and Sondhaus 2004).

Endogenous damage is sped up when we engage in aerobic exercise that increases oxygen metabolism and requires faster and deeper breathing. Thus, aerobic exercise speeds up the production of double-strand breaks, single-strand breaks, and mutations. Yet, as everyone knows, we benefit from moderate exercise, even though extreme exercise can result in permanent injury that is only partially healed. And this benefit derives from the accelerated repair of not only the newly exercise-inflicted damage but also some of the steady reservoir of endogenous damage. The phenomenon of benefit induced by damage is called "hormesis," from the Greek root meaning to stimulate, as in "hormone." Mental exercise can also be hormetic, since the brain normally accounts for some 20 percent of the body's oxygen consumption, which can be amplified by intense thought. It makes as much sense to call for minimizing radiation exposure in the radiology department or dental chair as to call for minimizing physical exercise and thinking.

Fox Chase Cancer Center radiation expert Mohan Doss (2013) explains:

> Concerns have been expressed recently regarding the observed increased DNA damage from activities such as thinking and exercise. Such concerns have arisen from an incomplete accounting of the full effects of the increased oxidative damage. When the effects of the induced adaptive protective responses such as increased antioxidants and DNA repair enzymes are taken into consideration, there would be less endogenous DNA damage during the subsequent period of enhanced defenses, resulting in improved health from the thinking and exercise activities. Low dose radiation (LDR), which causes oxidative stress and increased DNA damage, upregulates adaptive protection systems that may decrease diseases in an analogous manner.

You may ask, if DNA damage due to radiation is so much less than that due to endogenous oxygen metabolism, how can radiation even be detected by the cell given the much greater continual and fluctuating background of endogenous damage? That is, how can the cell "know" to exert its hormetic effect when exposed to much less frequent radiogenic damage?

There are two ways the cell can "know" that radiation is the cause of damage. One is that, as mentioned above, radiation acts through two different pathways, direct collisions between the radiation and DNA and indirect ROS production. And the direct collisions result in a more compact spatial arrangement of effects on the DNA. The second is that an acute dose of radiation delivered very rapidly, as from a CT scan, can significantly exceed the rate of endogenous damage during that brief interval, which can be detected and stimulate the hormetic response. Either the spatial or temporal signal can permit the organism to detect the difference between radiogenic and endogenous damage despite the former's generally less frequent occurrence.

Radiation then, particularly in low doses and dose rates, is a very weak toxin compared to oxygen. The body is of necessity more than capable of healing the destructive tendencies of both. If it were not, we would all develop cancer early in life and die in childhood, before we could reproduce, and our species would have long since become extinct. The fact that most of us in the US (about 60 percent, https://tinyurl.com/mpcjw4ch) will never develop clinical cancer—despite the continuous minute-by-minute initiation of a potential cancer by our metabolism throughout our bodies is evidence of effective repair/removal. This, combined with the pervasiveness of natural background radiation, and the fact that the 40 percent of us who *will* develop clinical cancer are bombarded by many more effective carcinogens than radiation, highlight the falsity of the downward extrapolation of linearity (proportionality) from high-dose radiation effects. These realities subvert the claim that damage to DNA always confers an increased risk of future cancer development.

The key missing ingredient in the LNT claim about low-dose radiation then is the fact of continual and adequate repair and/or removal of the damage. Exposure high enough to inhibit and/or overwhelm the protective response is far higher than that generally encountered by radiation workers or by the public, even from a nuclear power plant accident.

Many LNT adherents grant that evidence for their contention is next to impossible to obtain. They claim that cancers caused by low doses and dose rates of radiation are simply obscured by the US average 40 percent background incidence of cancer (with variation depending on geographical location), but they never doubt their existence. Here is an excerpt from the abstract to a paper by 31 authors (Cardis, Howe, et al. 2006):

26 April 2006 marks the 20th anniversary of the Chernobyl accident. On this occasion, the World Health Organization (WHO), within the UN Chernobyl Forum initiative, convened an Expert Group to evaluate the health impacts of Chernobyl. This paper summarises the findings relating to cancer. A dramatic increase in the incidence of thyroid cancer has been observed among those exposed to radioactive iodines in childhood and adolescence in the most contaminated territories. Iodine deficiency may have increased the risk of developing thyroid cancer following exposure to radioactive iodines, while prolonged stable iodine supplementation in the years after exposure may reduce this risk. Although increases in rates of other cancers have been

reported, much of these increases appear to be due to other factors, including improvements in registration, reporting, and diagnosis. Studies are few, however, and have methodological limitations. Further, because most radiation-related solid cancers continue to occur decades after exposure and because only 20 years have passed since the accident, it is too early to evaluate the full radiological impact of the accident. Apart from the large increase in thyroid cancer incidence in young people, there are at present no clearly demonstrated radiation-related increases in cancer risk. This should not, however, be interpreted to mean that no increase has in fact occurred: based on the experience of other populations exposed to ionising radiation, a small increase in the relative risk of cancer is expected, even at the low to moderate doses received. Although it is expected that epidemiological studies will have difficulty identifying such a risk, it may nevertheless translate into a substantial number of radiation-related cancer cases in the future, given the very large number of individuals exposed.

While granting that "there are at present no clearly demonstrated radiation-related increases in cancer risk," they assert, "This should not, however, be interpreted to mean that no increase has in fact occurred." They further admit that even after still more time, "it is expected that epidemiological studies will have difficulty identifying such a risk," but, they claim, this risk "may nevertheless translate into a substantial number of radiation-related cancer cases in the future, given the very large number of individuals exposed." Thus, expected statistical difficulty identifying such a risk does not dissuade them from affirming the cancer-causing effects of the radiation emitted from Chernobyl. Well, it is now 38 years since the accident, and still no cancers are found in excess of the normal background incidence, but the LNT adherents affirm that they are there nonetheless.[11]

The authors also claim that "prolonged stable iodine supplementation in the years after exposure may reduce this risk." But, while iodine supplementation—meaning nonradioactive iodine administered to keep the gland flooded and prevent uptake of its

---

[11] This type of reasoning exhibits a perverse analog of homeostasis. That is, every perturbation that tends to weaken the hold of the LNT concept is countered by its determined retention. Scientific investigation and reasoning, when done properly, is not homeostatic; rather it allows prior conclusions to be changed when evidence or its absence urges or demands such a change. Persistent belief in the admitted absence of direct evidence is a characteristic of religion, not science.

radioactive counterpart—would have prevented any excess thyroid cancers had it been administered at the outset, such a practice would have no further effect *on a prolonged basis.* I-131 decays, with a half-life of 8 days, to stable Xe-131 (xenon). Therefore, the radioactivity of any I-131 emitted by the accident, whether already in a gland or still on the ground, is reduced by half every 8 days. And in 80 days, after ten half-lives, the amount of remaining radioactivity will be reduced by a factor of 1,000 (1,024 to be exact, or 2^{10}). This steady reduction in radioactivity, even if a child continues to drink milk from cows that graze on the contaminated land, means the initial damage diminishes over time.

Not all LNT-endorsing scientists grant the inaccessibility of evidence for their favored hypothesis. In fact, there are published studies that the authors claim provide "strong" evidence for increased cancer risk at low doses, and for which they receive uncritical accolades in the popular scientific press (see, for example, https://tinyurl.com/cvvjs9b8). Prominent among these researchers are members of the WHO's International Agency for Research on Cancer (IARC),[12] who have published a series of articles using data from the INWORKS project (International Nuclear Workers Study) that includes over 300,000 workers. This is an impressive database, but these analyses have been shown to involve a significant number of blatant errors. These include circular reasoning (assuming what needs to be shown), violation of statistical rules (claiming support for findings that probably appeared by chance), improper statistical procedures (assigning the null to their favored hypothesis rather than to no-effect), ignoring possible confounders (such as age, which correlates to some degree with longevity on the job and with cancer incidence), confining individual exposures to the workplace and omitting those due to natural background and medical procedures (which are comparable to and often exceed the work-related exposures), thereby underestimating each worker's exposures by randomly varying amounts with the effect of falsely attributing to lower doses what are really the effects of higher doses (which in fact makes no difference, but from their point of view is an error), and then lumping this meaningless data into coarse bins that overwhelm and obscure finer details in the low-dose and low-dose-rate range, which, when recognized, carry

---

[12] Elisabeth Cardis, the lead author of the paper just quoted, is a member of the IARC group of researchers. She straddles both categories of LNT defense: admitting that evidence is hidden in the statistical noise yet claiming that evidence has been detected.

contrary implications. These many errors, which are only the most significant among them, are all explained in detail in a paper we published in 2016 (Sacks, Meyerson, and Siegel), and in further detail in a paper by Cardarelli and Ulsh (2018) that cites and expands on ours.[13]

Figure 8-1 compares dose rates of radiation in the Chernobyl region to natural background radiation in a number of European countries. The figure is taken from a commentary by the late Polish physician Zbigniew Jaworowski (2010), former chair of UNSCEAR.

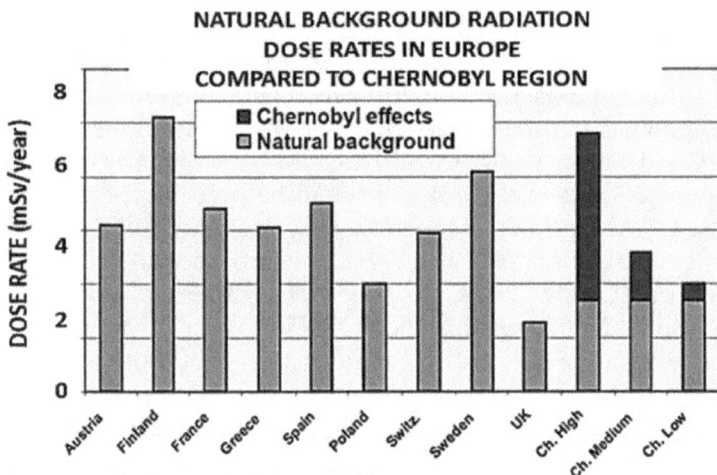

NATURAL BACKGROUND RADIATION DOSE RATES IN EUROPE COMPARED TO CHERNOBYL REGION

FIGURE 8-1

As you can see, even the highest dose rate around Chernobyl (natural plus added by the accident) is exceeded by natural background radiation in Finland. It is also exceeded in several other locations around the world, some much higher than in Finland. Furthermore, the portion added by the accident apart from natural background (the dark portion) is exceeded by natural background in several of the countries shown, even for "Ch. High."[14]

---

[13] John Cardarelli was then the president of the Health Physics Society (HPS) and Brant Ulsh is the current editor-in-chief of their *Health Physics Journal.*

[14] During the COVID pandemic, mounting evidence of the utility of low-dose radiation was found in clinical trials around the world testing its use to treat severe COVID pneumonia, including particularly the excessive immune response called cytokine storm. The trials showed a significant reduction in the death rate and a diminished need for ventilators (Hanekamp, Giordano, et al. 2020). However, the powerful influence of LNT on the medical profession, reinforced by fear of violating the "standard of care," has acted as a barrier to the rapid acceptance of this life-saving procedure. Even if LNT were valid, what is the prospect of causing a future cancer several decades in the future compared to the even higher prospect of allowing an unnecessary death in the next few days?

The campaign to ban nuclear energy is akin to a hypothetical movement to ban vitamins, on the grounds that they can cause diarrhea, stomach pains, hair loss, nerve damage, fatigue, and death when taken in large enough quantities. But banning vitamins (which, given their omnipresence in food, is no more possible than banning radiation or oxygen) would hypothetically spell the early death of the entire population. What we need with respect to vitamins, oxygen, water, and radiation is to discover the safe threshold doses and dose rates.[15]

* * * * *

Fortunately, growing opposition to the LNT dogma is indicated by the increasing attention being paid to hormesis. Figure 8-2 is a graph from the Web of Science showing the sudden onset about 27 years ago of a proliferation of scientific journal articles on the topic (https://tinyurl.com/8zp5n5vh). Naturally, these also include denials or underestimates of the hormetic effect, but if confirmatory studies were not mushrooming, opposition would be unnecessary.

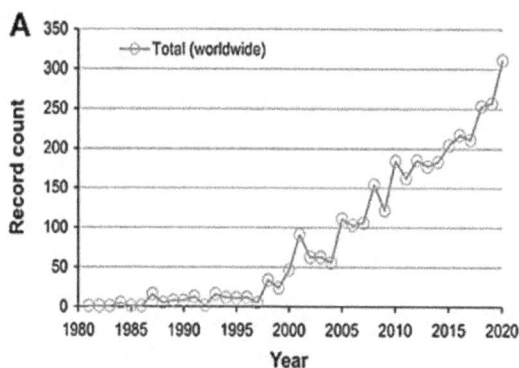

Annual Number Of Articles in Scientific Journals
Related To Hormesis

FIGURE 8-2

Reflecting this attention, the 2015 Nobel Prize in chemistry was awarded to three researchers for their several independent studies of

---

[15] In addition to diagnostic low-dose radiation (x-rays, CT scans, nuclear medicine), intermediate doses are used to mitigate or cure noncancerous diseases such as hyperthyroidism and high doses are used to treat cancer. As mentioned, the latter generally fail to provoke new cancers in the normal tissue unavoidably exposed. Thus, even high-dose radiation is a very weak carcinogen, dwarfed by everyday exposures to agents such as tobacco and benzene.

DNA repair in response to various causes of damage. From the Nobel committee's presentation (https://tinyurl.com/mphszjak):

> The Nobel Prize in Chemistry 2015 is awarded to Tomas Lindahl, Paul Modrich and Aziz Sancar for having mapped, at a molecular level, how cells repair damaged DNA and safeguard the genetic information. Their work has provided fundamental knowledge of how a living cell functions and is, for instance, used for the development of new cancer treatments.[16]

Harm has in fact been found experimentally from radiation doses that are too low. The harm is not simply an increased risk of cancer, but rather the total failure in experimental animals to develop an immune system.[17] This failure leaves the organism victim to all sorts of diseases in addition to cancer, prominently including infections. Since humans are never found in radiation-free environments, the immune system failure has been demonstrated in experiments using fetal and newborn rodents that are shielded from any form of ionizing radiation. In other words, a certain minimum amount of radiation exposure, during prenatal as well as postnatal development, is necessary for optimal health. This means that its deficiency can be just as harmful as its excess.

From LNT, in contrast, it follows that near or at zero exposure there is minimal or zero harm. It remains to be seen, however, whether this harm to rodents in the very low-dose-rate range is generalizable to humans. But one reason that rodents are commonly used in such experiments is that, being mammals, they often represent a useful surrogate for humans, on whom such experiments raise ethical questions. Thus, there is reason to believe that humans also suffer harm from insufficient radiation exposure.

It's instructive to consider the three dose ranges—too little, just right, and too much—as analogous to the three phases of, say, H_2O at various temperatures. Below freezing it is a solid (ice), between freezing and boiling it is a liquid (water), and above boiling it is a gas (steam). Such substances exhibit a triphasic response

---

[16] Ironically, the 2015 Nobel committee has in effect invalidated the 1946 committee's awarding of the prize to Muller, a denier of DNA repair. While there's no question that the science has advanced over the last three-quarters of a century, Muller, as we said above, was fabricating his conclusion even in 1946.

[17] Ongoing experiments in deep underground laboratories are described in such articles as the following: Liu, Ma, et al. 2018; Mitchel, Burchart, and Wyatt 2008; and Waltar and Feinendegen 2020, and in such videos as https://tinyurl.com/53ujtcwm.

to their environments (within the normally encountered range of conditions), and the three phases are separated from one another by two thresholds, at which phase changes occur—freezing/melting and condensing/boiling.[18]

The volume of H_2O does not maintain strict proportionality to its temperature as it passes through the three phases; neither is cancer risk proportional to radiation exposure from zero to high. For ice there is one relationship between volume and temperature, for water a different relationship, and for steam still another one. Similarly, in the three ranges of radiation exposure, there are three different relationships between cancer risk (or protection) and dose or dose rate.

US health physicist John Boice is a well-known and much-published scientist, who is the past president of the NCRP and a past or present member of several other advisory organizations. Boice says, with regard to acute doses, "In fact, the best fit in the range of < 2 Gy is 'linear quadratic' and not linear, but I'm challenged to see any practical difference."[19] He continues, "it is the current judgment by national and international scientific committees that no alternative dose-response relationship appears more plausible than the LNT model on the basis of present scientific knowledge" (https://tinyurl.com/42p58emm). However, neither the linear nor the linear-quadratic representation exhibits a negative slope anywhere in their domain, while hormesis exhibits an initial negative slope that turns upward toward a positive slope, eventually crossing a threshold from benefit back to harm (Figure 8-3).

Boice, in effect, appeals to Occam's razor, which prescribes that we base plausibility on simplicity, mathematical or otherwise. However, plausibility resides in correspondence to the predominance of evidence and possession of the best explanatory power. It enjoys mathematical simplicity only on occasion. Thus, Boice dismisses the understandings of numerous scientists who find no plausibility in a demonstrably inaccurate claim.[20]

---

[18] In addition, at extremely high temperatures, a fourth phase appears, which is a plasma with all molecules broken into their constituent atoms and atoms largely broken into their component nuclei and electrons. But here we're only interested in the three-phase analogy.

[19] "Gy" means gray, a unit of radiation exposure, roughly equivalent to a Sv, or sievert, but the latter refers to the biological response to a gray and accepts the validity of LNT. That is, each additional Gy allegedly causes an additional Sv (or some fixed multiple thereof) of biological damage in linear fashion, with no threshold.

[20] Boice has recently been part of a conspiracy within the HPS board of directors to oust then president John Cardarelli from his position. Cardarelli is not only a strong proponent of hormesis but he led in creating the 22-session video interview of Ed Calabrese,

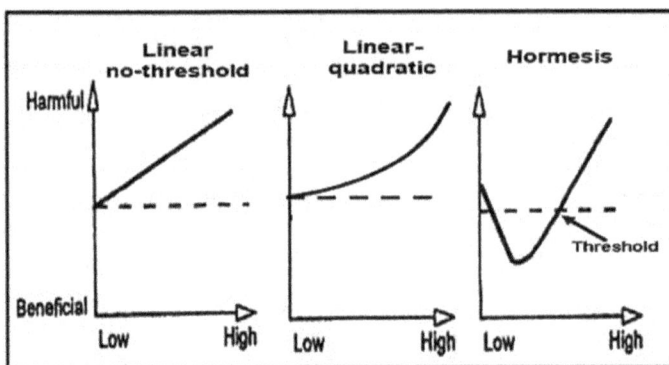

FIGURE 8-3

A revealing case of paradigm-associated blindness is exemplified by another much-cited LNT proponent, British-American health physicist David Brenner, along with his colleagues (Halm et al. 2014). As the paper relates, the authors measured blood concentrations of a marker of broken strands of DNA in three children preceding and following a CT scan. The authors cite a similar study nine years earlier (Löbrich et al. 2005), in which the blood concentrations of the markers in ten children were measured before the CT and following it at ½, 1 , 2½, 5, and 24 hours. Löbrich et al. found that the number of double-strand breaks (DSBs) unsurprisingly increased immediately following the CT and then began to decrease, with the final number at 24 hours even lower than before the exposure, though the small sample size made the difference between the starting and final counts statistically insignificant. Löbrich et al. concluded that DSBs were either repaired or removed over a number of hours, and somewhere between 5 and 24 hours the average number of DSBs per cell again approached values close or equal to the initial (pre-CT) number. In other words, CT scans produce initial DNA damage, which few scientists would deny, but no residual damage remains a day later. And this is aside from the undeniable diagnostic advantages that motivate CTs in the first place. Moreover, as suggested by Löbrich's study and by such studies and phenomena as those we

---

mentioned above, that exposes the fraudulent foundation of LNT (https://tinyurl.com/4secyrwh). The board settled for censuring Cardarelli over a letter he sent to Congress urging their involvement in halting the board's suppression of research countering LNT. However, the HPS membership rescinded the censure (https://tinyurl.com/398mkmbt, and personal communications).

review in the following section, the low-level radiation associated with CT scans seems to produce additional benefit directly from the radiation itself.

Peering through their LNT lens, Brenner and his coauthors were unable to perceive this suggested beneficial outcome. Rather than examining ten patients like Löbrich et al., they evaluated three. And rather than counting DSBs at several times following the CT, they focused on only one, at 1 hour. Despite citing Löbrich et al., they concluded two things from their own study. First, they surmised that had they counted the DSBs after ½ hour (as Löbrich had done), there probably would have been even more DSBs than at 1 hour. Second, completely ignoring the results of the earlier study, they postulated that the DSBs at 1 hour were permanent and would likely lead to cancer. And they recommended avoidance of CT scans if there were some other method of diagnosis, such as an ultrasound.

In arriving at this recommendation, they seem to have missed the implication of their first surmise that after the first ½ hour the number of DSBs was on the way down. They missed the further possibility that had they continued to measure at longer time intervals than 1 hour the number of DSBs might have continued to decline, perhaps all the way to baseline, and possibly beneath it—which was precisely the result found by Löbrich et al., whom it was their choice to cite.

One has to wonder whether their knowledge of the earlier findings prompted their failure to draw blood at later intervals for fear of undermining their preconceived conclusion. Or were they simply so blinded by their paradigm that it did not occur to them to do so?

The Evidence for Thresholds and the Direct Benefits of Low-Dose Radiation (Hormesis)

A dramatic demonstration of the falsity of LNT was found in an epic study led by the late University of Pittsburgh physicist Bernard Cohen (whom we quoted in Chapter 7). His team measured average home radon levels and rates of lung cancer mortality in 1,700 US counties that included some 90 percent of the US population and took several years of data gathering in the early 1990s (B. Cohen 1995; 2008).

Radon's most common isotope, Rn-222, is radioactive and a gas at the temperatures and pressures found at ground level. It is one of the decay products of uranium. Because radon is much heavier than the main constituents of air (nitrogen and oxygen), it tends to stay close to the ground and pervades houses and buildings, with the highest concentrations in basements and lower floors, varying widely from building to building and from moment to moment, particularly when windows are open.

Since uranium is abundant in the ground, so is radon; its radioactivity accounts for more than half the world's natural background radiation. Of the heaviest radioactive isotopes, radon is the only one that occurs naturally as a gas. Its chemical properties derive from the completed outer electron shell and prevent radon from forming molecules (compounds), even with other radon atoms. In acknowledgment of its aloofness, it is termed a "noble" gas, along with helium, neon, argon, krypton, and xenon; the noble elements occupy the righthand column in the periodic table.

Cohen investigated cancer of the lung because radon is inhaled and the lung is the first organ to be exposed. While radon decays with a half-life of less than 4 days, its decay products (daughters) are also radioactive, and can remain in the lungs for extended periods depending on their biological half-lives. The latter depend on the efficiency with which the cilia—waving brush-like extensions attached to the lining cells of the bronchi (airways)—are able to clear the bronchi and lungs. Tobacco products inhibit ciliary action, prolonging the biological half-lives of radon's daughters and other inhaled toxins. This inhibition largely accounts for the association of smoking with lung cancer and its synergistic combination with other carcinogenic respiratory pollutants.

Cohen embarked on his massive study having seen earlier smaller studies from Finland, Sweden, China, and the US, which suggested that radon levels were negatively correlated with lung cancer rates.[21] He and his co-investigators compared the lung cancer mortality rates county by county with the average home radon levels in each of those counties. They, too, found that in most cases the greater the average radon level in the county, the lower the lung cancer mortality rate in that county, and dramatically so (Figure 8-4).

The dashed line with the arrow pointing toward it represents the expected trend if LNT were true. Each black dot represents a group of counties with approximately the same average home/basement radon level.

Because the results ran counter to the expectations of many, and because his study was criticized by skeptics, and because the results (along with those of the earlier smaller studies) were counterintuitive

---

[21] Malcolm Browne, *New York Times*, September 28, 1988, Section B, Page 7 (https://tinyurl.com/u8p3ewf7).

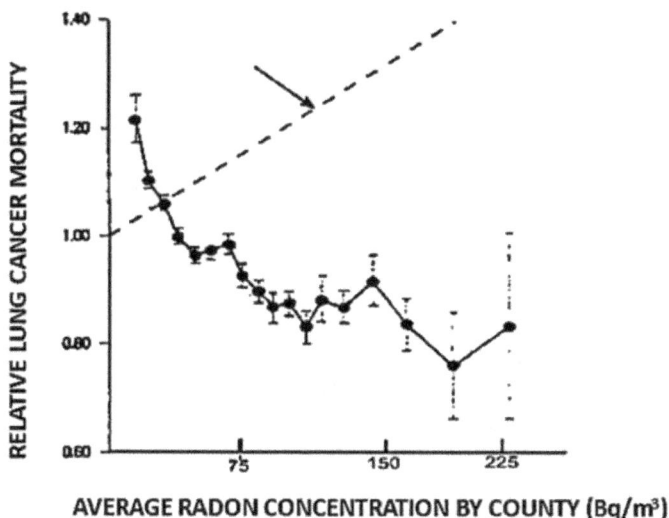

AVERAGE RADON CONCENTRATION BY COUNTY (Bq/m³)

FIGURE 8-4

at the time, Cohen sought the help of a statistician to search for possible confounding influences—correlations with features of the population, such as smoking and age, that might explain the findings. Together they examined hundreds of combinations of possible confounding effects but found none that could explain this great a divergence from LNT.

Naturally, Cohen's findings and conclusions occasioned many attempts by LNT defenders to find flaws in his study. One popular attempt at refutation was the charge that it was an ecological study—one that fails to examine person-by-person but instead groups people into large assemblages and averages the results for each group. One potential weakness of an ecological study is that it may examine population samples that are not representative of the population as a whole. But since Cohen's study covered almost the entire population of the contiguous 48 states and not just a small sample, this charge does not apply to his particular study.

A second potential weakness of an ecological study is that within each group there may be individual variation that follows a positive slope, regardless of the group's average. But if this were true, the likelihood is vanishingly small that the trend in population averages would be downward at higher and higher radon levels (within the range indicated in 8-4) and in a relatively comprehensive defense of his conclusion that LNT simply could not account for his results, Cohen pointed out that if the LNT relationship were true, it would imply the same result for average outcomes as that for each individual taken one-by-one—that is, no matter whether one were to consider individuals or groups, the slope would still have to be positive. And since it was negative, LNT could not be true. Moreover, lung cancer mortality rates are inherently a property of a population rather than an individual, each one of which has either died from lung cancer or has not.[22]

Cohen himself was reluctant to assert that his findings provided a degree of evidence in favor of hormesis and held only that they constituted a disproof of LNT. However, from a Bayesian point of view,[23] his findings provide relatively strong evidence for hormesis. Note that there are 18 dots on the graph, each representing a group of counties. Had Cohen combined his average county radon levels into only, say, two groups, and the group with the higher average radon level were found to contain a lower rate of lung cancer mortality, that could rightly be said to constitute extremely weak evidence for hormesis. With three groups, each one of which exhibited lower lung cancer mortality with increasing average radon levels, the data would provide slightly stronger evidence, and so on. The greater the number of groups (the smaller each group), given the predominant trend of falling mortality with greater radon levels, the stronger the evidence for hormesis. A glance at Figure 8-4 suggests that his findings earned the right to be labeled moderately strong evidence for hormesis. Indeed, Cohen pointed out that his data followed an averaged negative slope that was more than 20 standard deviations lower than the expected positive slope for LNT. That's at the very least extremely strong evidence that LNT is false, whether or not this study is regarded as providing relatively strong evidence for hormesis. His findings have since been corroborated more than once.[24]

However, the US Environmental Protection Agency (EPA) bases its radon-related recommendations on LNT. Figure 8-5 is taken from the their 2018 pamphlet titled "Home Buyer's and Seller's Guide to Radon" (https://tinyurl.com/ yc3nr552).

Ask yourself, "Which, if any, of these five bars doesn't belong?" While 90 percent of the battle is asking yourself the question in the

---

[22] Devanney (2023) has a useful history of the dispute over Cohen's study in Section 5.6.5.

[23] A statistical approach in which prior belief about probability of occurrence is continually updated as more evidence is acquired, whether it increases or decreases your prior belief.

[24] See, for example, https://tinyurl.com/yc33wuyv, which describes a similar inverse relationship in Worcester County, Massachusetts.

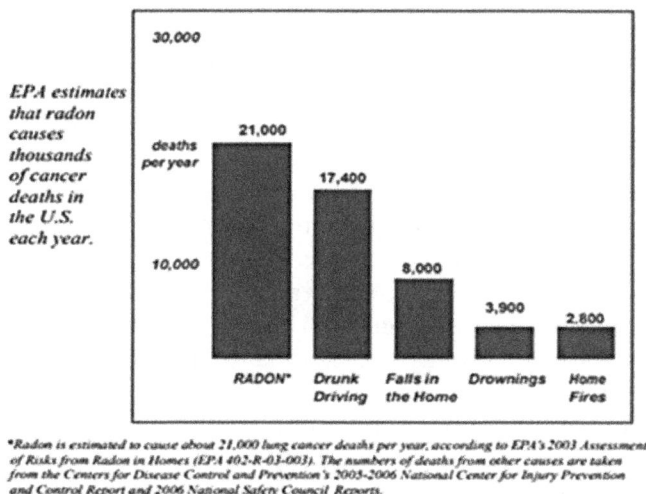

EPA estimates that radon causes thousands of cancer deaths in the U.S. each year.

Radon is estimated to cause about 21,000 lung cancer deaths per year, according to EPA's 2003 Assessment of Risks from Radon in Homes (EPA 402-R-03-003). The numbers of deaths from other causes are taken from the Centers for Disease Control and Prevention's 2005-2006 National Center for Injury Prevention and Control Report and 2006 National Safety Council Reports.

FIGURE 8-5

first place, if you find it difficult to answer you may be suffering from confirmation bias.[25]

Each of the categories besides "*RADON*" describes a type of discrete event involving an individual, one that can be confirmed or disconfirmed to have taken place, and which can be confirmed or disconfirmed to have caused a death. And because the events are discrete, they can be counted and summed to arrive at the numbers indicated above each bar. "*RADON*," on the other hand, is not a discrete event involving an individual, but rather a continuously varying concentration that, even according to the EPA's erroneous LNT-based reasoning, has to be weighted by the time spent by each person in each room, the varying time course of radon concentration in those rooms, and the proportion of time spent outside the house. This is virtually impossible to measure accurately, as it would require 24/7 monitoring of both people's (lots of people's) movements and the radon concentrations they encounter minute by minute around the year. Additionally, it would be impossible to know whether that exposure over time was the cause of the death, or for that matter the lung cancer, or even of the death rate in a

---

[25] Confirmation bias is seeing not necessarily what you want to see, but rather seeing what your beliefs prepare you to see, whether you want to or not. Propositions that confirm your beliefs tend unconsciously to be accepted uncritically, without a demand for evidential support, while those that disconfirm your beliefs are treated more skeptically. Indeed, disconfirming propositions often incur unwarranted rejection even in the face of evidential support, unless you are on your guard against confirmation bias and actively question results that are congenial to your point of view. We return to this issue in Chapter 11.

particular location, since possible confounding factors would have to be ruled out, as Cohen did with great effort. So the estimate of 21,000 radon deaths per year is just that, an estimate—one, moreover, that is based on false premises. The inclusion of the qualitatively different category "*RADON*" with the other four should alone raise a critical eyebrow or two.

The EPA pronounces radon "the second leading cause of lung cancer" next to smoking, and the first leading cause among non-smokers (https://tinyurl.com/4ranpjzv). This egregiously false pronouncement is widely repeated and acted upon. Compounding the EPA's irresponsibility in assigning guilt to an innocent agent, they recommend that you hire a company and pay hundreds of dollars to lower your basement radon concentration below 4 pCi/l (picocuries per liter, a measure of radioactivity; 4 pCi/l is equivalent to the point 148 Bq/m^3 on Cohen's x-axis, Figure 8-4). As you can see from Cohen's data, if followed (and some states and home buyers require it), the EPA's recommendation would *raise* the risk of lung cancer mortality, rather than lower it, because the protective effect of such low concentrations of radon is thereby lessened. In the end, lowering the protection against a risk is no different than raising that risk. So, you are advised, or required, to raise your, or your buyer's, risk of lung cancer at a cost in both monetary and health terms. It is as though the EPA were recommending that we decrease our regular exercise routine because aerobic exercise is known to produce microtears in muscle tissue. But such stress helps the body to repair and strengthen itself, and to diminish it would be to diminish our wellbeing.

This advice is the culmination of at least seven decades during which most of the world's radiation-related regulatory agencies have subscribed to the LNT myth. Extricating themselves from this condition is extremely difficult, but sticking with LNT represents a conflict of interest (careers and jobs depend on following the myth). The difficulty does not absolve them from the charge of scientific misconduct, and we should continue trying to induce them to adhere to a scientific approach rather than one they find convenient and professionally self-protective, both institutionally and individually. All cloaked by the rationalization that they are only following the "precautionary principle."[26]

---

[26] The precautionary principle dictates that we should always err on the safe side when the situation is uncertain. It is widely applicable under a broad set of circumstances, but only when the side assumed to be safer is in fact safer—when the consequences of erring in the direction of putative safety are less harmful than the outcome its advocates hope to avoid. This is decidedly not true of forced relocations, abstention from CT scans,

Sponsler and Cameron (2005) examined thousands of US shipyard workers who build nuclear submarines and aircraft carriers. They compared the health of workers who install the nuclear reactors to that of those who work elsewhere in the shipyard. On average the nuclear-related workers lived longer and had lower cancer rates, differences that were statistically significant.

Examining both nuclear and non-nuclear workers from the same shipyard, both groups performing equally stressful labor, was intended to eliminate the so-called healthy worker effect (HWE)—the idea that those with better outcomes were on average more fit to begin with. Among LNT adherents, the HWE is a popular way to explain away data that conflicts with their assumptions (Fornalski and Dobrzyński 2010). Sponsler and Cameron argued that if the nuclear workers had been chosen on the basis of greater fitness (as some critics speculated), then the differences in health status between them and the non-nuclear workers would have been apparent at the outset, but they were not. The differences evolved over time and were correlated with the time spent working with the nuclear reactors. This approach rendered the HWE inapplicable.

A striking phenomenon arises in Misasa, a spa town in the southwest part of Honshu, Japan's main island, where the radium-fed and radon-emitting natural hot spring baths are a major tourist attraction. Cancer mortality rates in Misasa are two-thirds those in the rest of Japan, and the mortality from stomach cancer, a common malignancy in Japan, is roughly half (Mifune, Sobue, et al. 1992).

Radon caves in various parts of the world are popular attractions where arthritis sufferers spend a few days to palliate their pain. In at least Germany and Austria, this therapy is covered by health insurance (https://tinyurl.com/37hnpdee). The sands at Guarapari Beach in Brazil are rich in thorium, making it a popular vacation spot where people bury themselves in the sand for hours at a time, thereby relieving their arthritic pain (https://tinyurl.com/3xer7kuf).

The most cited source on the effects of radiation is the Life Span Study (LSS) of survivors of the Hiroshima and Nagasaki atomic bombs. The LSS data was originally touted to be consistent with LNT, but more recent evaluations have rendered that conclusion questionable and have even come to opposite conclusions. Previously unrecognized methodological errors have emerged, one of which is coarse binning of dose ranges that obscures the initial dip in harmful

---

avoidance of nuclear energy, or many other radiophobic actions (Siegel, Sacks, and Greenspan 2021).

effects such as cancer incidence and shortened life expectancy. The initial dip suggests a hormetic response, but coarse binning renders it invisible. This is a common error in many LNT-supporting studies. Another methodological error is the failure to account for the unimaginable stress of such a cataclysmic event at the intentional hands of other human beings, leaving stress-related health impacts to be wrongly attributed to the radiation (Doss 2012).

A study by Japanese biologist Shizuyo Sutou (2020) indicates that the atom bomb survivors who were closer to the center of the blast, but not so close that they were subject to the blast and fires, had lower mortality and cancer incidence than those who lived farther. In other words, the closer they were to the blast and survived, the longer they have lived and the fewer solid cancers they have developed. Cuttler and Welsh (2015) found a lower incidence of leukemia among survivors who were between 2 and 3 km from the center of the blast than among those who were beyond 3 km.

This is a tiny sampling of a steadily growing number of studies (and phenomena) that support the conclusion that exposure to levels of radiation within a middle range of dose and dose rate actually enhances rather than diminishes health status, decreases age-specific cancer rates, and prolongs lives. So, radiation is no exception to the rule that agents that are toxic at high doses are generally benign and possibly health enhancing at lower but adequate doses. This literature consists of both epidemiological (observational) and experimental studies that search for the biological mechanisms through which cells and organisms defend themselves. Many more of these studies are listed in books by C.L. Sanders (2010) and T.D. Luckey (1991), even though these lists were compiled more than a decade ago. Brooks (2018) has a review of the scientific literature on mechanisms of action (at the cellular, tissue, and organismal levels) in response to low-dose radiation; see especially Chapter 5, "Paradigm Shifts in Low Dose Radiation Biology and Application of Data," and Chapter 7, "Mechanisms of Action." And for a concise review of studies disproving LNT and supporting hormesis, see Janiak and Waligórski (2023).

The fact of hormesis has been demonstrated repeatedly, but many questions remain. We need more information on mechanisms through which hormesis operates and better precision in the thresholds that separate the three phases associated with the different ranges of dose and dose rate. The latter will undoubtedly vary somewhat from person to person, from organ to organ, from radiation type to radiation type (photons versus particles, and which particles), whether the source is external or internal to the body, and so on. Research needs to be done to find additional applications for radiation that can heal or cure a variety of ailments, many utilized prior to World War II (Cuttler 2020). The duration of the hormetic effect in each of these settings is not well known at this point, and further investigation needs to be pursued. However, among its other detrimental effects, radiophobia obstructs funding for such research.

The Harmful Effects of Radiophobia

Radiophobia is the ideological foundation of the antinuclear movement. Not only has it permeated much of the public, but it has also affected the attitudes of many medical practitioners. Prominent among those, ironically, are diagnostic radiologists, whose training is immersed in the LNT paradigm.[27] The fear also governs the attitudes and actions of referring physicians in most other specialties. Following recommendations like those of Brenner and colleagues to avoid CT scans if possible, not only are some referring physicians hesitant to recommend radiological imaging, even when it is medically indicated and potentially life-saving,[28] but there is also a low level of refusal by adult patients, for themselves and their children, to accept such recommendations.

Additionally, there is a strong tendency among the miseducated majority of diagnostic radiologists to lessen radiation exposures for those CT scans that are consented to by patients/parents. There are even organizations among radiologists and health physicists, known as Image Gently (for children) and Image Wisely (for adults), that advocate this approach and urge radiologists to comply. As a result of excessively low exposures, a certain proportion of CTs have lacked sufficient diagnostic quality to avoid misdiagnosis, with all its attendant iatrogenic (doctor-caused) consequences (M. Cohen, 2012). The frequency of

---

[27] Indeed, it was more than two decades after one of us (Sacks) had begun his training in diagnostic radiology—while observing a debate between Eric Hall, coauthor of a radiobiology textbook for radiologists that takes LNT for granted, and John Cameron, coauthor of the study of nuclear shipyard workers mentioned above—that Sacks first thought critically about the issue, a reflection of his training and a manifestation of his own previous confirmation bias. What persuaded him immediately was Cameron's response to Hall (paraphrasing according to memory), "If what you are saying were true [that all radiation exposure causes permanent harm in proportion to the dose], it would mean that people are incapable of healing in response to illness or injury." Hall's position was tantamount to the assumption that a chef who cut herself numerous times over the years, and lost a little blood each time, would have eventually bled out completely.

[28] See, for example, https://tinyurl.com/bd2h2kad.

Sidebar: LNT's Dominance Confuses Honest Researchers

US radiobiologist Michael H. Fox, in a book defending nuclear energy and resisting radiophobia (2014), cites instances in which LNT does not fit the data, but he still hesitates to pronounce it false or even to cite interpretations of data leading to that conclusion. For example, he cites early research about Hiroshima and Nagasaki survivors, but he omits mention of more recent reinterpretations of the same data that refute LNT and the claim that the data strongly supports it.

Fox also cites experimental work in the late 1950s by William and Liane Russell, who examined the effects of radiation on mice. Yet a former graduate student of the Russells, Paul Selby, along with colleagues, reveals that his mentors had, in acquiescence to the Rockefeller genetics panel of which William was a member, knowingly suppressed data that refuted LNT (Calabrese, Agathokleous, et al. 2023). They show that the data cited by Fox involved flawed control groups and, when corrected, that data argues against LNT. Nor does Fox mention a reassessment, 21 years before his book came out, of the same data by William Russell himself, who this time, as part of an effort to defend the British nuclear industry, included his formerly suppressed 35-year-old data that refutes LNT.

---

such misdiagnoses would be difficult to discern, but it is highly unlikely that most radiologists would deny that a significant finding could be missed due to the noisiness (blurring) of the image caused by underexposure.

Even among radiologists and health physicists who recognize the harmlessness of low-dose exposures, some, thinking they are reassuring patients, nonetheless recommend advertising that they use exposures for diagnostic studies that are as low as reasonably achievable (ALARA). They also recommend the use of lead aprons to shield patients' body parts outside the x-ray beam. They fail to perceive that these efforts can't help but reinforce radiophobia. Furthermore, the exposure to other body parts occurs through internal scatter of the x-rays rather than through direct exposure. Direct exposure is precluded by collimators attached to the x-ray machines, and internal scatter renders the apron irrelevant for anything other than a pretense of protection.[29]

---

[29] On February 1, 2024, the American Dental Association finally announced that lead aprons need no longer be used on patients. But, pulling their punches, they still urge that every effort be made to hold the number of x-rays to a minimum. They urge parsimony not by way of following the general healthcare dictum to avoid unnecessary procedures of any sort, but because they still believe that every x-ray inflicts some harm on the patient, as limited as the harm and as valuable as the imaging information may be.

Still another major consequence of radiophobia has been the aforementioned (Chapter 7) government-sponsored forced relocations of populations in the vicinity of the nuclear power plant accidents at Chernobyl (Cravens 2007, pp. 91–99) and Fukushima (https://tinyurl.com/yc6dedkm). These relocations have killed thousands of people and otherwise negatively impacted hundreds of thousands of lives.

In the former case, some 350,000 persons were forcibly taken away from their homes, jobs, and communities and resettled in unfamiliar and inhospitable locations. This has produced a rise in alcoholism, heart attacks, strokes, and suicides among the evacuees. Radiophobia has also produced exhausting and debilitating mental stress among a far wider population. Throughout Europe, this secondary radiophobia resulted in a significant number of prophylactic abortions due to the unwarranted fear of malformed babies, backed by the advice of misinformed medical professionals. While the actual number may be difficult to obtain, the estimates are consistent with the degree of fear that was promoted, and even the lowest estimates, with some as high as 100,000, dwarf the recorded deaths due to the radiation.[30] The resulting panic is also reported to have spawned hundreds of suicides in Scandinavia due to unwarranted fear of the wind-driven plume of emitted radioactive isotopes—accuracy of the figures again difficult to confirm (Brooks 2018, pp. 226–27). These remote deaths were caused not by radiation, but by radiophobia.

In the case of Fukushima, more than 100,000 persons were forcibly relocated by the Japanese government, with similar results to those of the Chernobyl victims, including a government-admitted 2,300 or more immediate deaths, largely of elderly people hastily removed from hospitals and senior care facilities. But in addition to the many negative side effects that occurred to those evacuated from Chernobyl, there is a legacy in Japan (derived from the two atomic bombings) of regarding exposed survivors as being themselves radioactively poisonous to others. In an analog to racism, they are known collectively as the Hibakusha. They are often denied employment and shunned. This misunderstanding has unjustly tainted those adults, and particularly children, who were victimized by the relocations from Fukushima.

A spurt in divorces also resulted, as disagreements emerged over how to handle the stress of being moved. Spirits and families were broken by the radiophobia-based relocations. Boice (mentioned

---

[30] Imagine the added stress on prospective parents, who wanted the pregnancies, if the law had prohibited such fear-driven abortions.

above) has upbraided excessively fearful persons, seemingly oblivious to his role in creating the underlying phobia (https:// tinyurl.com/ ye2t27ea). Telling people that their exposure to low doses of radiation doesn't raise their cancer risk "very much" is of little comfort. The understanding, on the other hand, that their risk of cancer was not increased at all, and even lowered, might help to resolve anxiety.

Moreover, shunning would be obviated if the general public understood that external exposure to ionizing radiation dominated by gamma (photon) radiation does not cause anything to become radioactive, including humans. The immediate radiation from the atomic bombs was mainly gamma, while only 1 percent was from neutrons. And while absorbed neutrons can make their targets radioactive, were a person exposed to enough neutrons to make them a danger to those near them, they would have received a fatal dose and therefore would not have been among the survivors. And lets recall that we are all radioactive to begin with.[31]

Furthermore, the number of deaths due to the forced relocations from the Fukushima area exceeded the highest LNT-based predictions of future radiation-related deaths (https://tinyurl .com/58x8u2n7). And they were immediate, while any LNT-predicted deaths were expected to have been delayed by decades in the form of cancers. In fact, despite UNSCEAR's own past contributions to the radiophobia (over the objections of some of its members[32]), the organization now admits that not one single death, or any adverse health effect, has resulted from the radiation itself (https://tinyurl.com/bdeymr4u).

At Chernobyl, only the deaths of the 28 firemen who fought the 10-day fire can be definitively attributed to the radiation exposure—because leading to their deaths, within days, they each experienced the typical signs of radiation sickness. But due to mismanagement and lack of protective gear—from which those who managed the Fukushima accident learned, with complete success at preventing deaths[33]—they were unnecessarily exposed to

---

[31] When organic material is exposed to external gamma radiation, as is done to sterilize produce for supermarkets, the radiation enters and some of it performs its sterilizing function while the rest passes on through, leaving no residual radiation or newly created radioactive nuclei. Irradiation by sublethal neutron beams, on the other hand, can make someone radioactive for a few minutes, following which the radiation emissions fall to insignificance, and lower than that from most things in their surroundings. This phenomenon is relied upon by radiation oncologists who sometimes use neutron beams for cancer therapy.

[32] Jaworowski (2010) and personal communication from some others who objected.

[33] In addition to protective gear, the workers at Fukushima Daiichi were rotated during their management of the reactors after the diesel-powered cooling pumps on the lower level of the plant were inundated by the tsunami. In that way none of the workers was

very high doses, and the controversy only involves exposures to low doses, below the upper threshold for harm. Among the clean-up crew at Chernobyl, there have been, over the next several years, at most an additional 30 or so deaths that are possibly attributable to radiation exposure, but only 19 of the 30 are even likely to have been related to the radiation—for an upper limit of 47 deaths among firemen and clean-up workers. There are such conflicting reports about the more than half a million other clean-up workers over the following years that it may be reasonably considered to be unknown. However, no reliable report of an increase in radiation-caused cancer or associated deaths has emerged.

In addition to those among the firemen and clean-up workers, some deaths among children have been reported and attributed to thyroid cancer developed from drinking milk contaminated by accident-released I-131 (iodine). These would be the only deaths among the general public, and, if actual, they would be due less to radioactive material than to very specific and avoidable policy sequelae—paradoxically, due to *insufficient* fear of radiation (again, too little can be as harmful as too much). The estimated incidences of both thyroid cancers and associated fatalities vary widely. The literature includes counts assembled from reports by pediatricians that may be exaggerated and are likely unreliable. Part of the exaggeration derives from the discovery of occult cancers by new-onset screening, cancers that would never have come to light during the life of the bearer, and certainly would not result in death. Overdiagnosis is well-known for a number of cancers (including prostate and breast), indicated, among other things, by the prevalence of subclinical cancers found on autopsies of people who died for unrelated reasons.

Given these distorting influences, the highest estimate we can find for the number of radiation-caused thyroid cancers is about 6,000 more than the expected number had the accident never happened.[34] These were almost all in children under 18 years of age at the time of the accident. Thirty-six years after the accident (April 2022), the World Nuclear Association (WNA) assembled reports from a variety of sources and, citing a 2018 report from UNSCEAR, states that 15 deaths resulted from these

exposed to more radiation than was safe, much as radiation therapy for cancer is fractionated and delivered over several weeks.

[34] But there's no evidence that the incidence of clinically manifested thyroid cancer among children and youth prior to the accident was known or recorded. Without such information it could not be known whether there was an increase in incidence following it. And occult cancers found on screening after the accident do not provide such information.

cancers between 1991 and 2005 (https://tinyurl.com/2vdcrjc4), though the source for this estimate is not clear.

It turns out that thyroid cancer is one of the most successfully curable cancers and has a death rate of no more than 1 percent, and if treated properly even less. Indeed, UNSCEAR found a mortality rate of only 0.2 percent. Therefore, to take the highest number among the various estimates, of an excess of 6,000 childhood thyroid cancers attributable to radiation from the Chernobyl accident, at most 60 radiation-associated deaths among children may have occurred and as few as 15, or even 0. The true figure is more likely somewhere between 0 and UNSCEAR's 15, because the new onset of screening yielded many occult cancers (ones that would never have become manifest), some of whose victims may have died from some other cause that was then attributed to the thyroid cancer.

Even the highest estimate would not substantially change the total radiation-related death toll from the accident. Taken together with the 47 deaths among firemen and cleanup workers, the total number is still no more than around 100 and more likely somewhere between 47 and 62 (47 + 15). Even at 100, this is fewer than the number of passengers killed in hundreds of commercial airline crashes (the greatest number being 583 in the collision of two airliners in Tenerife, Canary Islands, in 1977), and not the 4,000 originally predicted by the WHO based on the demonstrably false LNT hypothesis, or the million(s) claimed by certain antinuclear authors. But even any actual deaths among bystanders (children) were preventable, had the respective governments of Belarus, Ukraine, and Russia disposed of the contaminated milk for a few months and given the children non-radioactive iodine to fill the glands and block uptake of I-131.

Despite the established latent period for radiogenic thyroid cancer of about 5 to 10 years, a period now already passed since the Fukushima Daiichi accident, no increase in thyroid cancers has been recorded in that region of Japan. This is in part because, unlike around Chernobyl, the environment in Japan provides ample dietary iodine, which floods the thyroid gland and greatly slows, if not prevents, uptake of newly released I-131. Nevertheless, to be on the safe side and learning from the experience of Chernobyl, the Japanese government quickly administered non-radioactive iodine to local children to help further guarantee blockage of I-131 uptake.

While Chernobyl may have resulted in real thyroid cancers, those claimed by the antinuclear crowd at Fukushima were invented. First, their claims were based on the numbers of thyroid nodules found at new-onset screening, though very few nodules are cancerous (generally about 5 percent). And a comparison of numbers of nodules found before and after the accident is useless, since screening had never been performed before the accident. Second, they fail to compare the number of nodules in Fukushima prefecture with those in other Japanese prefectures. In fact, there was a lower incidence of thyroid nodules around Fukushima than in surrounding prefectures, though this was mere statistical fluke. But had the fluke gone the other way, would they have omitted this information? Third, they omit mention of nodule size, which reflects the likelihood of subsequent cancer development (https://tinyurl.com/4savf93y).

While radiation may have played a role in causing the Chernobyl-related thyroid cancers and consequent fatalities, none of them was a necessary consequence of the escaped radiation from the nuclear plant accident. Nor were the deaths of the unprotected firemen and clean-up workers necessary. So the exaggerated number of deaths, even if they had been accurate, resulted from incompetence and political malfeasance and would not otherwise have resulted from the nuclear accident alone. The Fukushima Daiichi experience testifies to the fact that the deaths were all avoidable even in the face of the accident.

Antinuclear propaganda portrays such accidents as hazards that track you down and submit you to their horrors regardless of what you do. But just as the (imaginary) hazard from Fukushima's tritiated water would require that victims actively take it internally, so did any hazard associated with I-131-contaminated milk around Chernobyl require the same. The only innocent bystanders around Chernobyl who may have actually been harmed by the radiation were children who drank contaminated milk. The Soviet government's withholding information from parents about milk contamination victimized the parents as well (indirectly). But whatever harm children suffered depended on their action in drinking the milk, something that a warning against drinking this milk coupled with provision of an uncontaminated substitute could have prevented. And, while the antinuclear ideologues wrongly paint nuclear power plant accidents as imposing death on persons in spite of their own actions, it is fossil fuels that actually impose such passive mortality (Chapter 7).[35]

There was no call to abolish the use of petroleum in the wake of the 2010 explosion and fire on British Petroleum's Deepwater Horizon oil platform in the Gulf of Mexico—an accident that killed

---

[35] In fact, more than 7,000 pedestrians are killed each year in the US with an annual global total of well over a million road-related deaths; should we outlaw motor vehicles? The reason we don't is that walking near roads is unavoidable and motor vehicles are indispensable in the modern world. Well, so is nuclear energy.

11 workers and laid waste to the nearby shoreline, as it continued to spout nearly 5 million barrels of oil from beneath the gulf floor for 3 months. Nor was this event exceptional. The six worst deep-sea oil disasters of all time, involving either oil platforms or drilling ships, occurred in the 11 years from 1979 to 1989—a decade that included both the TMI (1979) and Chernobyl (1986) nuclear accidents. Each of these oil accidents killed numbers of workers comparable to those due to the Chernobyl event, some more than the Chernobyl total, some fewer. Yet it is likely that virtually no one unrelated to the victims either knows of or remembers any of them, simply because there is no anti-petroleum movement comparable to the well-funded antinuclear movement—at least no movement based on immediate oil-related fatalities.

Sidebar: The Six Worst Deep Sea Oil Accidents

These were, in chronological order, the 1979 sinking of China's Bohai 2 platform in a storm off China that killed **72** workers out of 74 aboard, the 1980 capsizing of Phillips's Alexander L. Kielland platform off Norway that killed **123** workers out of 212 aboard, the 1982 capsizing of Mobil's Ocean Ranger platform in a North Atlantic storm that killed all **84** workers, and the 1983 sinking of Arco's Glomar Java Sea Drillship in a South China Sea typhoon that killed all **81** workers—all prior to Chernobyl. Then in 1988, two years after the worst nuclear plant accident in history, the worst oil platform disaster of all time took place. It was a fire and explosion on Occidental's Piper Alpha platform in the North Atlantic that killed **167** (including 2 rescue workers) out of 227 on board, and was followed the next year, in 1989, by the sinking of Unocal's Seacrest Drillship in a typhoon off Thailand that killed **91** out of 97 workers on board. Each of these killed more workers than the Chernobyl accident (at most 47), with 618 deaths in all, averaging 103.

Reciting such a list of such disasters doesn't begin to do justice to the lost workers or their friends and families and may even seem a callous misuse of such tragedies. But the comparison is crucial to give perspective to the misuse of inflated figures when nuclear accidents are involved. Meanwhile the millions of pollution- and accident-related disease victims and deaths, among both coal and oil workers and the public around the world—from the extraction, refining, distribution, and burning of fossil fuels—go marching on year after year. Yet, besides any (preventable) deaths of children around Chernobyl, no member of the public has died from radiation due to commercial nuclear energy in the more than half a century of its use.

It's even more directly relevant to compare the numbers of deaths associated with AGW to those caused by nuclear energy. To pick just three outstanding examples, all from the richest country in the world, Hurricane Katrina in 2005 killed over 1,500 people (estimates vary), the 2021 winter cold snap in Texas killed almost 250,[36] and the recent wildfire in Maui (August 2023) killed 101. Each of these approximates or exceeds the (highest legitimately estimated) total death count of all three of the biggest nuclear accidents in the world over more than half a century of the technology.

Invented figures for deaths caused by the Chernobyl accident's radiation, numbering in the thousands to a million, or even two million (Helen Caldicott, https://tinyurl.com/m9fw6zzn), are sheer fiction, often thought up on the spot in the middle of a speech or interview (an action that is being raised to a fine art by a recently re-elected US president).[37] These exaggerated numbers are sometimes credulously accepted on the grounds that they are claimed, or at least appear, to be in defiance of governmental attempts to hide the reality.[38]

★ ★ ★ ★ ★

This completes the portions of the book on energy considerations. In the final three chapters we turn to an expanded critique of the Green narrative (other Green New Deals); to general geopolitical considerations—emphasizing the reasons why without a global cooperative system we will not likely be able to halt AGW, much less reverse it; and to an analysis of the impact of a fragmented competitive economic system on epistemic dysfunction.

---

[36] AGW causes not just excessively high local temperatures, but also unusually low local temperatures—both types of extreme weather events.

[37] British journalist George Monbiot takes Caldicott to task for citing references, only when asked, that do not say what she claims they say, some of them even written by herself. Here is Monbiot's confession of having been misled by the antinuclear movement and Caldicott in particular: https://tinyurl.com/3tzv5u8c.

[38] Two very useful and readable resources about radiobiology in general and radon in particular are written by Norwegian biophysicist Thormod Henriksen and his colleagues at the University of Oslo (2015, 2016).

IV

Capitalism:
Why It's an
Obstacle

9

Other Green New Deals: Ecosocialism versus Ecomodernism

In this book we analyze four positions on **energy systems** and **social systems**, all of which seek a solution to Anthropogenic Global Warming. These combine attitudes that favor or oppose nuclear energy and/or capitalism, as shown in the following schema. The chart summarizes the four positions and situates within the schema authors previously referred to and some to be discussed in the present chapter:

	CAPITALISM YES	CAPITALISM NO
NUCLEAR NO	JACOBSON, LOVINS	ECOSOCIALISTS: COX, AJL, ARONOFF, FOSTER & CLARK
NUCLEAR YES	ECOMODERNISTS: SHELLENBERGER, BRYCE, SMIL, HUBER & MILLS	SACKS & MEYERSON

We've tried to demonstrate that the upper left position is based on imaginary premises. The upper right and lower left positions each encompass an element that we consider valid and an element that we consider invalid. Our position combines what we consider to be the valid elements of these. We argue that the invalid elements undermine the valid, rendering them likely unworkable. However, if either could for a time achieve its goals, ecosocialism, because of its energetic infeasibility, would almost certainly fail long before ecomodernism, as energy is the foundation of life itself.[1]

Ecosocialists (upper right) reject capitalism but also reject nuclear energy in favor of inadequate and/or inefficient (low-EROI) energy, to be equitably distributed. Ecomodernists (lower left) favor nuclear energy but also favor capitalism or at least accept its permanence as an unquestioned default.

Our position is that to halt and reverse AGW the world needs both nuclear energy (technologically) and an alternative system to capitalism (socially), the latter based on universal cooperation for the egalitarian satisfaction of human needs, rather than profit for the few.

Two Key Ecosocialist Authors: Max Ajl and Stan Cox

Max Ajl's ecosocialist narrative (2021) simply rejects nuclear power without regard for evidence. First, he approvingly describes the 2010 Cochabamba agreement among largely indigenous peoples, centered on "the Rights of Mother Earth," including "the right to be free of contamination and pollution, free of toxic and radioactive waste." He summarizes (p. 10):

> Forms of development which dump detritus in reservoirs and fill the air with waste and carbon dioxide, nuclear power, willful genetic tampering ... and perhaps above all, industrial processes which spill out and over biomes' capacities for remediation would be banned.[2]

Ajl next quotes the well-known science fiction and fantasy writer, Ursula Le Guin's *Always Coming Home* (p. 18, Ajl's italics, presented as an epigraph):

> *The people of the valley did not conceive that such acts that they saw and felt much evidence of in their world—the permanent desolation of vast regions through release of radioactive or poisonous substances, the permanent genetic impairment from which they suffered most directly in the form of sterility, stillbirth, and congenital disease—had not been deliberate. In their view, human beings*

---

[1] The terms "ecosocialism" and "ecomodernism" were coined by adherents of the two outlooks.

[2] Note the call for banning biotechnology along with nuclear energy. In support, Ajl takes a swipe at the biotechnology-promoting Cornell Alliance for Science, simply because it's funded by Bill Gates (p. 36).

did not do things accidentally. . . . So these things human beings had done to the world must have been deliberate and conscious acts of evil, serving the purposes of wrong understanding, fear, and greed. The people who had done these things had done wrong mindfully. They had had their heads on wrong.

Third, Ajl mocks the ecomodernist call for high energy (p. 47):

While they nod at the notion that resource extraction is still occurring, . . . they argue that such remaining footprints can be lightened and erased through developing more efficient mechanisms for procuring needed resources: urbanization and nuclear power, desalinization and aquaculture, and agricultural intensification. . . . Finally, these demands are the ore with which a steel sacred calf is cast and at which all subsequent eco-modernist literature worships . . .

Fourth, as part of a discussion of the role of the US military in environmental destruction, he notes (p. 155):

White phosphorus and depleted uranium are left behind by US weaponry in Fallujah, Raqqa, and the Gaza Strip, with a dark bloom of birth defects. The Grants Mineral Belt in New Mexico remains one of the best endowed uranium deposits anywhere, providing the raw material for the world-killing weapons. The mining takes place close to homes and communities of the Diné people and those who have labored in the industry include Diné, Acoma, Laguna, Zuni, and Hispanic peoples, forcing upon them the social and ecological costs of US colonial-capitalist domination.

Fifth, he asserts: "Using solely domestic sources of renewable energy . . . would be preferable to emerging forms of green-energy colonialism, biofuel land-grabbing, nuclear, or oil and coal" (p. 168).

Finally, he approvingly cites an article by Stan Cox (https://tinyurl.com/a4tmx55b) that says (p. 70, endnote 37),

Beware of some Jacobson critics who, determined to save capitalism but cynically adopting the language of social justice, look at the inadequacy of the high-energy [*sic*] 100-percent renewable strategy and draw a suicidal conclusion: that the only acceptable alternative is a big rollout of nuclear power, carbon capture, and geoengineering.

Ajl relies on innuendo and guilt by association. In his fourth reference, he associates uranium mining strictly with production of nuclear weapons, familiarly conflating them with nuclear energy and thereby reinforcing nucleophobia. As we indicated in Chapter 8, no causal link has been found between the two atomic bombings and subsequent birth defects, nor are birth defects caused by the minimal radiation emanating from depleted uranium (DU, the byproduct of enrichment for nuclear reactors; see "Enrichment" in Glossary) which was used to make bunker buster shells during the two wars against Iraq.[3] Increased birth defects may be related, however, to other effects of the Iraq wars, including other pollutants and both mental and physical trauma. Health problems associated with exploded DU shells are related not to their extremely low radioactivity (lower than natural uranium), but rather to their chemical toxicity.

Ajl also singles out the minority of indigenous miners (15–20%) as special victims of radiation, as though the non-indigenous miners (80–85%), who also had slightly higher standard mortality rates, were irrelevant (http://tinyurl.com/3yw7nwtk). His appeal to emotion hides his neglect of scientific evidence. The elevated lung cancer rates found in the 1950s before the mines became ventilated (a reluctant response to strikes and other united actions by indigenous and non-indigenous miners) have declined with ventilation. Mines contain many types of dust other than uranium. In well-ventilated mines and in populations living near these mines, there is no valid evidence of elevated rates of lung cancer in any but smokers.

The abstract of a 2010 study (https://tinyurl.com/4pndvtrk) about Grants by Boice et al. says:

> Although etiological inferences cannot be drawn from these ecological data, the excesses of lung cancer among men seem likely to be due to previously reported risks among underground miners from exposure to radon gas and its decay products. Smoking, socioeconomic factors, or ethnicity may also have contributed to the lung cancer excesses observed in our study. The stomach cancer increase was highest before the uranium mill began operation and then decreased to normal levels. With the exception of male lung cancer, this study provides no clear or consistent evidence that the operation of uranium mills and mines adversely affected cancer incidence or mortality of county residents.[4]

---

[3] Uranium is the densest naturally occurring metal found on Earth, and therefore useful in shells. Elements heavier than uranium have much shorter half-lives and any naturally occurring quantities have long since become undetectable.

[4] The inclusion of "ethnicity" in a list of actual causal associations like "smoking" and "socioeconomic factors" reflects racist baggage, since ethnicity only operates through those accompanying factors. That is, if Native Americans have higher rates of some adverse medical condition—barring contrary evidence (which is not forthcoming)—it should be

Despite Boice's defense of LNT (Chapter 8), the article notes (p. 634) that uranium "has not been classified as a human carcinogen" because it is not "very radioactive" and "its chemical properties are often such that any inhaled or ingested uranium is excreted rather quickly from the body."

Ajl repeatedly touts indigenous peoples as sources of wisdom for ecological practices. While this attribution is not entirely without foundation, it is generally confined to the practices of small populations. Modern complex societies require largely different principles for ecological preservation. Indigenous populations as a whole are less familiar with such principles than modern scientists, including those with indigenous roots. But scientific practice is more important than roots. The use of indigenous populations and individuals as voices of authority, to which we return below, is an inverted variant of the ad hominem fallacy (focus on the messenger rather than the message).

Ecosocialist Association of Nuclear Energy with Capitalism and Renewables with Socialism

Arguing that technologies are socially branded, Ajl rejects the precept that "human-made tools can simply be repurposed by those in rebellion against the owners of the means of production" (p. 43). Mocking the claim that "[t]he issue isn't the knife itself, but who wields it," he says,

> Those who argue in an absolute way for technology's categorical social neutrality, especially from the left, forge one of the most dangerous, subtle, and effective instruments of ideological counterinsurgency.

Ajl believes that certain technologies are "welded onto the hands of the ruling class." He cites the late MIT historian David Noble's proof that "machine tools were designed to deskill workers and concentrate power"—though rather than design, deskilling was more likely a happy side effect of the profit-enhancing exchange of labor for machinery—but in citing it Ajl conflates the inherent properties of technologies with the social relations in which they are embedded. Thus, while the powerful owners of

---

assumed to be due solely to features of their subjugation to oppression and discrimination, not to their biological makeup.

industry did deskill workers with machine tools, the working class could easily turn them to their benefit were they in control of the economy and the state.

This is part of a popular argument about the relation between a "structure" and its "tools." For example, radical US feminist and civil rights activist Audre Lorde famously remarked, without specifying which tools, that "the master's tools will never dismantle the master's house." Applying this argument to language and colonialism, Kenyan Gikuyu writer Ngũgĩ wa Thiong'o for a time stopped writing in English. In contrast, Nigerian novelist/poet Chinua Achebe denied that languages have an essential character and continued to write in English. Similarly, James Joyce continued to write in English. Joyce and Achebe were among many colonized people who repurposed English as a tool against the British empire.

On the relationship between language and patriarchy, some French feminists argue that language itself is phallocentric. In a flight of exuberance, French theorist Roland Barthes even defined language itself as "fascist," on the grounds that it tends to "fix" meaning. Both Barthes and Bulgarian-French philosopher Julia Kristeva, influenced by and furthering a linguistic avant-garde, looked to areas of "non-meaning" to resist.

The very concept of guerilla warfare is predicated on using certain of "the master's tools" to dismantle "the master's house." It is, however, true that some tools cannot be so repurposed, for example, racism and sexism. But this does not apply to hammers, nails, guns, language, and most relevantly here, sources of energy.

There may very well be technologies that do not readily, if at all, lend themselves to repurposing (ICBMs come to mind). But such examples are inapplicable to most technologies, in particular nuclear energy for electrical generation. If everything associated with capitalism and exploitation were unrepurposable, humanity would be left with nothing upon a successful "rebellion against the owners of the means of production." The manifest absurdity of the generality forces us to examine each entity one by one. But even Ajl does not contend that all such entities suffer from this weakness.

Solar, for example, is included in Ajl's list of more democratic technologies, as it serves the many and, in his view, accords with his small-is-beautiful ideology—an ideology he adheres to inconsistently. Even solar, after all, is not small in either its requirements for land or battery backup (absent fossils and nuclear). And the machinery and techniques to obtain the raw materials, fashion them, and install the finished products in solar farms, along with their associated batteries, are dripping with capitalism's watermark.

Ajl favors small privately owned farms over large ones collectively owned, but he also favors low energy with low input, oblivious to both the high input and large farms required by wind/solar and the low input for nuclear.

His usual rejection of the big in favor of the small finds a parallel in his understanding of social systems. Capitalism tends to concentrate and centralize capital, so for Ajl big becomes a proxy for capitalism and small a proxy for anti-capitalism. However, there is nothing that saddles socialism (anti-capitalism) with smallness. Indeed, the working class, with control over the social and economic system (true socialism), would be able to increase efficient production of needed goods and services while eliminating production of the wasteful and unnecessary. It would be empowered to undo centuries of capitalist-induced deprivation that has left masses of people without adequate access to the means of survival—housing, healthy food, clean water, education, healthcare—while others bask in at least material comfort and some in luxury. Thus, even if Ajl's associations were valid, they constitute an illogical argument against nuclear energy and for socialism, bringing him closer to Proudhon (a champion of smallholdings) than to Marx.

Ajl shapes his theory to his prejudices with respect to nuclear energy, veganism, and meat. He regards nuclear as intrinsically authoritarian, a technology that could not be repurposed even if embedded in cooperative social relations (he is even mistaken about nuclear under capitalism). He interprets veganism in the same way, bound to forms of ecomodernism that he repudiates (pp. 35–36, 51–2).[5] In contrast, he grants that meat production can be repurposed away from factory farming that crams cows, pigs, and sheep so closely together that infectious illnesses become all but inevitable, and toward "agroecological" forms of production.

Ajl's exception for meat production is based on his concern for those pastoralists and peasants who produce it, which apparently dictates his view of the technology. And since he is repelled by meat-rejecting vegans and nuclear energy advocates, the objects of their affection are, for him, not "socially innocent." Like Gilbert and Sullivan's Mikado, who proposes to "let the punishment fit the

---

[5] Correctly differentiating between specific technologies and technology in general, however, Ajl denies that "because a technology is used to attack the people, the people should reject technology" (pp. 54–55). Thus, he exonerates himself from the charge of Luddism (a nineteenth-century movement attributing oppression to technology rather than to exploitation). But he ends up indicting himself anyway by claiming that biotechnology (GMOs) and nuclear energy are "welded" to ruling class power.

crime," Ajl lets the explanation fit the prejudice. To be consistent in his preference for low energy, he should, but does not, object to voracious energy hogs like the Internet and high-speed trains.

In sum, Ajl's advocacy with respect to technology and capitalism conflates low energy and low energy density (two different things) with low environmental impact and socialism. And it conflates high energy and high energy density with capitalism. And while he identifies what he considers to be small with socialism and what he considers to be big with capitalism, he has it partially the wrong way around: low energy density entails extensive repurposing of land, resulting in high environmental impact, and the reverse for nuclear. With regard to energy density, Cox is similar to Ajl, but his rejection of high-speed trains makes Cox more consistent than Ajl.

And now turning to Cox, his criticism of ecomodernism, whether in a capitalist or socialist/communist context, rests firmly on his rejection of nuclear power. He raises the usual objections around waste, danger, and proliferation, all of which we show to be misguided (Chapters 7 and 8). Like Ajl, Cox agrees that some technologies are not inherently wedded to certain social systems, but he refuses to exempt nuclear energy, which he contends is intrinsically capitalist and authoritarian. He maintains that nuclear's division of labor is elitist, and that nuclear engineers form a priestly caste, guardians of secret knowledge, and thus are authoritarian and antithetical to democracy. Consistency would dictate that he apply the same assessment to physicians, linguists, biologists—in short, any highly trained practitioner.

In fact, operating a nuclear power plant requires no more (or less) expertise than other engineers, scientists, airline pilots, master chefs, music composers, and so on. Cox's anti-elitism cloaks a kind of anti-intellectualism. Were the knowledge and expertise associated with nuclear engineering to be ruled out of a Green world order, the reasoning would tend to put a damper on all science. This would be to no one's advantage, as Cox would surely be quick to admit. He would likely confirm Greens' allegiance to science. The confirmation would probably include ecology and climate science, which are enormously complex endeavors that few people currently understand. Cox might object to the secrecy associated with nuclear weapons, but secrecy is foreign to the education programs for nuclear engineers.

Cox extends his list of intrinsically capitalist entities to include energy density, which he fears threatens to destroy biodiversity. He cites the observation of Wes Jackson, US agronomist and cofounder of the Land Institute, that "highly dense energy destroys information, both cultural and biological." Paraphrasing Jackson's evidence for this generalization, Cox says that energy-dense fossil fuels "have enabled the widespread cultivation of industrial crop monocultures, which harbor only a tiny fraction of biological information . . . that would be found in either natural ecosystems or pre-industrial farms" (p. 70).

On the contrary, high energy density, and the role it plays in greater power density, is precisely what enables us to leave large swaths of nature alone, even as we produce abundant energy. To blame monoculture on an energy source, and to conflate enabling with coercing, borders on parody, like blaming your hiking boots for getting you lost in the woods. Energy density is neither intrinsically capitalist nor agriculturally restrictive and lacks the ability to destroy biodiversity. Rather it is low-energy-and-power-density wind farms that have historically destroyed birds, bats, and insect species, though advanced turbine designs might be able to alleviate these hazards. And it is low-density solar farms that have displaced wildlife and required the destruction of neighboring ecosystems.

High energy density is a prerequisite for industrialization and mechanization, and these, in turn, for meeting real human needs—but it only succeeds if the social/economic system does not subordinate such needs to a competitive growth imperative.High-energy density is also a prerequisite for reducing our overall environmental impact. To believe it intensifies the exploitation of labor is to conflate the tools of production—energy, machinery, raw materials, and the like—with production relationships. Such conflation is a form of technological determinism. The relations of production differ in an exploitative system from those in a non-exploitative system, while the production tools are intrinsically independent of economic system, even if their employment is not.

The fear-driven rejection of high energy density necessitates the rejection of industrial society altogether. Playing on the failure of most ecomodernists to consider the need for system change, Cox's rejection of their manifesto is misaimed at their approval of technological progress. Paraphrasing a response from ecological economists to the ecomodernists, Cox notes, "Industrial modernity has certainly brought numerous benefits to humankind, but it has come at a heavy toll, and one that jeopardizes the possibility of creating a sustainable society (p. 61)."

It's ironic that some ecomodernists also regard high energy and technological progress as intrinsically capitalist. Yet they still correctly maintain that high energy density protects nature and that any alternative necessarily occupies and damages far more land. In particular, they expose the negative impact on biosystems due to the extensive gathering of biomass for fuel.

Others retreat from a necessary association between big technology and capitalism to affirm a mere tendency for technology to generate its own social relations. For example, Jim Thomas of Greenpeace says:

> So it's not like saying a technology will always go a certain way, I don't think it's as hardline as that, but I do think there's more likely a sort of drift towards certain outcomes and uses. And in technology, there's this idea of, there's a latency within a technology.[6]

Whether the probability is 100 percent or merely close, Thomas still locates the association in the realm of technology rather than the social system. Railing against biotechnology, he grants that a GMO product useful for small farmers may occasionally be produced, but he claims:

> Is it ultimately going to be centralising, is it always going to fall into requiring large centralised power, or does it go somewhere where it becomes something anybody could develop . . . in a free way? . . . looking at the last twenty years . . . [t]he . . . field has become controlled by large pharmaceuticals and agrochemical companies.[7]

Despite Thomas's downgrading determinism to a tendency, his formulation still contains the essential feature of technological determinism—namely, that the outcome is dominated by the technology rather than by exploitative relationships between owners and producers.

While some technologies require more training than others, this training is not intrinsically capitalist. Bullet trains, biotechnology, and nuclear energy are more complex than the Miracle Mop. While the Miracle Mop patents could be taken over by big capital—mass produced by Procter and Gamble and sold at Walmart—such developments are hardly to be discovered by examining the mop.

Big is not synonymous with bad, nor small with good. Thus, Thomas conflates large, centralized production with large, central-

---

[6] Quoted by British journalist Mark Lynas in his book *Seeds of Science* (2020, p. 220), about which more in Chapter 11.

[7] Lynas (p. 219).

ized power and concentrated capital. But capitalism itself is not a necessary or inescapable economic system, occupying as it has only the tiniest recent portion of humanity's 300,000 year existence.

Further Incoherence in the Green Arguments

Green arguments against nuclear are often internally incoherent. For example, while Cox argues that nuclear energy, like fossil fuels, is intrinsically wedded to capitalism, he also points out, along with Amory Lovins, that somehow nuclear fails in the capitalist market. Without his realizing the irony, this failure turns out to be largely due to the efforts of Greens like Cox. As we showed in Chapter 5, the high consumer price of nuclear power results from extrinsic manipulation. If the state were to stop subsidizing fossils and renewables and increase its subsidies to nuclear, if the grid operators were to stop giving priority to renewables when available, if the regulatory agencies were to eliminate arbitrary rules that boost the price of nuclear indefinitely,[8] were the nuclear industry to begin propagandizing in its own favor, and were the courts to reject phony litigations like they reject SLAPP suits, nuclear would easily succeed, as it has in France and several other countries.

Disregarding his charge of "suicidal" (as mentioned above), Cox gestures at the possible advantages of Generation IV reactors, in particular SMRs. But he cites one discouraging claim that they won't be available until 2050, too late to prevent further irreversible tipping points. Because it accords with his a priori rejection, Cox simply accepts the prediction of delay without citing predictions of earlier availability. NuScale gained NRC approval to introduce its SMR in the US in February 2023. Other SMRs include GE Hitachi's BWRX-300, a 300-MWe boiling water reactor, scheduled to be constructed in Ontario, Canada, by 2028, and ThorCon USA hopes to test its molten salt reactor (MSR), with its graphite moderator, by 2026 and commercialize it by 2028. There is also a pebble-bed research reactor in China (http://tinyurl .com/2n5h8hy9), and the DOE's EBR II ran from 1964 to 1994 without a hitch. See http://tinyurl.com/2bw777nu for advanced SMRs in the planning stage.

If Cox's approval of SMRs were not merely for show, why wouldn't he call for acceleration of this effort? After all, as we will

---

[8] In his book, Devanney (2023) demonstrates that the NRC, through their LNT/ALARA-based policy, has artificially boosted the consumer price of nuclear power out of the competitive market in the US.

see shortly, he calls for the elimination of capitalism, not eventually but by 2030—a preposterous timescale—with massive redistribution of wealth from rich to poor countries.[9]

In yet another example of technological determinism, Cox argues that nuclear can only supplement fossil fuels but cannot replace them. Yet as Qvist and Brook point out (http://tinyurl.com/4rmbtdka), nuclear certainly replaced fossils in France and Sweden. It is wind/solar that have not replaced fossil plants (without relying on nuclear or substantial hydro), because their inherent intermittency requires retention of reliable backup (mostly NG). The rejection of nuclear energy renders a Green future unattainable—for several reasons to be described below.

Intermittency as Viewed by Antinuclear Ecosocialists

Kate Aronoff and her coauthors (2019), attempting to defend renewables, trivialize the problem of intermittent electricity, as though the aversion to it were mere selfishness. They say (p. 115):

> Sun and wind are free but capricious. Moving energy around the grid counteracts their caprice but only up to a certain point. Conservatives love this: the right gloats that a windless night will shut off your T.V. in the middle of the NBA finals.

Having granted that blackouts born of capricious wind/solar can only be prevented "up to a certain point," they contradict themselves by confidently asserting that a TV cutoff "won't happen." They imply that in any case intermittency would only affect self-centered rightwing consumers. Ajl also trivializes the problem with some overlap with Aronoff et al.

The authors acknowledge daily and seasonal variation in wind/solar availability but reject Jacobson-type solutions through oversizing/overbuilding. They also reject reliance on battery storage, but for reasons different from Jacobson's – namely, to minimize overmining of poor countries. Their proposed solution is through "flexible demand," or more precisely demand management (by whom?), which involves smart meters and grids run by algorithms that allow "electricity flow" "to follow the elements." They cite US energy ana-

---

[9] Another author who extolls SMRs, in this case as a technology best suited to libertarian capitalism, is US art dealer and entrepreneur Reese Palley, in his book, *The Answer: Why Only Inherently Safe, Mini Nuclear Power Plants Can Save Our World* (2011).

lyst Jesse Jenkins, who, they say, suggests that battery storage requirements would be greatly lessened if flexible demand were employed.

Jenkins, however, is hardly their ally, as he is pronuclear, understanding that it is the only reliable substitute for fossils. And his proposed flexibility requires pairing wind/solar farms with nuclear plants in flexible operation mode (load following). This, however, is flexibility of supply, not of demand (http://tinyurl.com/ bdu6zkkn). While the main reason for the low EROI of renewables is the high energy input to manufacture and install extensive field apparatus (including batteries), their weather-dependent intermittency adds to the problem by reducing their energy output.

Ajl at least mentions the low EROI of wind/solar plus storage. He cites approvingly the work of Jacobson critic Ted Trainer (mentioned in Chapter 5) affirming that an all-RE economy, including storage, would lower the EROI to 3, or at best 4. Such EROIs, Ajl grants (quoting Trainer), are "well below the range of the thresholds identified in the literature as necessary to sustain high levels of development in current industrial and complex societies" (p. 66). Ajl goes further, giving voice to McGill University graduate student Tim Crownshaw's worry: "Will it be possible to run solar PV panel and wind turbine production lines using solar- and wind-generated electricity in the future [i.e., bootstrapping]? We don't know, but there are reasons to be skeptical" (p. 67). In other words, Crownshaw, with Ajl's apparent agreement, questions whether a wind and solar economy not only has too low an EROI to support a complex modern society but may not be able to reproduce itself (points we demonstrate in Chapter 5).

In introducing the Trainer comment, Ajl suggests that some of this lowering of EROI would be mitigated by the fact that wind/solar "often lose much less power [than the current energy system] on their way to being used" (p. 66).[10] But the basis of such a claim is not specified, and to the degree that it is true, it would only be for local uses, with short distances from source to appliances. While home rooftop solar might not lose much energy on its way into the house interior, and a privately owned wind turbine might not lose much energy on its way to the milking barn, electricity from wind turbines and solar panels sent across national and international long-distance lines would lose no less energy than that originating from fossils or nuclear. In fact, the losses may

---

[10] Similar energy-saving claims are made for EVs over ICE vehicles. Such claims are equally incomplete, as they neglect the immense energy needed to manufacture and repeatedly charge the batteries, and the accompanying heat losses.

be even greater, as an all-wind/solar electrical system would require a smart grid, and optimal sites would determine location rather than proximity to consuming populations.[11] While vast computing centers, like Google, and the growing fleet of EVs are already straining the capacity of the grid, the requirements of an all-RE system would further magnify the necessity for immense expansion—a problem that would be greatly lessened by reliable fossils and nuclear.

Ajl presents low-EROI as a problem but then (50 pages later) regards a low-EROI economy, or at least a low-technology economy, as a positive good because it is incompatible with capitalism. He argues that a low-energy, low-tech economy "would be an anti-capitalist qualitative advance: communal low-tech luxury" (p. 114). But to count on insufficient energy to push capitalism off the stage is like hoping to eliminate inequality by starving to death the residents of Palm Beach and Beverly Hills.

Alongside his approval of low energy, Ajl's assessment of the benign nature of intermittency follows in the footsteps of his tolerance of low EROI. He claims that intermittency is not a serious issue or is serious only for capitalists. He says, "current [capitalist] patterns of production are structured based on a certain kind of constantly available energy," as though blackouts were no problem for anyone else. Then he proposes to his fellow socialists that for planning purposes with intermittent energy sources, they could adopt "a more supple approach to manufacturing [that] might solve a large portion of the intermittency problem"—"supple" meaning, like Jacobson's "flexible," adjusting their production schedule to follow the unpredictable availability of electricity (pp. 114–15). In other words, adjusting demand to supply.

Ajl implicity contends that only under capitalism does the urgency of time make itself felt. While some production processes might not be negatively impacted by intermittency, meeting human needs in timely fashion would be, as we learned from supply chain bottlenecks during COVID.

---

[11] As briefly alluded to in Chapters 2 and 6, Bryce, citing the DOE, reports that the transition to wind/solar requires expansion of the grid by at least 57 percent, and possibly doubling it, over the next 10 to 15 years (http://tinyurl.com/38jah8h3). Since the US grid contains some 240,000 miles of high-voltage lines (the distance from the earth to the moon), this would require at least another 136,000 miles of long-distance wiring. At the current rate of expansion, this would take more than 80 years. So the expansion rate for high-voltage lines (never mind the millions of miles of low-voltage lines near the point of consumption) would have to speed up by more than a factor of 5 (80/15 = 5.3). At $4 million per mile, and rising, the cost for high-voltage alone would exceed half a trillion dollars over 15 years.

Ajl recruits support from others:

> Globally, industry uses around half of end use energy. Some of these processes rely on purely mechanical energy: turning, polishing, milling, hammering, crushing, sawing, and cutting. These could be run with intermittent power, as can great parts of food production . . . As Kris Decker [*sic*][12] notes, "intermittent energy input does not affect the quality of the production process, only the production speed . . . Running these processes on variable power sources has become a lot easier than it was in earlier times . . ." Factories could run on a mix of wind and solar. Such shifts do not mean reducing production or consumption. (p. 115)

In short, says Ajl, "Factories . . . can work when there is sufficient power and go idle when there is not" (pp. 72–73). "Goods can be transported by ships and trains, which move when there is power and stop when there is not. This is not . . . an obstacle to a complex society. It is an obstacle to a complex society that looks like the one in which we currently live" (p. 73).

Ajl cherry-picks situations that may tolerate intermittency, even if the owners of industry and consumers would prefer it otherwise. He and other low-energy enthusiasts, like Cox and Aronoff et al., seem blind to end-use applications that are dependent on reliable electricity, such as computers whose crashes are not simply disruptive but threaten to destroy data or damage the electronics, protective appliances like thermostats that run furnaces and air conditioners during severe cold snaps or heat waves, pumps to deliver fresh water (which many lost during the two-week 2021 Texas cold snap), electric subways stranding passengers threatened during rescue from a dark tunnel by the third rail's sudden revival, elevators trapping passengers between floors for indeterminate durations, electric surgical cautery interrupted while stemming a hemorrhage, life-or-death ventilators, dialysis machines, and so on. Auxiliary power sources, when present, as in hospitals, require a few seconds or more to become activated, but presuppose reliable electricity or fuel to produce or store energy. Such events transcend social system and negatively impact people's wellbeing. They are not reducible to irritating interruptions in the enjoyment of an overtime free throw.

Ecosocialists like Ajl and Cox would jettison energy-consuming heat pumps and air conditioning by substituting well-designed

---

[12] Kris De Decker, a Belgian theorist and blogger on technology and energy.

arrangements of shade and water. If this was ever effective at temperate latitudes, it is a fantasy dashed by AGW with summer temperatures over 110°F (43°C) that challenge the shedding of excessive body heat to survive another day.[13] Their vision represents nostalgia for a Tolkien Shire.

Ajl proposes that local sovereignty would prevail in a cooperative economy, especially over food, obviating long-distance transport and necessitated refrigeration. However, this would restrict desires and/or needs for non-local foods and shift labor back to agriculture. With nuclear-powered transport and refrigeration, unreliable and limited energy would cease to be problems, and food sovereignty would become unnecessary. Moreover, under Ajl's arrangement, crop failures could be fatal. In effect, Ajl endows his socialist/communist project with the authority to wave away necessity with a flick of the wrist. Whereas a cooperative system would enable humanity to recognize necessity and collectively find ways to accommodate it.

In endorsing energy intermittency, Ajl unwittingly undercuts such things as his desired high-speed trains and internet. The necessitated smart grid as well requires prodigious energy for its manufacture, construction, maintenance, and operation.

Ajl's thought gaps are paralleled by one of ecosocialism's founders, the late US author Joel Kovel (2007), who highlights the relatively low-energy egalitarian lifestyles of the NY Amish. While calling the patriarchal social organization of the Amish "egalitarian" is a stretch, Kovel, more to the point, fails to perceive that their low-energy lifestyle is largely maintained by selling high-quality artisanal products, which many members of the surrounding society can afford because they enjoy high-EROI energy. The low-EROI social order is largely dependent on the very society Kovel would have it displace, its matrix. This bears significant similarity in key respects to a central point developed throughout this book: that the ostensibly low, but no longer declining, costs of renewables, reflected in their only partially accounted low LCOE, is parasitic on the fossil/nuclear high-EROI matrix, and on the extrinsic pricing matrix. In both cases, greater penetration would progressively eliminate the supporting elements, slowly at first and then rapidly, resulting in disabling dyseconomies of scale.

* * * * *

Cox's approach overlaps Ajl's in some ways and differs in others. Along with Aronoff et al., Cox criticizes Jacobson on technical and

---

[13] See the discussion of Kim Stanley Robinson below.

equity grounds. As discussed in Chapter 4, Cox recognizes the tremendous material burden of an all-RE system, citing (in his footnotes 143 and 144) Heard, Brook, et al. 2017 and Clack, Quist, et al. 2017, respectively. While he acknowledges the futility of Jacobson's plan, Cox's rejection of nuclear leaves him without a workable alternative. In contrast, the critiques by the Heard and Clack teams are inseparable from their defense of nuclear energy. While in fact inseparable, these two elements are prised apart by Cox in order to say that nuclear "should be opposed for many reasons" (footnote 143).

Socialism Cannot Be Sustained Without High-EROI Energy

While low energy is implicit in an all-RE system, it is explicit in Cox's alternative proposal. He relies on consumption-side efficiency, calling for degrowth combined with massive redistribution of wealth in the "global North." He claims that sharing this energy can easily meet the needs for a good life in the "global South." This would, he says, only require 2 kW for every person in the world, not just electricity but total power. It would further require egalitarian distribution and the prioritizing of public goods over individual products. However, the 2 kW per capita would have to be reliable power, which in a post-fossil world would be rendered impossible by Cox's rejection of nuclear.

To persuade Americans that a decrease from 10 kW to 2 kW per capita is not unreasonable, he indicates that Europeans consume only 5 kW per capita. But even 5 kW is 2½ times what Cox claims is adequate, and it is largely reliable energy, dependent on the fossil/nuclear matrix. Of course, per capita is only a national average, and US energy consumption is greatly skewed toward industry, commerce, and their use of transportation (see Figure 2-1). The majority of individual US consumers, mostly working class, already enjoy far less than the average. A reduction in the average will further deprive members of the working class.

Finally, because so many people in the world lack adequate (or any) access to electricity and other forms of power, it turns out that 2 kW per capita is not much less than the current global average. As mentioned in Chapter 4, total global power consumption is 19.6 TW. Divided among 8 billion people, that's an average of 2.45 kW per capita. Despite the drastic consumption reductions that would be required in some countries, and by only some people in those countries, average global consumption would hardly

254	*Other Green New Deals: Ecosocialism versus Ecomodernism*

be reduced at all in Cox's proposal, and total consumption would presumably rise as population increases.

Cox argues that the rich countries, primarily in Europe plus the US, especially the latter, must cut carbon—not just CO_2, but methane (CH_4) as well—by 7.6 percent per year in order to meet IPCC guidelines and avoid catastrophic climate change. On equity grounds, he argues, other countries must be allowed to increase their standards of living to an amount where their needs are met. However, as we mentioned in Chapter 3, the rise in living standards is currently being underwritten by the rapid growth of coal consumption in China and India (http://tinyurl .com/ bdfhwd2z). Their growth in CO_2 emissions is far in excess of any projected reductions by the rich countries, and as we pointed out in Chapter 5, methane leakage alone prevents even these reductions from having any significant impact.

The argument is similar, on a global scale, to that of Aronoff et al.: fossils are necessary for catch-up development until RE can take over. Cox's futile proposal mandates massive direct cuts in carbon, something that even he points out cannot be done through market mechanisms and requires a rejection of the growth imperative. So for all practical purposes, cuts in carbon require a rejection of capitalism.

Given the presumed necessity of rationing, Cox is concerned that the rationing be equitable, and so his plan includes an ambitious redistribution program to move wealth, mainly capital, away from the top 33 percent to the rest of the population, all by 2030. While Cox's call for equitable redistribution seems laudable, it is astonishingly naive about what ruling classes do with their rule. Imagine a workers' delegation firing all plant managers and assuming control over production, letting the managers know that if they would like, they could roll up their sleeves and join in the production process.

The Conflation of Wealth with Capitalism

Cox's concept of political economy relies on the understanding of monopoly capitalism by US Marxist economists Paul Baran and Paul Sweezy (1966). In their view, monopoly capitalism has overcome the dynamics of competition but not the problem of underconsumption (relative to production). In Cox's reading of both *Monopoly Capitalism* and *The Limits to Growth* (Meadows et al., 1972), overproduction (relative to consumption) and overaccumulation of capital (relative to profitability) cause crises, in part because the surplus product cannot be fully consumed since the workers lack the purchasing power to soak up the surplus product, and therefore the surplus value cannot be realized by the capitalists. And monopoly capitalists need not reduce their prices since they enjoy the benefits of monopoly pricing.

While the problem of the unabsorbed surplus was partially and temporarily solved through debt-driven consumption, domestic monopoly pricing was undermined by global capitalist competition among Big Capitals (Shaikh, 2016). Baran and Sweezy's analysis was largely confined to the US, and prior to mature globalization. In any case, it has been seriously criticized by other Marxists. The inability of the consuming public to afford the surplus product in any isolated nation was perhaps only an earlier characteristic of US monopoly capitalism, though not its essential feature. A more relevant feature of contemporary global capitalism is that fierce competition has not been eliminated through monopoly; rather it has largely shifted from the province of small capitals within a nation to that of big capitals internationally.

Cox's solution to the surplus absorption problem focuses on individual consumption, instead of focusing on collectively-consumed products and on their repurposing. While Cox's egalitarian vision involves sharing wealth, that wealth, as noted, would be relatively short lived, since the material basis for reproducing wealth in a sustainable way is the plentiful energy that he proposes to largely banish. In other words, he proposes to overcome unequal distribution of social wealth through a progressive diminution of the wealth to be distributed.

Cox also tacitly assumes that the sole function of innovation in manufacturing technology (automation) is the amplification of profits. While this may be the intended purpose, and while enhanced automation may temporarily benefit the early innovators (or their employers), the technology soon becomes assimilated throughout the society and the rate of profit declines. Why? Because automation lowers the dependence of production on living labor, which, as Adam Smith, David Ricardo, and Karl Marx demonstrated, is ultimately the sole source of value and of profit. By assuming that technology is inseparable from profits, and thus from capitalism, Cox arrives at a parallel to his antinuclear argument. That is, advanced technologies, including nuclear reactors, are thought to be intrinsically capitalist, or intrinsically authoritarian, both amounting to the same thing. Thus, there is no possibility in Cox's world for repurposing energy sources, and who knows what else, that have arisen under capitalism.

Cox hopes to retain high wages while shortening the work week, both central goals of social justice, but to do so without reliable, plentiful net energy—something only nuclear can provide—to take the place of living labor. High wages emerge from, and can only be useful in the presence of, extensive surplus product, the very thing Cox aims to eliminate, because it depends on plentiful net energy. In short, without plentiful nuclear-powered net energy, a high-wage society with expanded leisure (shorter work week) cannot emerge, despite Cox's desire to see it happen (see Chapter 6).

*   *   *   *   *

As indicated by Michael Shellenberger (2020) and Leigh Phillips (2015), among others, much of the world needs more consumption, not less. And as they indicate, reducing labor productivity and perversely reducing the densities of both energy and power will lead to both less consumption and more work with more drudgery, thwarting the shortening of the work week. The poverty of Green philosophy here is at once material and epistemological, since social impoverishment and low consumption parade incoherently as higher wages and more leisure time.

Phillips clarifies aspects of this point with respect to Klein's writings, as we try to do here with respect to Cox's and Ajl's. Shellenberger repeatedly makes the same point to show that as serious as climate change may be, the low-energy solution is highly problematic and undesirable for labor. As an example, he opines that the sweatshop labor of a particular Indonesian factory worker, because of its embeddedness in a higher-EROI fossil-fueled economy, is far more preferable for both her and the planet compared to the agricultural labor rooted in a charcoal- and wood-fueled economy. Though Green ideology regards a lower-energy-density economy as closer to nature and therefore preferable, it is decidedly worse for both humanity and the environment.

In contrast, pro-capitalist ecomodernists, ignoring many amplifying and countering feedbacks of capitalism, argue that technological innovation requires competition as its source. They further argue that while capitalists seek what is to them cheap labor—but what is to workers drudge-laden, exhausting, and poorly paid labor—it facilitates an easy transition to factory work. The latter then produces the wealth to fund more education needed for the continued economic march toward imagined universal prosperity. Cheap labor becomes a step toward greater wealth leading to more investment, including more investment in education and higher wages. This ahistorically presupposes that cheap labor under capitalism is merely a steppingstone toward universal prosperity. Shellenberger, for example, argues that technological progress emerges without hindrance from free markets. And it is technological progress that leverages economic growth to reduce the human impact on nature and to create more job opportunities. While innovation-derived automation may reduce the number of jobs in one arena, it expands jobs in others, as though the expansion of job opportunities were automatically caused by higher unemployment (like pushing on a string). The contradictions of capitalism, as well as the role of chance, are thus erased.

In summary, while ecosocialist Cox wants what can only be had with the plentiful, clean, high-EROI energy that he rejects, ecomodernist Shellenberger, ignoring history, assumes that as long as there is plentiful high-EROI energy available, normally functioning markets will automatically meet human needs.

The striving for the cooperation necessary for global redistribution is undermined by insufficient energy. High energy and power densities enable economies, rather than dyseconomies, of scale, which in turn make possible coordination and cooperation, though guarantee neither. In other words, high energy and power densities are necessary but not sufficient for durable global coordination and cooperation—a goal that also requires an economic, political, and social transformation of all societies. For coordination and cooperation at large scales and in environments with today's large populations and high population densities, high energy and power densities that minimize materials and land are far more advantageous than low densities.

While economies of scale allow better coordination, dyseconomies undermine it. In particular, RE dyseconomies promote scarcity, increasing the tendency toward competition for survival, thereby rendering any attempts at egalitarianism in a highly populated world extremely fragile and in essence unsustainable. While scarcity could temporarily be compatible with cooperation and coordination, if the latter do not enable the abolition of scarcity, the arrangement will almost certainly break down in time. Economies of scale not only enable but are only derivable from cooperation and coordination. Just think of the cooperation and coordination logistics of far-flung (large scale) supply chains in a capitalist, or for that matter non-capitalist, world.[14]

---

[14] First, global coordination conflicts with Ajl's notions of sovereignty, even if some countries of the global South reject low-energy regimes. Second, a smart grid requires

The Ecosocialist Recruitment of Indigenous Authority and the Ecomodernist Refutation

Humane socialism requires scaled-up cooperation. And while cooperation is obviously possible under conditions of scarcity and privation, as exemplified by the beginnings of the twentieth century socialist enterprises in Russia and China, the goal of such cooperation was largely to end the privation and scarcity, not to sustain and merely share it equitably.

It may be objected that human beings lived under low-energy conditions for almost 300,000 years in relatively cooperative units. But these units were small, an option no longer available. Moreover, they had no other choice about energy. If we want a cooperative society functioning at local, regional, and global scales, we will need energy and lots of it. And it needs to be (relatively) clean, reliable, safe, and reproducible generation after generation. We can do this with nuclear energy, particularly Generations III and IV. We cannot do it without nuclear, barring unforeseen technologies. But such technologies would have to resemble or exceed nuclear in their energy and power densities and EROI. Wind/solar can't begin to satisfy these requirements.

Given the blind spots of the Green left's technical understanding, which in fact undermine their sociopolitical objectives, what remains is imagination and poetry. One form of this fantasizing involves recourse to indigenous peoples, as we saw with Ajl. This approach associates nuclear energy (for example uranium mining) with the oppression of indigenous peoples and rests on the unstated assumption that indigenous persons are generally intrinsically egalitarian. The former aims to provoke audience feelings of guilt, while the egalitarian assumption romanticizes historically marginalized groups. The approach parades as antiracist and prejudges defenders of nuclear energy as racist. However, the appeal to indigeneity is, in its own way, racist, essentializing groups that, in fact, are divided by class in the modern world. Essentializing any population stereotypes individual members, a central feature of all forms of racism, even if the stereotype is admiring rather than disparaging. Offering a podium to a member of an indigenous group

---

economies of scale, but its champions, like Aronoff et al., reject the energy and power density that enable the economies of scale required to build a smart grid in the first place. Meanwhile Cox (consistently, on this point) rejects smart grids as part of rejecting economies of scale, high energy density, and high power density. Instead he favors the local and the small, but he incoherently couples it with a proposal for massive global planning for redistribution, a proposal that we argue is undermined by low energy availability.

in the hopes that their status will induce audience acceptance of the message is a subtle form of ad hominem—in which the identity of the speaker takes precedence over what is spoken.

There is no question that different cultures have much to learn from each other, since different angles of approach to any subject virtually always contain germs of truth and good advice—a point we return to in Chapter 11.[15] But they also contain germs of untruth and bad advice that other cultures may have already recognized and rejected.

Using this uncritical appeal to indigenous cultures, Cox turns to the Indigenous Climate Network based in Canada to make his argument for "non-market solutions." In doing so, he seems to rely on an argument from authority, citing the Climate Network's claim that "[w]e have proven our peoples' expertise and knowledge in developing successful non-market solutions that surpass current carbon market mechanisms" (pp. 117–18).[16]

Leaving aside the reference to non-market solutions, if the solution is to be scaled and sustained we will need nuclear energy and cannot rely on the wisdom of Anishinaabe prophets to choose the Green path. Cox's principal reference in this regard (footnote 281) is to a speech by indigenous antinuclear activist Winona LaDuke given to the E.F. Schumacher Center in 2017 (see http://tinyurl.com/m7vvecbz for the video form and a briefer transcription at http://tinyurl.com/bde3hpfm).[17] Her speech calls for opposition to what she calls the Wiindigo economy (there

---

[15] As just one example of what the dominant US farming culture has to learn from indigenous peoples, Hopi dry farming is a method developed through trial and error over the last two millennia, in what is now northwest Arizona, to produce crops that require minimal amounts of water and hence are resistant to endemic drought. Necessity gave birth to invention (http://tinyurl.com/4w7838we). Unfortunately, much of the dominant farming culture is prevented from absorbing such lessons, due to the unstated assumption that its adherents have nothing to learn from indigenous peoples.

On the other side, many of those who celebrate and look to indigenous cultures for wisdom miss out on lessons as well. It turns out that these crops are, in fact, genetically modified organisms (GMOs). Yet GMOs are the victims of a blanket rejection by many of the very people who celebrate indigenous invention. This unwitting inconsistency in the anti-GMO movement is similar to that surrounding nuclear energy. Indeed, the false distinction between GMOs developed in the field (regarded as good) and those developed through scientific methods in the laboratory (regarded as bad) is analogous to the false distinction made by antinuclear guru Helen Caldicott between "natural" radiation (good) and that resulting from human effort (bad).

[16] Ironically, there have been recent agreements between indigenous peoples and the nuclear industry. See, for example, agreements between the Canadian nuclear industry, especially the SMR producers, and the First Nations of Canada. Such agreements will be good for the planet and will weaken the poetics of nuclear fear based on the false opposition of all things nuclear to all things indigenous (http://tinyurl.com/4nktj34m).

[17] The late German-British economist E.F. Schumacher's book *Small is Beautiful: Economics as if People Mattered*, first published in 1973 with multiple subsequent editions,

are various spellings). Wiindigo is a creature, very often though not necessarily taking human form, that preys on the living, sucks their blood or eats them materially and spiritually. This is LaDuke's indigenous term for capitalism, which she thereby associates with vampires and cannibalism. This is a reasonable image; Marx used similar metaphors to describe capitalism. However, her solution, following Schumacher, is a kind of smallholder, low-energy egalitarianism, with some modern features, like infrastructure, and of course an uncritical reliance on wind/solar power. To illustrate the Wiindigo economy, she makes a brief mention of radiation and the uranium mining in Grants, New Mexico, discussed above. She also approvingly refers to Elon Musk's declaration of intent to solarize Puerto Rico to overcome its dependence on mainland aid.

Related to her smallholder vision, she emphasizes the development of self-reliance, understood primarily as a world made by hand, rather than by, as she puts it, "fossil slaves," meaning people who are slaves to fossil fuels. This is an important part of a power-down world view that makes power-down sound egalitarian by referring to labor-saving energy and machines as enslavers. While under certain social conditions such machines are tools in the hands of enslavers, under other conditions (outside her purview) they could free people from back-breaking drudge-ridden labor to enjoy more leisure time. Yet at one point she praises the importance of efficiency. "Who wants to be inefficient?" she asks rhetorically, not realizing that her own mixture of assumptions guarantees inefficiency, and that her favoring of renewables directly undermines her desire to protect land from human abuse.

She also seems unaware of the Osage Nation's years-long legal struggle (recently successful) to drive Enel from their land in Oklahoma, where this originally European company was building, not a fossil fuel or nuclear plant, but a massive wind farm over the objections of the tribal members (a group brought to public attention through the Oscar-nominated movie *Killers of the Flower Moon*). Which indigenous groups, then, deserve the title "voice of authority"?

LaDuke is most incoherent in her discussion of mining, in opposition to which she has been actively involved. While promoting renewables, she (like Jacobson) ignores the fact that scaled-up renewables require a far greater extent of mining than nuclear. She

---

is one of the ideological signposts of the Green left, with its call for low-energy-dense, decentralized power, both in the energetic and political senses. Schumacher, incidentally, was a coal industry advocate opposed to nuclear (Visscher 2025).

largely reprises Cox's mixture of power-down assumptions, from the promotion of low energy density and self-reliance to making things by hand. She makes incoherent references to "industrial Green energy," touting Musk as the savior.

The left has largely abandoned the critique of power-down proposals to the libertarian right—particularly Bryce, who comments on the bowing to indigenous wisdom (2014, p. 55):

> The Green Left's romanticization of the past along with its continual claim that renewable energy and organic agriculture are the only way forward ignores the deprivation, lack of social, intellectual and economic mobility and short life spans that dominated pre-industrial societies.

Bryce cites two anthropologists who determined through an analysis of 12,500 Native American skeletons dating from the pre-Columbian era that "in the healthiest cultures in the 1,000 years before Columbus, a life span of no more than 35 years might be usual" with few surviving past the age of 50 (p. 59).

Shellenberger points out that much Green ideology is in thrall to the back-to-nature fallacy, including the poetics of indigenism and the preference for metaphor and association over analysis and explanatory power. The nature-worshipping embrace of wind/solar is fine if electricity is not your goal. If it is, then the neglect of industrial-scale conversion apparatus is at best careless. While Greens point out that nuclear power plants are unnatural, they miss that millions of PV panels and 60-story wind turbines are not to be discovered hidden in the forest.

Cox's and LaDuke's admiring attitude extends to locavorism—the favoring of locally grown food—on the grounds that remotely grown food requires long-distance transportation, leading, they claim, to greater environmental impact. The degrowth tradition favors small, local, and low energy, and romanticizes the smallholder as skilled, smart, and democratic. In response to this kind of argument, US journalist Will Boisvert, in his article "An Environmentalist on the Lie of Locavorism" (http://tinyurl.com/5d79dyf3),[18] shows that the economies of scale afforded by the high urban density lauded by locavores require high-volume, long-distance transport by trucks and freight trains. And again, only if we have high-EROI, energy-dense, clean sources for electricity generation and for the manufacture of clean synthetic fuels can we take

---

[18] The "Lie" in Boisvert's title might better be termed "Fallacy," since most locavorists have likely not thought this through and are innocent of lying.

advantage of such economies of scale, the very hallmark of industrial society, including the much-maligned industrial agriculture.

> Boisvert summarizes the locavorist world view almost satirically: by minimizing the food-miles from field to fork, urban farming reduces fuel consumption, shrinks the city's carbon footprint, cuts out corporate middlemen and composts sinful food waste into life-giving fertilizer. But it's a cure-all for just about every other discontent as well, according to UDL [Columbia University's Urban Design Laboratory], a "synergistic solution" that alleviates food insecurity, fights diabetes, rehabilitates vacant lots, provides job skills, sparks economic development, and promotes "community empowerment," "environmental justice" and "social interaction." It will even soothe our neuroses, letting New Yorkers "reconnect with a food system that many feel is somehow out of their grasp."

He shows that the fuel to move a ton of food is an order of magnitude less per mile for long distances than for short laps within the city. Smaller numbers of large trucks can carry huge amounts of food long distances, while it takes many more small trucks to carry the same amount of food a few urban miles—trucks driven by many more drivers who require still more fuel for the trips between home and work, more than the food itself requires. And farming on large parcels of cropland is far more efficient than on the numerous smaller urban gardens.

Long-distance trucking and rail hauling, including bullet trains, can take advantage of even greater economies of scale in a nuclear world. Boisvert explains:

> That's because the high-volume, long-haul food transportation perfected by industrial agriculture is fantastically more energy-efficient than the low-volume, short-haul shipments of locavore distribution systems.

US food and agriculture expert Robert Paarlberg (2021) offers a similar critique of the limits of locavorism (p. 100):

> One calculation found that UK consumers who drive six miles in a personal car to buy green beans flown in from Africa will actually emit more carbon dioxide per bean than the aircraft bringing them in 4,200 miles from Nairobi.[19]

---

[19] Paarlberg also shows that food shipped over long-distances enjoys the added safety of refrigeration.

And while Boisvert and Paarlberg both speak of fossil-fueled vehicles, these insights would be even more striking for synthetic fuels or hydrogen fuel cells, efficiently produced in a high-EROI economy with economies of scale.

Before the present authors began to study energy a decade and a half ago, we largely fell into this localist misconception. While we never conflated bigness and economies of scale with capitalism, our preference for the local, with its assumed greater fuel economy, and our downgrading of industrial-scale farming, thought to devour fossil fuels, seemed to be necessitated by the imminent prospect of peak oil (minus the layered-on spiritual additives). This concern ballooned into the assumption of peak everything, including peak minerals. Because industrial society, the argument went, was powered by, and was inseparable from, fossil fuels, degrowth appeared again to be a necessity. The degrowth philosophy was then transferred to the climate crisis, and the focus turned to changes in social relations designed to make low consumption feel like home.

US journalist Michael Pollan (2006) argued for bringing big agriculture back to local and small scale. His book featured Polyface Farms in rural Virginia, which raises grass to feed chickens and pigs in a sustainable way and eliminates cutthroat competition and capital accumulation. Pollan aimed to share best practices in a cooperative way, and to encourage emulation by local, necessarily small, farmers. The model featured the nutritional benefits of eggs and meat from more humanely treated free-range chickens—humaneness amplified by, and inseparable from, the smaller scale and the rejection of capitalism. This rejection followed Proudhon's smallholdings rather than Marx, evoking the yeoman farmer so dear to American mythology. But it was underwritten by the necessity of powering down and going local in response, first, to peak oil and second, to climate change. Peak oil seemed to put paid to big capital and Polyface was offered as a small scale, sustainable, and egalitarian alternative. Paarlberg, however, reveals that Polyface is necessarily reliant on relatively affluent customers who can afford the high prices for the eggs and chickens, reprising Kovel and the Amish. In other words, the low-energy, less efficient Polyface is parasitic on the high-energy matrix that provides the affluence of its customers.

Our study of renewables and nuclear revealed to us the erroneous attraction to smallness when taken out of context. Powering down turns out to be unnecessary, not merely because we had the timescale of peaking wrong, but because nuclear doesn't face the constraints and destructive externalities of fossil fuels. The seductive low-energy vision, weaving misaimed metaphor with poetry and divorced from science, needs to be soundly rejected. The more so since population growth without nuclear energy promises a Malthusian future in which perhaps hundreds of millions will suffer early deaths (more about Malthus below, also see Glossary).

To sum up, high energy density (and low to zero emissions) are better for our environment (which includes each other) than low energy density. All things being equal, remediation and preservation of the environment cannot be served with low energy density, as it spells high environmental impact and stymies the energy-requiring remediation. This is irrespective of the social relations surrounding production.

Given high energy and power densities, there is reason to believe that a cooperative economy rooted in managed, rational growth—or, where justified, a steady state—will prove itself more workable than a capitalist economy characterized by repeated crises and a growth imperative, both deriving from fragmented competition-driven accumulation (Chapter 10).

*   *   *   *   *

This book has argued for an energy source with a number of indispensable characteristics: **clean, carbon-free, sustainable, low environmental impact, plentiful, high EROI** and **relatively safe**. Today the only energy source with all these is **nuclear fission**. We also indicate in the following chapter why we think that the cessation and reversal of AGW may well require a different social system, one based on a **cooperative** global economy. In this section, we describe in more detail the critique of ecosocialism by ecomodernists Shellenberger, Bryce, and US energy experts Huber and Mills (2006).

A key ecosocialist contention is that a system whose inherent mechanisms force it to grow despite anyone's desire or intent cannot be sustainable, as it eventually collides with finite resources (peaking). And efforts to improve efficiency in end-use consumer applications are often negated by Jevons's paradox. Additionally, efficiency improvements in general encounter obstacles in capitalism's fragmentation and the resulting competitive imperative. While competition fosters many efficiency gains, it also prohibits the scale of cooperation necessary for timely sharing of the most efficient technologies and often functionally, or even legally, prohibits their development. One of the most glaring examples is interference with the further development of nuclear energy technology by fossil fuel interests.

Shellenberger [20] (2020) describes big-fossil funding of the Green movement and synergies between big fossils and big renewables. He notes (p. 219):

> During the last decade, Stanford's Mark Z. Jacobson eclipsed Amory Lovins, EDF [Environmental Defense Fund], NRDC [Natural Resources Defense Council] and all other groups in supposedly proving that renewables alone can power the planet and that nuclear energy isn't needed to replace fossil fuels. When I debated him . . . in 2016, Jacobson claimed it would be cheaper to shut down existing nuclear plants, including Diablo Canyon, and replace them with renewables. He did so as a senior fellow at the Precourt Institute for Energy, which was named after an oil and gas magnate and board member of Halliburton. . . . Precourt's board is comprised of [*sic*] leading oil, gas and renewables investors. It is hard to imagine a more direct conflict of interest.

While ecomodernists don't believe that capitalism's competition and growth imperative will undermine sustainability, the greater efficiency that they defend is preferable to the low energy advocated by ecosocialists. Their pro-capitalist, pro-plentiful energy, pronuclear path would necessarily prevail, because low energy is deadly. The termination of rational (rather than imperative) energy expansion in the face of population growth would make humanity poorer, even if access were equally distributed.

However, capitalism's competition-driven accumulation imperative propels unlimited economic growth on a finite planet. For example, with 3 percent annual compound growth, the global economy would double every 24 years.[21] Ecosocialists commonsensically argue that a planet with finite resources cannot be subject to a growth imperative without running into shortages at some point. In Richard Powers's Pulitzer Prize-winning novel, *The Overstory* (2018, p. 321), one of the radical defenders of old-growth forests, Maidenhair, asks the logger about to cut down the majestic tree that Maidenhair is defending, "Do you believe human beings are using resources faster than the world can replace

---

[20] While writing very useful pages in his more liberal days, Shellenberger has more recently joined the extreme rightwing, pampering Trump's big lie about the 2020 election and defending deliberate disinformation as a fundamental right that is being infringed by Big Tech autocrats. Since the recent presidential election, those autocrats have prohibited fact-checking of Trump's fabrications and perversely justified this prohibition as a defense of free speech.

[21] Using the handy rule of 72: 72/3 percent => double in 24 years, 72/4 percent => 18 years, and so on.

them?" She is pleased with the affirmative answer and points out that the rate of use is rising, not falling. She sums it up, "It's so simple. So obvious. Exponential growth inside a finite system leads to collapse. But people don't see it." Farther down the same page, we are told, "We'll soon be eating two-thirds of the planet's net productivity. Demand for wood has tripled in our lifetime."

This idea is expressed by US Marxist ecosocialists John Bellamy Foster and Brett Clark (2020, p. 245, note their focus on energy consumption rather than production):

> Technological change under the present system routinely brings about relative efficiency gains in energy use, reducing the energy and raw material input per unit of output. Yet this seldom results in absolute decreases in environmental throughput at the aggregate level; rather the tendency is toward the ever-greater aggregate use of energy and materials. This is captured by the well-known Jevons Paradox, named after the nineteenth-century economist William Stanley Jevons who pointed out that gains in energy efficiency almost invariably increase the absolute amount of energy used, since such efficiency feeds economic expansion.

For these ecosocialist authors (who reject nuclear), such a society cannot be sustained, since "ecological analysis points irrefutably to the fact that we are up against the earth's limits." They go on:

> Not only is continued exponential economic growth no longer possible for any length of time, it is also necessary to reduce the ecological footprint of the world economy. And since there is no such thing as an absolute decoupling of the economic growth from the throughput of energy and materials taken as a whole, this means that exponential economic growth, particularly in the rich countries, must cease while economics must be aimed at the rational satisfaction of human needs. As an immediate objective, the world economy must wean itself entirely from fossil fuels as an energy source—well before the one trillion metric tons of carbon is emitted into the atmosphere.

The "one trillion metric tons" to which they refer is ostensibly the amount that would raise the global temperature by 2°C, which is commonly taken to be the threshold for irreversible climate disaster (though there is no single threshold, given numerous irreversible tipping points).

An alternative, optimistic view by ecomodernists is the confidence that human ingenuity, driven to innovate by the profit motive, can overcome any limits imposed by nature. That is, the capitalist growth imperative fosters the impetus to either find new reserves, seek substitute materials, or invent more efficient technologies and thereby conserve known resources.

Huber and Mills sum up what happens when efficiencies are gained (p. 121):

> Rising efficiency has led to more energy consumption, not less. From wood to coal to oil to uranium, new fuels packing more energy into less space could be distributed more quickly and conveniently, and burned faster, and so they were. Altogether new forms of demand materialized around the new fuels. Coal extracted initially to replace wood for heating found much larger markets in steam engines and then electric power plants. Oil initially used to fuel cars found other large markets in ships, jets and chemical industries. Electricity initially supplied to power Edison's new bulb was soon tapped to power electrical motors, then compressors in refrigerators, then air conditioners, and then microprocessors.

Innovations "multiply in new niches," and make possible a "limitless range of new interventions." Again (p. 117):

> LEDs are already displacing incandescent bulbs in cars, inside the passenger cabin and in the brake lights, to begin with – and certainly saving energy as they do. But . . .the new light bulbs . . . won't just emit light . . . they already project millimeter waves for active cruise control or sense infrared light for enhanced nighttime vision . . . The new solid-state light bulb doesn't just supply a more frugal desk lamp, it does laser surgery and makes it affordable for the masses.

These examples of technological advance under capitalism are part of Huber and Mills's breathless encomiums to capitalist progress. While we differ with their faith in capitalist markets (Chapter 10), we appreciate their insights about energy efficiency and its implications. Marxists and other ecosocialists need to capture and assimilate, rather than reject, these grains of truth.

For many ecosocialists, the Jevons paradox of efficiency is viewed as capitalism's Achilles heel, preventing conservation of resources. In contrast, pro-capitalist ecomodernists regard the paradox as a boon rather than a drawback, because it encourages an outpouring of human ingenuity, which, they believe, always responds to pressures on resources. Huber and Mills believe this ingenuity turns our planet into "The Bottomless Well" (their title), such that "We Will Never Run Out of Energy" (their subtitle).[22]

In effect, the ecomodernist response to Jevons's paradox is, "So what?" Humans will transition to cleaner energy and will, they assume, find ways to mitigate the effects on climate. Ecomodernists see free markets (transparent and fairly regulated) as necessary for the proper investment climate in which nuclear's advantages will win out, leading to its rapid deployment, especially Generation IV. However, they neglect or underestimate the effect of stranded capital in preserving and reinforcing the established fossil competition.[23]

Pro-capitalist, pro-market ecomodernists recognize something that escapes the attention of ecosocialists, namely, that technological innovations—foremost among them, the efficiencies offered by advanced nuclear energy—would not only expand employment, but would expand interesting and meaningful work, and, under different conditions, leisure. Ironically, this has always been a major goal of socialists and communists.

Huber and Mills argue that if the "technologies of power were now at a standstill, the future of employment in the U.S. would be bleak." After noting that in modern industrializing society, "cataracts of power are extracted from coal, oil, uranium and natural gas,"[24] they assert that "the historical correlation between rising employment and rising consumption of these now conventional fuels is simply too strong to deny." (See their pp. 126–27, including the EIA graph of the relationship.)

They then comment (p. 127):

Power is one of the three fundamental inputs that determine the productivity of labor in every sector of the economy; the other two

---

[22] If the "well" is, for all practical purposes, "bottomless," it's because human ingenuity enjoys the earth's "bottomless" fund of uranium and thorium, given the finite life expectancy of the solar system.

[23] See for example, Hargraves (2012) and his associate Devanney (2023). Devanney believes that when capitalism works the way it is supposed to, nuclear in fair competition with fossils (and renewables) will bring lower costs and higher wages. Marxists like ourselves recognize that competition leads inexorably to increased concentration and centralization of capital. Both barriers to entry and resistance to such barriers are intrinsic properties of capital, such that markets often don't yield the best outcome for consumers. Devanney calls for the government to intervene in nuclear's favor by imposing strict rules of competition, without selective subsidies, which, as we have said, would destroy the fossil and renewables industries. In particular, he calls on the government to end its reliance on LNT and ALARA, which the NRC currently uses to boost nuclear's costs. See also Conca's 18-minute talk at https://tinyurl.com/27hkj2k8. Our attitude toward competition aside, we have learned a great deal from both Hargraves and Devanney about nuclear energy and its barriers.

[24] Throwing uranium in with that set of oxygen-requiring fossils is like saying that a trip from Sacramento to Savannah is achievable by walking, bicycle, or pogo stick . . . oh, or jet.

are material and information. Capital constructs intelligent configurations of concrete, steel and silicon; power makes these structures run. Power moves the worker to the increasingly distant workplace, drives the intelligent machines that surround him [*sic*] there and processes the materials and information that he works with.

The strength of this argument is that we need power, more not less, not to fuel a growth imperative but to meet humanity's needs. A weakness, which we consider below, is that for Huber and Mills, it is capital that runs the world. Labor follows. For us, it's labor that constructs these "intelligent configurations" of materials and capital that expropriates the wealth embodied in and enabled by these configurations. In a classic example of capitalist ideology, Huber and Mills perceive capital to be the constructive agent, rather than the labor that creates that capital.

Growth itself is not the problem; rather the problem is the mindless and irrational growth imperative connected to a system in which each individually owned parcel of capital must expand or succumb to the competition. This then results in the expansion of the economy as a whole. Similarly, automation is not the problem; rather the problem is its employment under capitalism, where it is used primarily to expand individually owned parcels of capital. It does so in part by shrinking the portion of the labor force controlled by that capital, though not necessarily the labor force as a whole. In other words, automation is necessary to increase the efficiency of labor exploitation and diminish the owners' cost relative to that of other capitals. That it may also diminish the price of the product to the consumer is a byproduct of the interaction between supply and demand, which is out of the hands of the individual owners (unless they engage in price collusion, which risks disciplining by the legal system when it clashes with the interests of other powerful capitals).

As Sci Fi author Kim Stanley Robinson (about whom more below) has noted, not all efficiency is good efficiency. Or to put it more precisely, improvements in efficiency may be of a type that is good for some, but not necessarily for everyone, and in some cases not even for most.

Ecosocialists Fear the Growth Imperative Necessarily Leads to Peaking of Materials

Two comments before we continue: one regarding the weakness of the ecosocialist position and the second that of the ecomodernists. In opposition to our assertion—supported by nuclear scientists and engineers as well as by ecomodernist energy analysts—that the earth contains plentiful energy that will last for the lifetime of the sun and planet, ecosocialists assume there are inescapable "limits to growth." Their fear of material peaking suffers from tunnel vision, since the all-RE program threatens to exhaust the Earth's material reserves—possibly sooner than fossil fuels.[25]

The ecomodernists' grasp of energy is not the problem; rather, it's their political/economic adherence to capitalism. They attribute indisputable historical innovation, accelerating productivity, and amplifying prosperity to the competitive quest for profits. Their position maps presumed past onto imagined future, denying that a cooperative system based on human needs could also motivate continual and accelerating innovation.

In Chapter 10 we explore the ways that, in addition to the incentive to innovate, the anarchy of capitalist production at the same time vitiates the spread of those practices needed to meet the accelerating cataclysm of AGW.

Another example of ecosocialist reasoning comes from Chinese-American political economist Minqi Li (2008), especially his chapter titled "The End of the Endless Accumulation."[26] Li opines about the three main energy sources in turn. Taking nuclear first and citing a 2006 study by the US Geological Survey (USGS), Li says that to fuel the current number of reactors in the world (about 440), known uranium reserves contain only 30 years' worth, potentially 70 years if new reserves are discovered.[27] Either way, the world, he says, is soon to exhaust its supply of uranium.

Li thinks supplies are better for renewable energy, but not by much. First, the emptying of mineral reserves accelerates as renew-

---

[25] The "limits to growth" position tends to regard both economic and population growth as a form of disease, like cancer. But not all growth is predatory, or uncontrolled, or usefully associated with disease. After all, plants and animals are mainly healthy as they grow.

Human ingenuity and high productivity can restore living nature and don't necessarily deplete it, even if dead organisms (fossils) and mineral resources are fixed and finite. The exhaustion of fixed mineral resources can be greatly delayed if not evaded through technological improvements in mining and refining, coupled with recycling and reuse, as well as with the elimination of the growth imperative. Fossil fuels, in contrast, cannot be recycled and will eventually be exhausted, even if the timescale is only vaguely knowable. In that sense there is indeed a limit to growth, but one that can be avoided through nuclear energy.

[26] The publishers (including Foster) also put out the socialist magazine *Monthly Review* and generally censor pronuclear articles. Like the editor of *CounterPunch* (Chapter 7), Foster refused to reply to an inquiring email from us about their antinuclear practices some years ago.

[27] As though such a number could even be guessed at. One of the chief characteristics of the unknown is that it is unknown—a point the present authors forgot when we thought the world had reached peak oil some years ago.

ables penetrate more deeply. He says, based on a 2006 USGS study (pp. 164–6):

At current rates of production, all the probably recoverable resources for 14 out of the 32 basic metals would be exhausted in less than a hundred years. If the world's resources consumption keeps growing at 2 percent a year, then all the probably recoverable resources for 25 basic metals would be exhausted in less than a hundred years and all the potentially recoverable resources for 17 basic metals would be exhausted in less than 150 years.

Li's 2 percent expansion rate is independent of Jacobson's proposed rapid build rates before 2050, and of the bootstrapping period following 2050 (our Chapter 5). Nor does Li's estimate account for accelerating wind/solar penetration, the required oversizing/overbuilding, batteries to stabilize the grid, EV batteries, amplified electrification, or expanded applications. Were Li's early USGS estimates to prove correct (there is good reason to doubt them), updating these estimates would render Jacobson-type plans absurd.

Finally, Li was not alone in expressing relative certainty in 2008 about the imminent arrival of peak oil. Those who, around that time, predicted the imminent exhaustion of petroleum argued that oil production had peaked at around 85 million barrels per day (MBD) and that tar sands, oil shale, undersea deposits, etc., requiring more energy-intensive extraction and refining methods, would be unable to profitably compensate for the decline of light sweet crude. However, oil production has blown past 85 MBD (Chapter 10).

This by no means falsifies the peak argument but should give us pause in predicting timescales. The concept of reserves is ambiguous: it is generally taken to refer to deposits that are already known or guessed at, and only those that are profitable given current extraction techniques. Technological innovation leads to discovery of additional reserves, while those previously thought unobtainable can become recoverable. Moreover, substitutes can often be found and the day of reckoning postponed—especially when reuse and recycling are also available.

USGS's updated report in 2013 still estimated that profitably extractable uranium would last the same 30 years they had estimated 7 years earlier. But at least the two figures here are in the same ballpark. Further revealing the fragility of timescale predictions are advances in uranium extraction from seawater (by the US DOE) and advances in reactor design (by private entities) that use thorium (thought to be three to four times as abundant on Earth as uranium). According to Conca, reporting in 2016 on DOE research (https://tinyurl.com/3t52yueb):

> Nuclear fuel made with uranium extracted from seawater makes nuclear power completely renewable. It's not just that the 4 billion tons of uranium in seawater now would fuel a thousand 1,000-MW nuclear power plants for a 100,000 years. It's that uranium extracted from seawater is replenished continuously, so nuclear becomes as endless as solar, hydro and wind.

And in 2018 DOE's Pacific Northwest National Laboratory (PNNL) estimated that the oceans currently contain 500 times as much uranium as there is on land, with a far lower ocean-to-land ratio for thorium (https://tinyurl.com/3jrwkesj).[28]

While the cost of seawater extraction of uranium circa 2016 was twice the cost of on-land extraction, the cost is coming down. Indeed, a March 2020 article in *Applied Materials Today*, just four years later, reports on technological advances with adsorbent materials (https://tinyurl.com/5esh9z2e). Conca had said in 2016, "marine testing shows that these new fibers had the capacity to hold 6 grams of uranium per kilogram of adsorbent in only about 50 days in natural seawater." The 2020 article reports that with a new adsorbent, "[t]he used PVDPA-modified fibers showed a uranium uptake capacity of 392 mg/g and reached the adsorption equilibrium in less than 3h." To express these in the same units, in 2016, the prior adsorbent could extract 6 gm/kg while taking 50 days, whereas in 2020 the new adsorbent could extract 392 gm/kg while taking only 3 hours. This represents a gain of over 65 times as much uranium in only 1/400 the amount of time, or an improvement by a factor of more than 26,000. And this extraction technology is in its early development.

For all practical purposes, there is no foreseeable peaking of nuclear fuel, any more than there is foreseeable peaking of moving air or sunlight. All preferences for wind/solar over nuclear, on the basis of renewability and implied inexhaustibility, are without foundation. And as we explained in the Sdebar headed "What's Meant by 'Renewable'?" in Chapter 2, nuclear fuel can be renewed in both the senses of reinfusing and refreshing. In addition, there have been impressive advances in the fabrication of uranium/tho-

---

[28] Thorium, in contrast, is relatively insoluble in water, so there is less in the oceans than in the ground; some beaches, however, have high concentrations of thorium in the sand, as exemplified by Guarapari Beach in Brazil (See Chapter 8).

rium mixed fuels and in reactor designs that can utilize the combination (https://tinyurl.com/2rmdwm6x and https://tinyurl .com/
4rv46m73).

Australian socialist Simon Butler, in his 2021 "rebuttal" on the
ecosocialist website *Climate and Capitalism* (https://tinyurl.com/
2h32ev3w) of US socialist Bhaskar Sunkara's [29] earlier pronuclear
article in *The Guardian* (https://tinyurl.com/2p9hhth9), claimed
that if 70 percent of global energy production were to be provided
by nuclear, there would be only 6 years of uranium left on Earth.
Butler embedded this evidence-free assertion in nine other equally
invalid antinuclear arguments, virtually all of which we have
addressed in this book.

We could ask those who are worried about the growth imperative's tendency to speed up the arrival of whatever peaks await,
"Do you oppose mining in general in order to avoid the exhaustion of any minerals?" Many Green advocates might answer in the
affirmative, such as the makers of the documentary film *Planet of
the Humans,* Jeff Gibbs and Ozzie Zehner, as well as Cox and Ajl.
But they should realize that this would be tantamount to opposing
the development of all those industries and technological developments that depend on mined resources. So, if Gibbs and Zehner
were to decide to amend their answer to say that they are not
opposed in principle to all mining, our next question would be,
"Would you want to reduce the impact of mining?"

Recall that Cox is quite aware of, and opposes, the mining burden of an all-RE economy. In Chapter 4, we discussed the problem
of lithium mining in an all-RE economy but lithium is just one of
many important minerals whose supply would almost surely continually be under stress, either for physical or geopolitical reasons
or both, in a scaled-up all-RE economy. More generally, "the mineral requirements among the metals alone" include "lithium,
cobalt, silver, copper, aluminum, nickel, iron and the exotic rare
earth elements cerium, lanthanum, dysprosium, neodymium and
praseodynum [*sic*]" (p. 68).

One study Cox cites concludes that an all-RE (capitalist) world
(p. 68):

> would probably run out of cobalt and lithium well before 2050.
> Even with recycling of metals, the quantity of cobalt for battery
> manufacture would equal 120 to 210 percent of all known

---

[29] Sunkara is the founding editor of the magazine *Jacobin*, a columnist for *The
Guardian*, and that rarity, a pronuclear socialist.

reserves on Earth, while 90 to 160 percent of all lithium reserves would be required.

If Gibbs and Zehner were to answer our second question affirmatively, then they should be vehemently pronuclear, which they are decidedly not. Low-energy socialism is at least as inimical to the satisfaction of human needs as growth-imperative capitalism. But the former can't possibly sustain itself and thereby defeats its own progressive egalitarian social goals, while the latter has sustained itself for a few centuries, albeit with people's needs subordinated to the profit imperative.

The two biggest takeaways when looking at these two antithetical narratives (ecosocialism versus ecomodernism) would be, first, an answer to the question, which narrative, if those were our only two options, could better meet human needs while preserving the planet's resources? And, second, if we are looking only at peak arguments, it really comes down to the relationship between the technological efficiencies that a pronuclear capitalism can capture in the face of a growth imperative versus the very serious technological inefficiencies of an antinuclear commitment to all-RE in a hypothetical low-energy socialism. Granted, low-energy socialism would escape the pressures of a growth imperative, but it is purely hypothetical as it is unsustainable and would be unacceptable to virtually everyone, regardless of class.

As we have seen, the resource intensity of an all-RE energy system is more than an order of magnitude greater than that of an all-nuclear energy system, and future improvements also look to be on the nuclear side. In addition, the avoidance of peaking favors the pronuclear supporters of capitalism insofar as many important innovations generally require plentiful energy.

But the biggest takeaway is the first, namely, that between ecosocialism and ecomodernism, one is vastly more plausible—the pro-capitalist pronuclear position, with its capacity for plentiful energy. Because, if our thesis is correct, low-energy socialism cannot be reliably reproduced, and as we have pointed out, while it may be a starting point it can never be an endpoint or social goal. A low-energy society, one giving up on reliability of energy supply, could not even approximate a socialist society, except in the fiery imaginations of low-energy socialists, and would likely require a relatively rapid and massive population reduction—a Malthusian nightmare that no socialist would sanction.

The fear that capitalism's growth imperative will soon exhaust the Earth's fixed resources, while perhaps seductive philosophically, looks to be wildly wrong empirically. And yet we see both Foster/Clark and Li making exhortations that mindless self-propelled growth cannot be allowed to continue. There's still plenty of room for growth, and if there were not, it would still be the most paradoxical form of magical wishful thinking to believe that human needs could be met through material-intensive, low-EROI wind/solar.

So, if the consequences of the growth imperative are to be evaded, it can only be by following neither the ecosocialist nor the ecomodernist solution, but rather by following the few-and-far-between pronuclear socialists. That said, the growth-imperative argument is subordinate to the broader argument about the intrinsic contradictions of capitalism, since it is competition-driven profit maximization that fuels the growth imperative. Therefore, to base an anti-capitalist argument on the critique of the growth imperative will be less persuasive than basing it on an analysis of the overriding contradictions that, in turn, entail the exploitation of the many by the few. Instead of, or at least in addition to, focusing on capitalism's uncontrolled growth, Marxists (and everyone else) should focus on the barriers thrown up by a fragmented competitive profit system to the large-scale coordination needed to institute the cleanest, most efficient, and safest technologies throughout the world—energy-producing and otherwise.

Foster and Clark (2020, p. 247) quote fellow US Marxists Harry Magdoff and Paul Sweezy, who cogently make the argument about the limits to coordination and cooperation under capitalism (quotation marks in the original):

"Suppose we forget about trying to control growth and instead focus on abating the effects of growth by reducing pollution and arranging for a more rational use of raw materials and energy. Such an approach would entail a high degree of social planning: nothing less than a wholesale redirection of the economy involving among other things changes in population distribution, methods of transportation and plant locations—none of which can be subject to real social planning without violating the rights of private property in land, factories, stocks and bonds, etc.

"From whichever side the problem is approached—controlling growth or restructuring existing production . . . , we come up against antagonisms and conflicts of interest that capitalists and those charged with protecting capitalist society cannot, in the very nature of the case, face up to. In the final analysis, what stands in the way of any effective action is the contradiction between the social potential of present-day technology and the anti-social results of private ownership of the means of production."

Much has changed since 1974 when Magdoff and Sweezy wrote this, but the underlying antagonisms that render the capitalist mode of production inimical to the cooperation required to meet the challenges of climate change have not. This is the strength of the anti-capitalist argument, and it's routinely employed in discussions of imperialist rivalry over resources, leading to resource wars driven by capitalist competition.

Cox (p. 69) quotes a 2014 statement by Don Fitz, one of Cox's fellow editors of the socialist journal *Green Social Thought*, about wars over renewable energy resources (nuclear excluded):

"Would the Green World Order mean that Venezuela might have less reason to fear an invasion aimed at gaining access to its heavy oils? Or, would it mean an additional invasion of Bolivia to grab its lithium for green batteries? Would northern Africa no longer need to fear attacks to secure Libyan oil? Or, would new green armies to secure solar collectors for European energy be added to existing armies? Across the globe, those marching with the red, white, and blue banner of the War for Oil would continue to invade. But they could be joined with those marching with a green banner."

Cox also quotes the Winnipeg-based International Institute for Sustainable Development (pp. 69–70):

"the extraction and trade of mineral resources can fuel grievances, tensions and conflict, particularly when they happen in a context defined by weak governance, multidimensional poverty, human rights violations and youth unemployment."

These points about geopolitical, interimperialist rivalry are real insights but not on the radar screen of pro-capitalist ecomodernists like Bryce (2010, 2020) and Huber/Mills (2006). For them, competition leads intrinsically and inevitably to innovation. Shellenberger also treats the capitalist market as intrinsically innovative and effectively exonerates it from obstructing nuclear expansion. Instead he locates the barriers to the spread of nuclear energy in the political realm, beyond the market. He is correct that external forces do indeed play a role in this obstruction, but they do so mainly through the market, supplemented by antinuclear lobbying and propaganda. To imagine that the market could be left to operate without outside interference is as will-o'-the-wispish as the attainment and sustainability of an all-RE system.

In contrast, there are those who appreciate that competition in its various forms has as easily blocked innovations as it has encouraged them, depending on context (Ridley 2020). And to the extent that competition does produce an important innovation in either energy or pharmaceuticals or virtually anything else, the proprietary constraints (patents, copyrights, proprietary information), essential to encouraging innovation in the context of competition and profit, block its rapid spread. In other words, competition forces securing profits against the interests of competitors rather than motivating mutual help. This impediment, in turn, hinders, without entirely preventing, further innovation, as it impedes learning from the mistakes and successes of others.[30] These barriers to cooperation and the sharing of the benefits of innovation apply to any technology, including nuclear.

Competitive barriers to cooperation also include opposition by vast capitals heavily invested in rival fossils and renewables. Furthermore, once commercialized in a regulatory environment favorable to the technology, production and diffusion would be dominated by the most powerful capitalist countries. For example, ANEEL (Advanced Nuclear Energy for Enriched Life) fuel, as with all things nuclear, would be fabricated and controlled by imperialist powers in their own interests. Conca (https://tinyurl.com/2rmdwm6x), for example, recommends the use of the fuel for energy-poor countries side by side with his advertising the geopolitical advantages of developing the fuel in the US to restore America's "competitive nuclear energy advantage." That is, he sells it as a way for US capital to retain control of the fuel cycle, including price and distribution.

---

[30] In fact, a cogent example was the 1979 Three Mile Island accident, the culmination of a series of events that had earlier been experienced at the Davis-Besse plant in Oak Harbor, Ohio, near Lake Erie. There the engineers had found the problem quickly and prevented a meltdown, so that almost no one ever heard about the incident. Had the engineers been allowed to share information immediately throughout the country (and the world), the TMI meltdown would also have been preventable. The NRC had no plan to widely purvey information about such experiences, virtually nullifying a learning curve. While in the immediate sense, the TMI accident hurt no one other than the power company's investors, it had a decidedly negative effect on the general acceptance and expansion of nuclear energy, which hurt us all. Indeed, to this day, 45 years later, TMI's ghost is still being exploited to oppose nuclear.

278   *Other Green New Deals: Ecosocialism versus Ecomodernism*

Because their insights are undermined by their antinuclear commitments, the ecosocialists have inadvertently trapped themselves and their followers. In a capitalist world shaped by international competition, an all-RE world could not come into being because economies and societies rooted in higher-EROI energy technologies would outcompete those rooted in lower-EROI technology. But the problem is more severe. We have demonstrated that wind/solar can only operate relatively efficiently and (in capitalist terms) profitably at low penetration, and only with government subsidies. Without such a matrix, they cannot even sustain themselves, for technical reasons and regardless of the social/economic system.

Default Malthusianism in the Ecosocialist Position

Ecosocialists for the most part strenuously keep their distance from both Malthusianism and Luddism, positions with which certain members of the Green left nevertheless flirt, like US journalist and energy specialist Richard Heinberg. Both Ajl and Cox, however, reject any argument that their low-energy commitments bring Malthusianism into the picture. But the ecosocialist position inevitably invites its nemesis.

Malthus's original argument in the late 1700s was based on the premise that population grows geometrically (exponentially) while the food supply could only grow arithmetically, and people would soon outstrip food. Therefore, continued Malthus, those who could not feed themselves should be left to starve. Generalizing his argument to incorporate all resources was a position based, as Marx and others recognized, on the false assumption that the productive forces would remain stagnant. Malthus's critics respond that technological innovation has allowed resources to win the race with population size.

Ecosocialists like Ajl and Cox avoid focusing on population curtailment. Their focus is instead on curtailing the growth imperative under capitalism. These appear to them to be very different. Yet once you commit to the low-energy road, which does in fact involve the willful curtailment, even destruction, of the productive forces, the degrowth position, while nominally different from an anti-population position, is nevertheless Malthusianism by default.

Malthus harbors pessimistic assumptions about the human capacity for innovation and creativity that could solve the resource shortage problem. His assumption about the exponential rise of population has also proved misguided in the modern world, particularly in the more affluent (imperialist) countries where population growth is slowing, or reversing to the point that soon there will not be enough young workers to generate the Social Security (or equivalent) fund to help support retirees. In other words, it's the slowing of population growth that has become a problem, as life expectancy and retirement durations continue to expand, and as long as capitalism is in charge.

The only thing that would matter in a cooperative world is whether the young workers throughout the world were generating enough product to satisfy the needs of the retirees, along with their own needs and those of others not engaged in manufacturing, including service providers, children, and those with disabilities. Increasing productivity, which necessitates adequate energy, would solve that problem. Otherwise, many retirees would have to postpone retirement or return to work. In fact, in the US 10–30 percent of retirement age persons still work or have returned to work for whatever reason (https://tinyurl .com/2h4pyzuj).

The ecosocialists unwittingly arrive at a position similar to that of Malthus, but by a different route: their antinuclear narrative would constrain the further development of the productive forces, preventing precisely the development that would refute Malthus. This in turn would constrain the production of the social wealth needed to meet human needs in an egalitarian fashion, and would therefore require, and eventually result in, a drastic curtailment of population. While they reject the Malthusian recommendation to let the poor starve, and instead insist on egalitarian redistribution, their energy proposal would ineluctably stymie their objective.

Both the ecosocialist and Malthusian positions entail an underdevelopment of the productive forces—the former by intentional policy, the latter by limited vision. Ajl would get nothing like his desired "low tech luxury," and Cox nothing like his desired 2 kW per capita.

The self-negation in the ecosocialist plan is not that different from the general Green opposition to fracking. While ethically opposing fracking and, in general, the production of fossil fuels, the Greens fail to see that, without nuclear, wind/solar can only operate reliably through continual support from the fossil matrix (mostly NG). That only through such support can the overall EROI of the electrical system attain a practicable and socially acceptable level.

Two Novels Exhibiting the Incoherence of Both Ecosocialism and Ecomodernism

US Sci Fi writer Kim Stanley Robinson's novel *The Ministry for the Future* (2020) contains both the strengths and weaknesses of the Green left analysis of climate, energy, and capitalism, though he subtly points the way toward nuclear energy.

The novel features American scientist Frank May and opens in India in the late 2020s during an intense heat wave. The demand for air conditioning has overwhelmed and shut down the grid. People, mostly poor and middle class—the rich, it is implied, are safe—struggle to keep cool enough to survive. Many have window air conditioners running on NG-powered electrical generators, though they are in short supply. The implication is that governance has largely broken down.

Frank has brought many people into his office building, where a generator runs upper story window air conditioners. Despite the generator's placement on the roof to prevent its theft, a gang hears it running and, keeping Frank at gun point, takes the generator. To keep himself cool he hikes to a medium-size lake filled with people trying to stay alive. But the water is hot under the baking sun and cannot cool at night—both day and night becoming hotter as climate change worsens. Frank passes out in the lake and awakens in the hospital to learn that 20 million Indians have died. The near-death experience leaves him with PTSD.

The scene shifts to a meeting in Zurich of the organizations connected to the Paris Climate Accords. Paraphrasing the Indian climate representative, Robinson's narrative voice explains (pp. 23–24):

> More people had died in this heat wave than in the entirety of the First World War, and all in a single week and in a single region of the world. The stain of such a crime would never go away, it would remain forever.
>
> . . . In dealing with the poverty that still plagued so much of the Indian populace, the Indian Government had had to create electricity as fast as they could, and also, since they existed in a world run by the market, as cheaply as they could. Otherwise outside investors would not invest, because the rate of return would not be high enough. So they had burned coal, yes. Like everyone else had up until just a few years before. Now India was being told not to burn coal, when everyone else had finished burning enough of it to build up the capital to afford to shift to cleaner sources of power. India had been told to get better without any financial help to do so

whatsoever. Told to tighten the belt and embrace austerity, and be the working class for the bourgeoisie of the developed world, and suffer in silence until better times came—but the better times could never come, that plan was shot. The deck had been stacked, the game was over. And now twenty million people were dead.

The people listening sit in shame and silence. Robinson writes (p. 25):

So when the funerals and the gestures of deep sympathy were done with, many people around the world, and their governments, went back to business as usual. And all around the world, the CO_2 emissions continued.

For a while, therefore, it looked like the great heat wave would be like mass shootings in the United States—mourned by all, deplored by all, and then immediately forgotten or superseded by the next one, until they came in a daily drumbeat and became the new normal.

Taking a line from Slovenian philosopher Slavoj Žižek, Robinson sums it up, "Easier to imagine the end of the world than the end of capitalism" (p. 25).

In response, the fictional Indian voters repudiate the ruling party, the BJP (currently ruling in fact), and the entire ruling elite, replacing them with an "inchoate and fractured resistance of victims" in a new party whose name means "survival" in Sanskrit. The new party begins dismantling its caste system, nationalizes the power companies, and (p. 25):

a vast force was put to work shutting down coal-fired power plants and building wind and solar plants, and free-river hydro, and non-battery electrical storage systems to supplement the growing power of battery storage.

As we soon learn, the new Indian government has had enough and breaks with the "world community" (global capitalism), pursuing the geoengineering strategy of spreading aerosols in the stratosphere to create a Mount Pinatubo-shielding effect to reflect sunlight back into outer space and cool the planet.[31]

---

[31] In 1991, the Philippines' Mt. Pinatubo erupted in the second most intense volcanic eruption of the twentieth century, spewing a haze of sulfate aerosols that girdled the earth and lowered the global temperature for three years, countering the rising heat trapping by the continually emitted GHGs.

Later in the novel, a reference is made to the failure to solve the "storage problem" and to successful efforts to desalinate water on a global scale in order to address the fresh-water shortage. The energy for desalination comes from the "tremendous amounts" of excess "clean renewable" energy the world has produced. Robinson does not quantify the "tremendous amounts," relying instead on strong adjectives. He joins antinuclear propagandists in evading estimates of what it would take to generate so much net energy.

A comment first on Robinson's political economy and second on his technical analysis. Burning fossil fuels does not lead smoothly to more wealth followed by clean technological innovation as the ecomodernists would predict. Instead, we get austerity, the standard neoliberal capitalist recipe. Due to its underdevelopment (the result of deliberate policies by the British empire), India is prohibited from following the path of the "developed" world, where fossil fuels allow for the addition of renewable energy.

Here we offer a Marxist rebuttal to a confused mixture of class analysis, anti-imperialism, and nationalism. India is not the working class to the "bourgeoisie of the developed world." Such an analysis converts what ought to be nuanced class analysis under advanced globalization into a simple dichotomy of global North and South, where the South becomes the working class and the North is equated with the exploiters. Neither is valid, as both hemispheres include both exploiters and working class.

Robinson's political economy merges into his technical analysis, as he speaks of the developed countries, in which capital is accrued through the burning of fossil fuels. In his view, a fossil fuel bridge allows the developed world to reach a threshold where wind/solar can take over. However, as we have indicated, humanity cannot afford to restrict either fossils or nuclear merely to serve as a bridge on the way to all-RE, and wind/solar will never be able to do away with whatever bridge is proposed.

The fictional reference to "excess" renewable energy, flowing from what would have to be massive oversizing and overbuilding of wind/solar farms, is very misleading. In real life, it would be no match for the plentiful energy from nuclear. And since that fictional "excess" would have to be stored, the self-negating energetics of storage are also glossed over by Robinson, who substitutes feel-good references to "free-river hydro" and "the growing power of battery storage."

Interestingly, the novel repeatedly references immense geoengineering strategies to reverse AGW, like the Indian aerosol seeding.

Some of these involve reforestation and rewilding, thereby restoring carbon sinks and biodiversity, and some are aimed at countering the rising sea levels. One involves pumping the meltwater from under the gigantic Antarctic glacier and spraying it to refreeze on the exposed upper surface. Easily imagined by a novelist, this strategy would require amounts of energy taken for granted and unquantified.

In his earlier novel *New York 2140* (2018) about the future of AGW, Robinson had already acknowledged the necessity of geo-engineering (including biotechnology) and taken the Green opponents of such alleged hubris to task for their purity. We return to this shortly.

In the later novel Robinson's description of AGW is devastatingly cogent, but in it he looks solely to wind/solar to solve the problem. His failure to recognize the shortcomings of these renewables traps him into following in the footsteps of the various Green New Deals. Ironically, however, in one out of 106 chapters, he offers a laudatory paean to nuclear energy, apparently unaware of the contradiction. Three-quarters of the way through, in chapter 76, his narrator, an anonymous woman in the US Navy, possibly a nuclear engineer, notes (p. 381):

> Among other things, we've run 83 nuclear-powered ships for 5,700 reactor-years, and 134 million miles of travel, all without nuclear accidents of any kind. I lived within a few feet of a nuclear reactor for three years, no harm no foul. My dosimeter showed just the same as yours would, maybe better. How can that be? Because the system was engineered and built for safety, whatever the cost. No cutting corners to make a buck. Done that way, it can work. Probably the Navy should run the country's electricity system. I'm just saying.

Robinson's view of nuclear energy over time has been decidedly mixed. In his Mars trilogy, written in the 1990s, the planet is powered without incident by Integral Fast Reactors (IFRs), the successor to the US DOE's EBR II, the world's first fast breeder reactor. But more than 20 years later, while *Ministry* unequivocally praises nuclear energy in just one offhand chapter, the rest of the novel essentially writes it off in favor of renewables.

Cracks in Robinson's Green Narrative

As we said, in *New York 2140* Robinson had not only elaborated approvingly on geoengineering, he had also harnessed it to a searing critique of the Green left's rejection, sharply exposing their conceptual poverty and its devastating consequences. This novel might have been subtitled (in a nod to Naomi Klein) *Disaster Capitalism Meets Catastrophic Climate Change*. Robinson makes clear that while AGW is catastrophic, it is not apocalyptic. After all, humans have adapted to catastrophic conditions many times over, but so has capitalism (which he views as unfortunate). Melting of land ice has produced a 50-foot sea level rise, completely submerging lower Manhattan and rendering upper Manhattan as the new Venice. Between them is a zone of buildings falling into ruin, an area ripe for speculators and gentrification.

The character Amelia hosts a popular TV show that takes wildlife tours in a blimp around the climate-challenged planet. However, she is unavoidably compromised by capitalism—a media star who wears sexy outfits to attract internet tourism and who tries to do good by featuring endangered animals and trying to save entire species.

Backed by ecologists, she tranquilizes polar bears and transports them in her blimp from the North Pole to Antarctica, all televised. Despite the danger, she pulls it off, only to have the Antarctic Defense League, fashioned after Greenpeace, drop a neutron bomb on Antarctica to keep it pure of polar-bear contamination.

Amelia's response (pp. 318–320):

> Look, we're in the sixth mass extinction. . . . We caused it. Fifty thousand species have gone extinct and we're in danger of losing most of the amphibians and the mammals and all kinds of birds and fish and reptiles. . . . It's . . . a fucking disaster. So we have to nurse the world back to health. . . . There's genetically modified food grown organically. There's European animals saving the situation in Japan. . . . We've been mixing things up for thousands of years now, poisoning some creatures and feeding others, and moving everything around. . . . So when people start to get upset about this, when they begin to insist on the purity of some place or some time, it makes me crazy. . . . It is fucking crazy to hold on to one moment and say that's the moment that was pure and sacred, and it can only be like that, and I'll kill you if you try to change anything.
>
> . . . I've met some of these people, because they come to meetings and they throw things at me. Eggs, tomatoes—rocks. They shout ugly hateful things . . . and they think everyone else is wrong because they aren't as pure as they are. . . . I hate their self-righteousness about their so called purity. . . . There is no such thing as purity. It's an idea in the heads of religious fanatics, the kind of people who kill because they are so good and righteous. . . .

> So now there's a group claiming to defend the purity of Antarctica. . . . It's nice but it's no more pure or sacred than anywhere else. . . . It's part of the world. There were beech forests there once, there were dinosaurs and ferns, there were fucking jungles there. There will be again someday.
>
> Meanwhile, if that island can serve as a home to keep the polar bears from going extinct, then that's what it should be.

Okay. Quite a rant and also quite in line with our own critique of Green philosophy, with all of its dyseconomies and boomerangs. And it's certainly a sound rebuttal to Naomi Klein's self-righteous assertion (2019, p. 100), "Nuclear power and geoengineering are not solutions to the ecological crisis; they are a doubling down on exactly the kind of short-term hubristic thinking that got us into this mess."[32]

One may object that Robinson's work is fiction and so exaggerates. However, in our final chapter we discuss the real life-and-death dangers (minus the neutron bomb) of self-righteous Greens at work. They are driven by an anti-corporate activism—at once well-meant but utterly manipulative and infused with confirmation bias, itself derived from a Green culture that necessarily bears the marks of the cutthroat competitive society it aims to oppose.

We note in closing that the purity Amelia hates reflects a crushing blindness on the part of certain Greens to the extreme conflicts that pervade the natural world, a world characterized poetically by Tennyson as "red in tooth and claw." Long before our species emerged, nature was both beautiful (in the future eyes of humans) and brutal; the arrival of humans was not the event that introduced destruction to the world. However, supplementing the animal kingdom's widespread ability to feel and in some cases to think, humans have the ability to reason beyond the most elementary level, and we are capable collectively, under suitable social circumstances, of mitigating at least some of the destruction all around

---

[32] Geoengineering is a vast category of human endeavor. It includes such beneficial things as the construction of sewage systems, roads, bridges, tunnels, canals, reservoirs, hydroelectric dams, the electrical grid, broadcasting towers, and underwater communication cables. It includes the building of cities, the clearing of brush, the planting of crops, the restoration of beaches and wetlands, not to mention the building of wind/solar farms that Klein supports. The list is endless. To condemn geoengineering as inherently and solely mess-creating, based on the unintended side effects of some projects, is like damning healthcare on the grounds that the annual number of iatrogenic (doctor-caused) deaths in the US is not far behind those due to heart disease and cancer. The need is to choose appropriate forms of geoengineering while trying to avoid unintended harmful side effects, many of which are predictable.

us, whether unwittingly initiated by our species or not. Therefore, organizations which hope to build a genuinely egalitarian world that can meet the needs of the world's people, including the need for species diversity, must find a way to fight the ruling classes without destroying each other.

10

Competition and the Dyseconomies of Capitalism

Having discussed the dyseconomies of scale encountered by renewables in the early chapters, and having briefly alluded to a number of contradictory features of capitalism in the previous chapter, we now turn to those dyseconomies erected by capitalism that render the system a barrier to halting AGW and to the general satisfaction of human needs. In particular, capitalism's fragmentation and competition for profit, and the resulting accumulation and growth imperatives, are barriers to the replacement of fossil fuels by the only energy technology that on its own can interrupt the growing climate crisis and provide abundant energy.

Because profit is the goal of capitalist investment, there is powerful resistance, first, to the stranding, both actual and potential, of fossil and wind/solar assets, and second, to the sharing of technologies among competing firms and nations. These two objects of resistance are, as we will see, entangled manifestations of the tendency of the rate of profit to diminish over time—a direct consequence of competition under capitalism.

If competition is unleashed in a free market, the conventional story goes, it promotes innovation and its spread globally. Consumer items like cell phones or cars lend intuitive appeal to this concept. But while taking the universality of capitalist competition for granted, even libertarians and liberals see certain competitive obstacles to the unleashing of innovation. Libertarians, paralleling the *neo*liberal view, believe that innovation is hampered by an *extrinsic* entity, government and its intervention in the market. In contrast, left liberals, following British economist John Maynard Keynes, find the impediment in an *intrinsic* entity, the inevitable tendency toward monopoly derived from competition. The Keynesian view does not disparage competition but holds that a global economy can only grow equitably if competition is modulated by a fair, extrinsically imposed, regulatory framework that includes progressive taxation, and in the case of AGW, a global system of carbon taxes.

The Keynesian celebration of regulated competition spawns a widespread advocacy for a global Green economy, based primarily on wind and solar. Liberals maintain that global Keynesianism can overcome the processes that lead inexorably to increasing inequality—poverty in the face of exploding riches—and recurrent economic crises. In this view, firms compete not to cheapen the labor of producers (through automation and layoffs), but to pay it better so it can consume (raising effective demand) and thereby contribute to global growth. To Keynesians, competition is magically turned from a race to the bottom for workers, as both producers and consumers, into a race to the top. The concern of libertarians, in contrast, is for workers only as consumers, while their situation as producers drops out of sight.

In contrast to both, Marxists see inequality and economic crises as unavoidable results of a fragmented and competitive economy, producing the so-called contradictions (inherent conflicts) of capitalism. These contradictions, we argue, undercut both the economic and epistemic cooperation necessary to, among other things, the rapid expansion of nuclear energy to replace fossil fuels.

In summary, in this chapter we explore the ways the global capitalist economy obstructs the cooperation necessary to spreading the most advantageous energy technologies rapidly and equitably. Our analysis is rooted in the Marxian diagnosis of the contradictions of capitalism, contradictions whose existence the various pro-competition advocates, from free marketeers to Keynesians, flatly deny. They both believe in capitalism's ability to address climate change and celebrate competition as the sole, or at least greatest, source of innovation—a view that captures part of the truth but only part.

Real Competition

Standing opposed to the libertarian and Keynesian concepts of what we might call "happy competition" is real competition, as explained, for example, by Pakistani-American Marxist economist Anwar Shaikh (2016). Real competition is a battle and an inescapable imperative, a form of warfare. Rejecting the idealized imaginings in the conventional economic concept of perfect competition, Shaikh focuses on the conflicting, rather than shared, interests of competitors. Real competition, he shows, rules out any stable Keynesian race to the top, in which all parties, both sellers and buyers, can be satisfied.

As Shaikh puts it (p. 259, italics in the original):

> Capital is a particular social form of wealth driven by the profit motive. With this incentive comes a corresponding drive for expansion, for the conversion of capital into more capital, of profit into more profit. Each individual capital operates under this imperative, colliding with others trying to do the same, sometimes succeeding, sometimes just surviving, and sometimes failing altogether. This is *real competition*, antagonistic by nature and turbulent in operation. It is as different from so-called perfect competition as war is from ballet.

While newer startups can always come along to replenish the declining number of contending firms, late entrants are disadvantaged, and those that catch up are exceptions. Novel technologies like the internet or AI (better termed IA, for intelligence augmentation), however, may face no "older winners" with which to catch up, but once established, "real competition" sets in.

Shaikh shows that cost cutting, which enables price reduction and expanded market share at the expense of competitors, is crucial to capitalist success and survival. "Cost-cutting can take place through wage reduction, increases in the length or intensity of the working day, and through technical change [automation]. The latter becomes the central means over the long run" (p. 259). Automation is also central to Marx's explanation of the falling rate of profit, a concept we describe in a Sidebar at the end of the following section.

Marx versus Keynes

French-American economists Emmanuel Saez and Gabriel Zucman (2019) argue that fair taxation would make capitalism more inclusive. They propose a global tax plan to counter "international tax competition," in which small and relatively poor countries increase their income by enticing big firms to locate their corporate headquarters within their boundaries. The bait is tax rates lower than at home.

Saez and Zucman propose that, to begin with, a tax schedule be imposed both nationally and globally to reduce income inequality among nations and among citizens. It would begin with a minimum global tax of 25 percent on all multinational production, to be enforced and received by the home country, regardless of where the company headquarters or production process may have been moved. This would cut the legs out from under the use of havens for tax evasion, but some smaller, poorer countries would lose this part of their income—a normal result for the weak in capitalist competition. What seems fair to some in the regulated (or unregulated) market is unfair to others, but for it to work most countries have to generally agree to participate.

Once this international coordination were in place, other taxes could be imposed in the service of equity, from a wealth tax on the superrich to a national income tax of 6 percent to cover healthcare and substitute for the premiums paid directly by workers or taken out of their wages for health insurance, a burden that Saez and Zucman, with some justification, regard as a hidden tax. The national taxation method of healthcare funding is common to many European countries.

They further argue that while a progressive income tax could be fine-tuned to maximize economic output, higher confiscatory taxes should be levied in the interests of equity. With such tax rates in place, they maintain, competition can be transformed from a race to the bottom for the majority to a race to the top.

With a high enough tax floor the logic of international competition, they say, would be turned on its head. Companies would go where the workforce is productive, where infrastructure is high quality, and where consumers have enough purchasing power to buy their products—that is, to countries, or back to countries, that are industrialized and more fully developed economically. Instead of competing by slashing wage rates, companies would, they maintain, compete by boosting spending on infrastructure, by investing in wider worker access to education, and by putting their money into research. Instead of improving the bottom line of shareholders, international competition would contribute to more equality within countries (p. 126).

Saez and Zucman's foundational assumption is that, even under capitalism, profit can take a back seat to other desiderata. They point to the fact that income taxes in the US used to be far more progressive and remained so until the government chose the "low road" of neoliberalism, minimizing its intervention in the market and lowering taxation. In fact, the highest marginal tax rate, to fund World War II, topped out at 94 percent, versus 37 percent today.[1]

Because of US dominance following the war, which saw the decimation of the economies of Europe, Japan, and the Soviet Union, the US government was able to continue with Keynesian measures for a period, the so-called golden age of capitalism. In the

---

[1] Capitalists often approve of government regulations imposed on them, particularly regulations that are more burdensome to smaller firms, spurring those firms' elimination and an increase in the survivors' market share.

1970s, the US produced nearly 50 percent of all global manufacturing. It is now down to 13 percent, with China the world's leading manufacturer. As both result and enabling contributor to its present and growing power, China's electricity grid is more than twice the size of the US grid. In other words, in the world of competition, with its inevitable unevenness of advance, US economic dominance is waning. This condition is mitigated for the moment by the US's continuing military and financial dominance. But contingent conditions cannot be generalized, and conditions keep changing.

As British Marxist economist Chris Harman (2009) put it during the 2008 Great Recession (p. 303):

> Today the sheer scale of integration of national economies means that serious implementation of state capitalist solutions would cause enormous disruption to the system as a whole. Yet for national states simply to sit back and leave giant firms to go bust in the hope of crises liquidating themselves . . . would do even greater damage. The two long-term tendencies pointed to by Marx—for the rate of profit to fall on the one hand and for the concentration and centralisation of capital on the other—combine to put the whole system in a noose. The attempts of capitals and the states in which they are based to wriggle out of it can only increase the tensions between them—and the pain they inflict on those whose labour sustains them.

FDR's New Deal, designed to put an end to the Great Depression (and inspiration for the modern phrase "Green New Deal"), was a major application of Keynesianism. The New Deal became the leading model for social democratic forms of the good capitalism envisioned by Saez and Zucman. But while the golden age of capitalism that followed was non-generalizable, neither was it brought about by Keynesian policies. It was primarily the result of the US's participation in World War II and its emergence as the dominant capitalist economy, displacing the British Empire. The war put vast portions of the working class back to work at home and on the battlefields abroad and happened to spare US infrastructure from destruction. Keynesian policies played a decidedly secondary role.

British Marxist economic geographer David Harvey (1982) explains (p. 444):

> In the end the war was fought to contain autarky and to open up the whole world to the potentialities of geographical expansion and unlimited uneven development. That solution, pursued single-mindedly under United States's hegemony after 1945, had the advantage of being super-imposed upon one of the most savage bouts of

devaluation and destruction ever recorded in capitalism's violent history. And signal benefits accrued not simply from the immense destruction of capital, but also from the uneven geographical distribution of that destruction. The world was saved from the terrors of the great depression not by some glorious 'new deal' or the magic touch of Keynesian economics in the treasuries of the world, but by the destruction and death of global war.

A recent book edited by two Marxists—Italian-Dutch economist Guglielmo Carchedi and British economist Michael Roberts (2018)—directly compares Keynesian with Marxian assumptions. They and their coauthors demonstrate with abundant data that the Keynesian narrative fails to explain the crisis-laden character of capitalism, while the Marxian law of the falling rate of profit explains it well. According to the latter, a global capitalist system cannot sustain a global Green New Deal. While an all-RE electricity system would collapse of its own weight for physical reasons regardless of economic system, the competitive, profit-driven character of capitalism, even with an energetically feasible all-nuclear system, would necessarily lead to crises. Escape from crises would still require massive destruction of certain capitals, whether through economic means (for example the demise of weaker firms whose equipment is bought up by the stronger firms for reduced prices) or military means.

Real competition disallows a "race to the top"—in which competition for profit could hypothetically serve to pay labor better, to share innovations for the good of all, and to maximize human flourishing.[2]

Again, there are many examples where big capital approves its being regulated, but only when regulation gives those capitalists a competitive advantage. There are also many examples of firms paying workers better, but only to gain that same competitive advantage. Bryce (https://tinyurl.com/rxjbwc46) mentions the shortage of trained high-voltage linemen (all but a few percent are men) as one obstacle to electrifying everything in the US (and the rest of the world). As a result of the shortage, Investor Operated Utilities (IOUs) engage in poaching, as they can afford to pay line-

---

[2] The world of sports appears to demonstrate that competition inherently produces excellence. However, it is competition combined with the cooperative draft and sharing of TV revenues that allows professional baseball, basketball, and football to flourish. This cooperation is a feature entirely absent, not even possible, within the capitalist war of each firm on all.

men $10–20 more per hour than smaller electrical cooperatives can.[3] IOUs enjoy superior economies of scale and therefore cheaper unit costs in areas other than labor. So the strong poach the weak, yielding the strong a net gain in the competition.

But far from being a race to the top—the win-win claim of Keynesianism—the objective is to destroy the competition. At the same time, the poaching threatens the well-being of the millions of consumers served by cooperatives trying to survive in a profit-maximizing matrix. One work-around would be increased social spending to train more linemen and pay all equivalently well. But that would be a "race to the top" through cooperation rather than competition. If, on the other hand, the cooperatives raise wages, they would be forced to raise prices, putting them at odds with both individual and business consumers. A refusal to raise prices would risk their going out of business, perhaps to be bought by IOUs, which could then lower linemen's wages. Cooperation is prevented from scaling by a competitive profit-maximizing system.

Keynesians like Saez and Zucman (and like Van Jones; see Chapter 6) deny that the contradictions of capitalism are inherent in the system, claiming that we can choose between good and bad capitalism. They also believe that investment, rather than profit, is the independent, and explanatory, variable and can boost effective demand. Investment is thought to call the tune, and if encouraged through government spending, economic growth and rising income will follow. For Polish economist Michal Kalecki, "the determination is from rising investment to rising profits, and in the relation there is little room, if any, for reverse causation" (Carchedi and Roberts, p. 92). And for US economist Hyman Minsky, causality "'runs from investment and government spending to taxes and profits,' and in recessions, 'Big Government, with all its inefficiencies, stabilizes income and profits'" (p. 95).

An alternative outlook, still within a capitalist framework, is held by the Austrian austerity school of neoliberalism. Contrary to Keynesianism, this school holds that government spending is the problem, not the solution. While Keynes emphasizes that low wages and underconsumption lead to a realization crisis—in which products cannot be sold so profits cannot be realized—the austerity proponents emphasize that high wages lead to a profit crisis—both capturing a part of the truth but failing to see the inherent

---

[3] There are some 900 electrical co-ops in the US, which together serve some 42 million people.

Scylla-and-Charybdis nature of capitalism. Keynesians see austerity itself as the cause of crises while the austerians see pump priming as the cause. To Keynesians, austerity in response to crises is both unnecessary and crisis enhancing, as is financial speculation. So in the Keynesian view the distinction between good and bad capitalism is the distinction between governmental pump priming on the demand side versus austerity, or between productive versus financial capitalism.

But far more important than the austerians in opposing Keynes are the Marxists, for whom profitability is the short-term independent variable that produces changes in investment and for whom the falling rate of profit causes repeated profitability crises over the long term.

One-sided analysis pervades both the "race to the top" argument and the claim that free-market capitalism is an unmitigated source, indeed the only source, of innovation. Rising wages can be sustained by firms only as long as those raises follow an increase in the firm's productivity—meaning, among other things, that the higher wages are shared by fewer workers, and as long as the firm enjoys an increase in profits relative to rivals. But, as we explain in the Sidebar below, once competitors catch up with such changes, and the competitive advantage of the initiating firm is lost, the profit rate of all competitors ends up lower than before. As Chris Harman (2009) notes: "the competitive drive of capitalists to keep ahead of other capitalists results in a massive scale of new investment which cannot be sustained by the falling average rate of profit."

As a result of this fall, countertendencies kick in, functioning in part to cheapen labor, either directly through wage cuts or through further automation. From the latter, the productivity of labor keeps rising, and its continued growth spells further layoffs and unemployment. Sometimes relieving the unemployment is the chance emergence of new competitors or new industries, or the emergence of, or access to, a new offshore market. The 1991 dissolution of the Soviet Union and the Eastern bloc enabled the emergence of such a new offshore market for Western capitalists. Such chance and contingent eventualities are offered by apologists as if they were a necessary result of the automation-driven rise in unemployment, or at least an inevitable development. While a bigger reserve army of labor (as Marx termed the pool of unemployed workers) offers opportunities for new industries to emerge, other conditions determine whether those opportunities are seized. That is, even if unemployment were a necessary condition for such emergence, it would hardly be sufficient.

As Michael Roberts (2018) makes clear, Keynesianism appears to carve out a "third way between the status quo of rapacious markets, austerity, inequality, poverty and crises and the alternative of social revolution . . ." (p. 108). To which illusion he responds (p. 109):

> What the Marxist analysis of the capitalist mode of production reveals is that there is no 'third way' as Keynes and his followers would have it. Capitalism cannot deliver an end to inequality, poverty, war [while delivering] a world of abundance for the common weal globally, and indeed avoid the catastrophe of environmental disaster, over the long run.

The contradictions of capitalism, which include periodic and unplanned crises, cannot be "regulated out" by a global form of Keynesianism envisioned by Saez and Zucman and US politicians like Elizabeth Warren and Bernie Sanders. Every external input to the economy invites a feedback response that either negates the change or exchanges one problem for another, or both.

The Keynesian prediction is that increases in investment should restore both profit rate and economic growth while increasing equity. But not only does investment have no positive predictive power over profitability, there is evidence that it has negative predictive power—that an increase in investment leads to a decrease in profits. In his essay in Carchedi and Roberts (Chapter 3), Spanish-American Marxist professor of politics José Tapia notes that the Keynesian position entails that there be a "*positive* dependence of profits on investment." Tapia demonstrates that not only is this "inconsistent with the data" (which he provides), but the data also shows "substantial evidence of a *negative* causal dependence of present profits on past investment, as well as very strong evidence of a *positive* causal dependence of investments on past profits" (p. 115, Tapia's emphases). He says (p. 111), "regression analysis provides further evidence that changes in profits are followed by changes in investment *in the same direction*" while also providing "evidence of movements in investment being followed by movements in profits *in the opposite direction*, a fact that is completely at odds with the expected causal link inspired by the Keynesian scheme" (Tapia's emphases).

Using another way of expressing the same idea, Carchedi (Chapter 2) demonstrates the fallacy of the "Keynesian multiplier"—a theory that government borrowing to finance civilian investment in public works would "spur other investments in the private sector, and these would spur still other investments, thus multiplying employment and profits" (p. 65). According to this narrative, rising wages should prevent or help exit crises by soaking up inventories (unsold commodities), thus solving the realization problem. But rising wages cut into profitability, all other things being equal, and Carchedi shows that of the "twelve post-World War II crises, eleven have been preceded by rising wages and only one by falling wages (the 1991 crisis)" (p. 65). Whether in the form of distribution to labor or investment, Carchedi shows that government expenditure doesn't correlate with rising profitability, nor does it reverse declining profitability: "the correlation between government expenditures and the ARP [average rate of profit] is negative in ten out of twelve cases [crises]. Government expenditures do not reboot the economy" (p. 68). He concludes that "empirical data showing the non-existence of the Keynesian multiplier are overwhelming" (p. 68). Finally, he notes, in concert with Harvey (quoted above), that the recovery of profitability arises from neither rising nor falling wages (the neoliberal view), but only from the destruction of capital (p. 69):

> If both Keynesian and neoliberal policies fail, the only way out of the crisis is that generated spontaneously by capital itself: its rejuvenation through a massive destruction of its less efficient units. The world economies exited the 1929-33 crisis with World War II, not through state-induced investments.

Sidebar: The Inexorable Tendency of the Average Profit Rate to Fall

If there is one feature of capitalism that thwarts the Keynesian Green goal of state intervention to regulate the market for the benefit of all, over the long term, it's the inherent falling rate of profit. That is, capitalist competition produces an inexorable downward trend in the average global rate of profit. Some nations and segments within nations can experience a reversal of this tendency, but only at the expense of competitors and only temporarily.

An unavoidable consequence of real competition is automation. Automation serves two purposes: it restores profits by reducing the wage bill and the efficiency gains permit the poaching from less efficient capitals, thereby transferring value. Automation raises what Marx called the organic composition of capital—roughly speaking, the ratio of the value of machinery, plant, and raw materials that goes into the product to the value of living labor necessary to the same product. The rising organic composition causes the profit rate to fall because living labor is the sole source of surplus (exchange) value, namely, the portion that constitutes *new* profit. This is the famed labor theory of value, discovered first by Adam Smith in the 1700s, elaborated by David

Ricardo in the early 1800s, and explained in greater detail by Marx in the later 1800s.

Only living labor can produce more value than is required to obtain and maintain it, maintenance to be understood as the value required to sustain a worker and her/his family (which Marx called labor power as opposed to labor time). That is, only living labor time produces more value than it costs the capitalist to purchase the labor power (wages). Machinery, plant, and raw materials, in contrast, do not produce profit, because a prorated portion of their embedded value—considered as the value of dead labor (labor already done)—is simply transferred unchanged to the product. The transfer from machinery and plant happens through wear and tear (accounted for by depreciating the machines and plant over time), while the raw materials are transferred in toto. The value of living labor in a product is the number of person-hours required to produce it at the average productivity of labor. Competition relentlessly forces a rise in the organic composition and an accompanying reduction in the average profit rate—profit divided by total investment in machinery, plant, raw materials, and labor power.

The escalating process of automation and layoffs increases labor's productivity (more output per worker), but as a means of enhancing profits it only works for the first capitalists who automate, as they can at first lower their costs while continuing to charge the going price. But the others in that industry are forced to do the same to defend their positions. As they all automate and lay off workers, the latecomers, having now also lowered their production costs, are freed to compete for market share by lowering their prices, and the scramble causes the profit rate to decline for all.[4] The new average profit rate is then lower than that prior to the onset of that round of automation. And it continues to spiral downward, temporarily interrupted, as we have said, only by the destruction of some capitals—a process appropriately termed "creative destruction" by Austrian economist Joseph Schumpeter.

The periodic economic crises resulting from this falling profit rate never prove fatal to the system—since the bigger capitalists always have ways of recovering—but they have negative, sometimes devastating, consequences for smaller capitals and most people. The crises undermine any possibility of hoped-for equity, and competition pushes beyond reach the hoped-for cooperation that underlies the Keynesian dream at the heart of any Green New Deal.

Contradictions Following from Real Competition: Innovation and Stranded Assets

Capitalist competition both stimulates and hampers innovation. What it stimulates can be profitable, but for the public it may be

---

[4] Unless there is collusion, which the government frowns on and attempts to reverse in behalf of capitalists in other industries who must purchase from the colluders, either directly or indirectly.

either useless or harmful. Competition forces corporations that innovate to protect their innovations and investments against rivals. They do so through proprietary secrecy, patent protections, intellectual property laws, and, on an international level, ultimately by military means through their state. Otherwise, rivals might improve on the innovation and render the initial innovation obsolete, thus stranding capital. Despite such attempts at prevention, however, this happens every day. Innovators will only welcome improvements made by their own hand, but they generally delay introduction into the market until their initial investment has been largely amortized.[5] For innovations that happen to be generally beneficial, the public may suffer from such delays—for instance, vehicle safety devices. But since capitalist innovations are intended mainly to enhance profits, they may or may not be beneficial for consumers and may even be detrimental to human needs—examples include opioids, fast foods, or semi-automatic rifles.

Defenders of capitalism claim that competition fosters a race to produce, for example, the best and cheapest wind turbines and solar panels (though best and cheapest are often incompatible characteristics). The proposal by renewables advocates that logically follows from the (mis)estimated relative costs is to unleash competition and thus innovation internationally. For example, in an early Democratic Party debate in 2020, US presidential candidate New York Senator Kirsten Gillibrand called for promoting competition between China and the US to ramp up the production of solar panels and wind turbines, thereby unleashing a sort of renewables "arms" race. Similarly, the market-oriented nuclear hope is that SMRs, designed for mass production on ships, such as Thorcon's reactor, can be produced rapidly, locally, and cheaply and thereby compete with coal—even absent price-boosting state control over coal's GHG emissions. Then electricity could be sold at something around 3 cents per kWh, and profitably at that.[6]

But this view of competition over an energy source, which appears to lead to a virtuous circle of Green energy production, gives an incomplete picture of capitalism's self-negating nature. Thorcon's inventors and engineers explain that by using thorium (Th-232) they will eventually be able to deliver nuclear energy more cheaply than coal. But for the time being they are developing a transitional

---

[5] Software companies are something of an exception. Their capital is less vulnerable to stranding, which leaves them free to innovate continually, rendering their prior innovation obsolete and forcing customers to purchase the updates.

[6] It's too soon to know how much of this is commercial hype.

SMR that employs slightly enriched (5.95 percent) U-235. It turns out that Th-232 works most efficiently combined with U-235 at a greater enrichment of close to 20 percent (called high-assay low-enriched uranium, or HALEU). But HALEU is in scarce supply, and Thorcon's competitors cannot operate without it, which forces them to pay more to secure their supply—a price that Thorcon cannot now afford, anticipating that this will not change until the late 2030s. So competition among buyers forces up the price, even though competition among sellers can force down the price. That is, competition cannot serve the interests of both buyers and sellers. Thus, the claim that it is an unmitigated good is one-sided, neglecting the fact that competition always produces losers as well as winners. In contrast, in a non-competitive (cooperative) world all could work together to produce the HALEU needed to optimize the performance of thorium-fueled reactors.

Fossil fuel entrenchment offers an additional powerful example involving massive profits and what would become stranded capital and infrastructure, which elicits determined resistance by the fossil-related capitalists and others to their elimination. The related infrastructure consists of such things as mines and equipment, oil wells and oceanic platforms, refineries, pipelines, fracking equipment, oil and liquid natural gas (LNG) tankers, combustion chambers, and gas stations, as well as hundreds of millions of private ICE vehicles and modes of public transportation. Green New Deals put ICE vehicle owners in a contradictory position—AGW constitutes a threat to them and their descendants in the long run, while the expense of replacing the vehicle could be prohibitive for many in the short run.

A recent article in *The Guardian* suggests that by 2036 a successful net-zero energy transition could make half the world's fossil fuel assets worthless, though to the authors net-zero means renewables with nuclear eliminated (https://tinyurl.com/mr78rbvv). Something like $11–14 trillion worth of assets, they report, could be lost in such a transition, resulting in another Great Recession like the 2008 downturn. They note that those countries with the greatest fossil assets, including Russia, Brazil, and the US, would be the most vulnerable to stranding and as a result would resist the transition, or at least the related businesses would. Meanwhile, net fossil fuel importers like the EU would benefit the most. And China, with its numerous coal mines accounting for some 60 percent of its electricity, but also with its expanding production of wind/solar apparatus and nuclear reactors (not mentioned in the article), would see both stranding and benefit. The "global proliferation of renewables" for electricity has hardly affected oil-dependent transportation. EVs at this point are only being accessed by the relatively affluent, even with government tax incentives (and their batteries are mainly charged by NG or coal, defeating the intended purpose). So for the moment, stranding of the oil industry awaits its turn.

The authors of the *Guardian* article assume that an all-RE electricity system could eliminate fossil fuels from the grid and predict that fossil investments should begin to dwindle. They are oblivious to the eagerness with which fossil companies promote renewables.[7] Besides, the article's projected stranding of $11–14 trillion in assets might be a serious underestimate, since annual fossil fuel revenues alone are around $5 trillion (https://tinyurl.com/yszz3h2f).

In 2008, it was easy to think that US oil production had reached its pinnacle more than three decades earlier at around 9–10 million barrels per day (MBD), as it had since plummeted to only 5 MBD (https://tinyurl.com/5ctc7dhb). Fifteen years later, the US leads the world in production of crude oil at 13 MBD. And when biofuels and LNG are included, the total is around 20 MBD (https://tinyurl.com/2us4eb5m)—almost double second-place Saudi Arabia's 12 MBD and third-place Russia's 11 MBD (rounded off). Additionally, the US nearly doubled its NG output in the same interval (https://tinyurl.com/muu7zs2u). The *Guardian* authors are naively optimistic to believe that the US oil and NG industries will happily ride off into the sunset, while PV panels welcome the sunrise, over the next 12 years. Smil sees the global net-zero transition as practically impossible (https://tinyurl .com/ycy2yt3e). And Bryce cites a University of Michigan study concluding that copper shortages alone render practically impossible a transport system based on EVs (https://tinyurl .com/4unr5pmz).

Jacobson ignores the problem of stranded assets and claims that the huge global investment in wind/solar would save enough money to pay for itself in under 6 years (https://tinyurl.com/jyesxf53). For an elucidation of the speed of his projected all-RE

---

7 Fossil interests had long engaged in climate denialism until their strategy changed to greenwashing around 2008. They not only support "clean" wind/solar, but actually partner with it sometimes. Intermittent sources offer little threat, as they depend for now on fossil fuels. Both the fossil industry and now the rising wind/solar interests handsomely fund NGOs to campaign against nuclear, which is capable of replacing them both (https://tinyurl.com/3fdj6eza). And numerous governments parade their virtue by pursuing "clean" fossil-approved renewables. Meanwhile, pronuclear countries sometimes use nuclear fearmongering against rivals, for example, the hypocritical complaints by China and (to a lesser extent) South Korea about Japan's release of Fukushima's tritiated water (footnote 16 in Chapter 7). These all exemplify real competition as a barrier to real solutions.

build rate, we refer readers to the excellent analyses by Maloney (2016) and Conley and Maloney (2017), in which they show just how unrealistic Jacobson's projection of speed really is. But in order to lend as much plausibility as they could, they indicate that, contra Jacobson, providing the energy for construction (assuming no nuclear assistance) would require the building of some 500–600 new 1GW NG plants. Then, upon completion of his projected all-RE buildout in 2050, these plants would no longer be needed and would become stranded assets.

Only war has produced such extensive destruction of capital assets historically, and only those of rivals (turned overnight from rivals into enemies) have been targeted. As discussed above, destruction of capital is the required pathway out of endogenously produced economic crises, but such periodic and chaotic destruction of capital is not the same as preplanned stranding of temporary NG facilities.

Aside from the stranding, Jacobson's supporters have not completely ignored the criticism of his intended build rate. To indicate that humanity can rise to the occasion, some refer to the colossal speeds of the Manhattan Project and the Space Race. But these two monumental efforts were driven by self-preserving imperialist military considerations that united virtually the entire US capitalist class (such as it was prior to the Trump-promoted antagonistic fracturing). In this case, the snail-like progress toward 100 percent penetration by renewables (only 5–6 percent of world primary energy consumption so far) indicates an absence of such unity of purpose—an absence inseparable from the competitive system.

Differing from the Manhattan Project and the Space Race, the Roadmap would require *anti*-imperialist *global* cooperation. A major energy transition requires local to regional to national to global cooperative planning, a feat to which the competitive nature of capitalism is entirely inimical.

Capitalist firms cannot, by opting for a lower-EROI technology, reduce their unit costs to improve their competitive position. Lower EROI requires greater labor, because it calls for an expansion of EI if it is to reach the desired EN (Chapters 5 and 6).

Josh Floyd penned a prescient warning in 2014 about any project akin to Jacobson's (https://tinyurl.com/ns964ahp):

> the harder we push a transition away from conventional energy sources, the more we starve other economic activity of its lifeblood. This is the energy trap. Under conditions of energy scarcity, a

crash program to shift to alternative sources—particularly where
their exploitation involves relatively high up-front energy costs—
makes that scarcity worse in the short term. The political implications . . . are immediately apparent: leaving aside all other issues
relating to the prospects for renewables to directly substitute for
fossil fuels, anyone championing a long-term vision for renewable energy predicated on short-term pain must contend with the
rather significant political problem that abandoning the program
at any point will bring immediate relief. This presents proposals
for transitioning energy supplies to 100 percent renewable
sources at the global (or even national) scale with a formidable
obstacle. Overcoming it would entail a remarkable—and likely
unprecedented—degree of collective discipline at population-
wide scales.

In other words, in light of the competitive imperative, were
such a plan to even be initiated, eventually the strongest capitalists would be tremendously tempted to retain the much
cheaper and more energy-efficient NG bridge, utilizing it in a
relatively conventional manner. This is especially the case given
the capability of greater speed of construction for an NG plant
compared to that of a wind farm with an equivalent amount of
actual, rather than nameplate, power production (which could
at best be only an average of rapidly fluctuating and unreliable
power). So the stranded NG bridge assets would then turn
into a primary energy source and the wind and solar farms
built to date would become the stranded assets. But either
way, massive asset stranding would become inevitable. Were
our erstwhile Greens somehow able to carry the all-RE plan
out further, we would face the aforementioned discipline
called for by Jacobson in order to align demand with supply—
including hours of work, lighting, and any other usage of
electricity.

And this discipline would have to be imposed by not just a
nation state but by a global authority, since there is no reason for
the working class, let alone the capitalist classes, to voluntarily
and collectively subject themselves to such austerity. Furthermore, were the project technically sustainable, this externally
imposed discipline would extend at least from the date of completion into the indeterminate future. There is no good reason to
think such a global state of affairs—operating against the interests of all socio-economic classes, in the service of a plan technically incapable of being brought to fruition and predicated

almost solely on an irrational and uninformed fear of nuclear power—could ever come into being.[8]

The problem of stranded assets is amplified by the progressive concentration and centralization of capital and the relentless declining of the rate of profit. The competition among capitals, each necessarily seeking greater market share—both defensively and aggressively—leads periodically to overcapacity and overproduction (both relative to demand), and eventually to economic downturn for the time being. As the late American-Canadian Marxist historian Ellen Meiksins Wood (2012) puts it, summarizing the argument of American Marxist economic historian Robert Brenner (p. 275, italics in the original):

> Investment in fixed capital allows producers to stay in the market even when lower-cost competitors enter the fray, and they can stay in even at a lower rate of profit. But the point is also that the same heavy investment means they *must* stay in, even just to recoup their costs, or, at least, it is hard to get out at the right time. So manufacturers hang on to surplus-plant instead of closing it. The end result is a declining rate of aggregate-profit across the industry, with wider effects throughout the whole economy.

Such dynamics undermine the Greens' Keynesian narrative.

The stranding of fossil and renewable assets is delaying the development of nuclear. To assist in this delay, the capital interests who dread the stranding are foisting a different kind of fear on the public. But sooner or later other capitalists—as well as a growing portion of the public—will be forced to realize that a nuclear-based electrical system is what they need. And sooner or later

---

[8] Palmer (2014) cites another proposal from 2012 for an almost all-RE system, but which assumed the necessity and permanence of a parallel fossil fuel grid to smooth the intermittency. This was a

> study of the PJM [Pennsylvania, Jersey, Maryland] regional grid in the USA, [which] assumed that 28,300 MW of fossil fuel plant [capacity] would be retained in their "99.9% scenario," equal to just under the average grid demand, but accounting for a miniscule [*sic*] 0.02% of system energy and hence operated at well under 1% capacity factor.

Palmer notes that the plan's authors fail to address the incompatibility, within a capitalist context, between the creation of so much excess capacity to "reliably meet demand" and (at least in Australia) removal of "the conventional revenue stream required to allow the [NG] generator to remain profitable"—because it is based on actual energy produced rather than capacity. On the other hand, Angwin (2020) refers to payments to the NG producers (in the US) based on capacity rather than production. Such a high "revenue stream" to NG producers would undercut any carbon tax and delay the intended end to fossil fuels.

government officials will be encouraged, lobbied, and funded to agree. More favorable regulation, at least in some places, is likely to follow as newer reactor designs are created and become commercialized. Absent the impediments encouraged and imposed by stranded capitals, nuclear may eventually be able to compete successfully in the market. This is a necessary precondition for nuclear to achieve global predominance.

But the cooperative and planned phase-in of nuclear and phase-out of fossils that will be required to avoid dangerous gaps in energy supply will be stymied by competition among nations, if not within nations. Among other things, the requisite expansion of skills and equipment necessary for creating and operating nuclear plants is also applicable in part to the creation and operation of nuclear weapons. Without international cooperation, nations will be unable to interdict a nuclear arms race. As it is, imperialist rivals already must agree to monitor each other as well as countries desiring access to nuclear technologies like uranium enrichment. Less powerful countries are often motivated to develop weapons by a desire for national sovereignty and defense against more powerful nations, but they are often not allowed to do so. The cases of Iraq and Iran suggest that the coordination said to be possible according to the Keynesian "happy competition" outlook is an illusion. The pretext for the 2003 US invasion of Iraq was its alleged efforts to develop nuclear weapons, from purchasing yellowcake (raw uranium) to producing centrifuges. The Bush administration mendaciously claimed that the Iraqis had already produced nuclear weapons (weapons of mass destruction, WMDs), which were never found. With regard to the monitoring of Iran, the international cooperation necessary and possible under the Green Keynesian view was not forthcoming.

Offered as a rebuttal to this pessimistic view is the megatons to megawatts agreement in which the US purchased decommissioned warheads from Russia to fuel US domestic reactors (Chapter 7). But that exceptional example took place between established nuclear powers. Keynesian coordination to share nuclear technologies would require the abolition of serious geopolitical rivalry. This in turn would require the abolition of capitalism's inherent competition on a global scale. Or the establishment of a fantasized stable "super imperialism," in which the strongest country would benignly dominate weaker countries ostensibly for their own good. Such an arrangement, however, is not stable, nor could it be, if only for the uneven development in a capitalist world where hegemony is never assured. And what would restrain a dominant country to remain benign toward all others when its own growth imperative is thwarted (under real competition)?

Imperialist rivalry creates a global energy dilemma. Renewables, if they were technically feasible, might avoid the weapons threat, but they would produce geopolitical conflicts over limited raw materials. Nuclear is technically feasible but is associated with direct military threat in a competitive world. Thus, to halt AGW requires a solution to both the absence of agreement on the technical issues and the competition-derived geopolitical threat.

Finally, the stranding of tens of trillions of dollars in assets, in order to engage in a global nuclear energy build-out, would necessitate a comprehensive international agreement and coordinated plan to phase out fossils and renewables as nuclear is phased in. The timed phase-in/phase-out plan would have to avoid even a temporary deprivation of life-preserving energy. This agreement would have to entail pledges, vigorously monitored and faithfully carried out, to refrain from establishing any new coal mines, oil wells, or fracking sites and from building any more fossil fuel plants or commercial-size wind/solar installations, in favor of building only nuclear facilities—possibly tolerating the temporary establishment of renewables installations if local conditions require them, understanding that they will become stranded for energetic reasons.

However, capitalism's inherent competition exerts extreme headwinds against both the emergence and the stability of such an agreement—particularly since it would need to include every country in the world. Capitalism has seen no such development, but this is what would be required to halt AGW. Given the nature of the system's imperative and inescapable competition, halting AGW may well require its replacement with a different and cooperative economic system. It remains to be seen whether the necessity of halting AGW will eventually prove sufficient to mobilize enough of the world's population, particularly the working class, who, to put it mildly, have the least stake in the system that exploits them and are the most vulnerable to the destructive effects of accelerating AGW.

The Hope that a Carbon Tax Will Foster a Popular Rejection of Fossil Fuels

Many advocates of an all-RE electrical transition call for a global agreement on carbon taxes to incentivize a switch to clean energy. A carbon tax would, they propose, be set low initially and gradually raised, giving each country's energy capital time to transition to lower carbon sources. The tax revenues could either go to research and development for clean energy or be distributed to the nation's consumers in a fee-and-dividend plan.

A less-favored plan relies on carbon trading, in which a cap would be imposed on CO_2 emissions, and businesses exceeding the cap would be required to purchase permits from companies remaining below the cap. This would, it is hoped, provide a profit incentive to go Green, since the higher-emitting companies would want to charge higher prices to cover their permits and would thereby lose market share to the lower-emitting companies in the same industry—all other things being equal.

Goldstein and Qvist (2019) approvingly quote one physicist, who, seeing Sweden's partial success in diminishing its CO_2 emissions, says, "solving global warming does not require us to 'tear down capitalism.' The world just needs to be a bit more like Sweden" (p. 207). As they note, the Swedish government was able, without destroying economic growth, to put a hefty tax of $150 per ton of emitted CO_2, the highest such tax in the world and at the high end of considered proposals. In addition to Sweden, Goldstein and Qvist point to France and the Canadian province of Ontario for successful mitigations of CO_2 emissions, suggesting that the whole world could follow their example.

But even in Sweden the resulting emissions reduction has been modest, with a 25 percent reduction in total annual CO_2 emissions since the onset of the program in 1991 and no ability to affect the global atmospheric concentration. At that, each of these reductions was aided by specific special features. Sweden and Ontario have abundant hydro and insufficient political opposition to their nuclear facilities to constitute an obstacle. France, having virtually no domestic access to fossil fuels, is forced to depend either on outside sources or on nuclear, and has mainly chosen the latter.[9] As it turns out, those countries that did switch energy sources did not do so to avoid higher prices or to lower GHG emissions, but to gain energy independence. And their switch was not hindered by the potential stranding of domestic

---

[9] As they further point out, an even greater reduction was achieved prior to 1991 by the expansion of Sweden's nuclear contribution. In short, carbon pricing is less successful in reducing emissions than the increased penetration of nuclear. However, despite pressure against fossils, the market is not allowed to foster nuclear over renewables because of the several forms of government interference in the market described above. That choice has to be made deliberately by the various states. Also, fee and dividend is circular and self-negating, since, first, the carbon-taxed entities can pass on their added costs to the now dividend-enriched consumers without much opposition, canceling their incentive to lower their emissions. And second, even the ostensible benefit for a company that moves to low emissions depends upon the continued existence of high-emissions producers.

fossil fuel-associated assets, a contingent and nongeneralizable advantage.

Moreover, even where the transition has occurred, it has affected only the electricity sector (roughly 18 percent of total energy consumption in the US) and not transportation (37 percent or so, see Figure 2-1), which is everywhere almost entirely dependent on oil, with EVs still a marginal contributor (and mainly dependent on another fossil fuel at the charging plant, NG). And as long as electricity is mainly produced by fossil fuels, EVs may even *increase* carbon emissions. The reason is that transmission of battery-charging electricity over the grid incurs heat losses and therefore, for a more remote power plant, more fossil fuel will need to be burned at the plant than would be burned by an ICE vehicle (even allowing for the lower emission of carbon by NG than by oil for the same energy output). Second, it may take more energy to produce EVs and batteries than ICE vehicles (https://tinyurl.com/yhhsuv4u).[10] The transportation sector exemplifies the problem of stranded assets under capitalism.

Competitive pressures interfere with the long-term global cooperation demanded by the halting of AGW. Neither the carbon tax nor cap and trade have worked everywhere, and GHG emissions are still rising, particularly in China and India, for reasons inseparable from the nature of a competitive system governed by profit—namely, the need to outcompete other nations, the only way to escape external dominance. This international competition militates against a universal agreement to impose a carbon tax on the whole world. Success could brook no violators. As Goldstein and Qvist point out (p. 197):

> Ideally, a carbon price should apply to the whole world, but no practical mechanism exists to set such a price and implement it. When

---

[10] Furthermore, for longer-established auto makers the manufacturing cost of EVs is so much greater than that of ICE vehicles that the automobile manufacturers are forced to take a loss on every vehicle to make them affordable to consumers. Indeed, Ford's loss of over $65,000 per vehicle sold resulted in a net loss to the company in 2023 of several billion dollars. Moreover, the market has been confined to relatively affluent people concerned with climate change, and sales of EVs are declining (https://tinyurl.com/bdhuda9a). Relative newcomer Tesla's profitability is apparently largely related to their greater efficiencies in several areas—first, car design (weight and aerodynamics), second, automated manufacturing process, and third, minimizing of intermediary auto dealers (https://tinyurl.com/4vr456ze). While US government subsidies and tax credits have also helped Tesla transition from net loss to profitability, they only hastened that transition rather than accounting for it altogether. But even at that, Tesla's fortunes are beginning to recede with saturation of the EV market, accompanied now by the abandonment of Tesla by potential customers appalled at Musk's role in Trump's slashing of US government services.

only some countries or regions charge for carbon pollution, there is an incentive for dirty industries to simply move to places where they can compete for free. Then the places with a carbon price have to impose a tariff (border adjustment) to account for the difference, but this is a difficult system to implement.

For any individual corporation or nation, the large upfront investment for an energy transition—to renewables or to nuclear—will generally lead to an intolerable competitive disadvantage, since competitors will not permit the time to amortize that investment as they invest in shorter term changes that will pay for themselves in shorter order. This allows them to play a more dominant role in the world market, not just the energy-related market but all other products, since all extraction and manufacturing rests on the foundation of energy. Only if all nations could cooperate to permit that transitional time to pass could such a massive changeover succeed everywhere, but this would require the rejection of capitalism's competitive imperative. Just as the hypothetical extrapolation from low to complete RE penetration encounters self-undermining *technical* features rooted in low EROI and intermittency, extrapolation of the local to the global encounters self-undermining *geopolitical* features rooted in capitalist competition.

While competition presents a barrier to worldwide cooperation, agreements among subsets of nations (and companies) are common: for example, the Axis and the Allied Powers during World War II, and NATO, the Warsaw Pact, and the Southeast Asia Treaty Organization (SEATO) following the war. Also cooperative agreements to limit enrichment facilities in Iran or START treaties between the US and Russia to limit nuclear weapons, agreements that require monitoring because long-term trust is beyond reach when competition prevails. In each case, such alliances among a few nations have been aimed at gaining economic and/or political advantage in the broader competitive world rather than at replacing global competition and are never aimed at satisfying the needs of humanity as a whole. In short, in the face of a global system of competition for profit, a global carbon tax encounters dyseconomies of scale.

An Example of Real Competition's Inhibition of Real Innovation

We close out our critique of the argument that competition (and only competition) fosters innovation with a brief look at the pharmaceutical industry, an industry that most often produces mere pseudo-innovation. It suggests that in an industry intended to meet human needs rather than elective pleasures, pseudo-innovation acts as an obstacle to real innovation.

The US FDA regulates the development and marketing of medically indicated drugs, among other needed products. The agency only requires, despite some internal objection, that clinical trials demonstrate that more people are helped by the new drug than by a placebo. Trials are not required to show superior efficacy to drugs already on the market. This approach encourages imitation rather than innovation and fosters the proliferation of drugs with the same, or even less, efficacy.

Drug companies are thereby encouraged to enhance their profits without necessarily benefiting the public. And they can do so by either adding a slight chemical modification to one of their products or discovering a second use for it, and either way marketing it as a new drug. Among other detriments, this policy tends to waste the scientific skills involved in research and experimentation.

In this process, any benefit to the public is incidental; profits prevail, and without them no drugs would be developed at all in a capitalist system. Indeed, there are conflicting interests around drug prices: company profits versus affordability to the public. Government officials, whether recipients of campaign funds or not, are generally loath to challenge high prices lest they inhibit new drug development. This reticence leaves the profit makers in control, holding those who may need the drugs hostage. To lend their reputations a patina of respectability, pharmaceutical houses claim that the high prices are needed to recoup their investments in the drug's development, though much of the bill is often paid by government funding of university research (Angell 2005).

Besides being able to charge whatever the market will bear (for more affluent patients), the drug companies are allowed by the government to protect their profits from competitors with 20-year patents, thus foiling the free market. Some drugs are happily found to have a side effect that happens to be desirable and for which it was not designed. The FDA then allows the drug to be marketed with a fresh name and a fresh patent, thereby extending the profit protection for another couple of decades. An example is minoxidil (brand name Rogaine), which was originally developed to treat high blood pressure and was found to slow hair loss, for which it is now marketed under a renewed patent. The inefficiency imposed on the market, by retarding improvements in products and by multiplying a bevy of old drugs with slightly modified form, is a reflection of the competitive system applied to products necessary to our well-being.[11]

A second example is semaglutide (brand name Ozempic), introduced to treat diabetes, but found to help with weight loss. For weight loss the drug comes in a higher dose and is sold with the new name, Wegovy, and has a new 20-year patent. However, given that obesity is coming to be recognized as a precondition for many serious illnesses, it is clear that the newly discovered use for an established drug can sometimes have significant benefits for many patients. Since such fortunate discoveries are less motivated by therapeutic utility than by profit, these benefits are largely a matter of chance rather than planning.

British physician Ben Goldacre (2012) gives some detail on the way clinical trials can amplify profits without benefiting patients. If a trial looks for more than one possible benefit, there's an elevated probability that at least one of the benefits, simply by chance, will appear to be significant during that one trial. So, for example, if the trial looks for, say, 20 different possible benefits, even if the drug has none of them it is likely that one will seem to be present merely by chance. When this occurs, if the trial were to be repeated looking for that benefit alone, it is likely that it would fail, but if run 20 more times, with the same number of new enrollees looking for the same benefit, it would likely succeed again once out of the 20 runs. The 5 percent definition of success (one out of 20) could also be set at 10 percent or any other arbitrary small percentage. This way, if a drug produced and marketed for one purpose turns out not to do much good for the intended purpose, it may be able to be sold as a treatment for a different condition.

Another example of conflicting interests is AstraZeneca's omeprazole, a proton pump inhibitor (PPI) for stomach pain and heartburn. The patent was about to expire early this century, and the company was in danger of losing $5 billion a year, a third of its revenue, as other manufacturers would be freed to manufacture it as a generic at reduced price (p. 145). So AstraZeneca introduced the "me again" drug, esomeprazole (brand name Nexium, "the

---

[11] An example of the "bait and switch to enhance profits" game involving an elective, rather than necessary, product is the Converse Chuck Taylor All Star gym shoe. This product was transformed from a basketball shoe to a classic example of "cool" when gym shoes became an accompaniment to dress clothes. The old cheap canvas black and white shoe now comes in a variety of colors and at higher prices (https://tinyurl .com/2d82kw3a).

purple pill"). This is a twist on the "me too" drug, which consists of "entirely new molecules that work in a similar way to the old ones," while the "me again" drug consists of the same molecule with "one clever difference." In this case, the drug can exist in right and lefthanded forms called enantiomers of the same molecule. The protein chain of omeprazole twists in one direction while that of esomeprazole twists in the opposite direction. Goldacre (2012) says, "[t]he companies claim this as a new drug, and so add a whole new patent lifetime to their profits" (p. 146). He says later, "these two drugs are almost identical, and esomeprazole is basically no better than omeprazole, just much more expensive. The advertising campaign was highly effective, so we waste money on drugs that are no better than those that already exist" (p. 250).

Thus can "innovation" driven by competition in the capitalist market enhance profits without really innovating or benefiting the public—a pseudo-innovation. Risk analyst and epidemiologist Geoffrey Kabat (2017) carries over this critique of competition to biomedicine in his preface:

> it is widely recognized that there is a crisis in the field of biomedicine, characterized by a "culture of hyper-competitiveness." In this environment, scientists may feel the need to overstate the importance of their work in order to attract attention and obtain funding. Other symptoms of this climate are a "lack of transparent reporting of results" and an increasing frequency of published results that cannot be replicated.

Goldacre notes that companies write academic papers claiming efficacy of their drug and then search for a prominent academic to claim authorship, thereby giving the paper more academic weight; this constitutes "covert promotional literature." The practice both promotes the drug and enhances the reputation of the academic, edging out "people studying social factors or lifestyle changes, or side effects, or medicines that are out of patent" (p. 299).

The Fallback Claim—Regardless of Its Dyseconomies, There Is No Alternative to Capitalism

An extension of the "happy competition" narrative is the claim that there is no workable alternative to capitalism. Known by its initials, TINA (there is no alternative) implies that a planned economy (whether authoritarian or cooperative) cannot work. Associated prominently with the Austrian-British libertarian economist F.A. von Hayek, TINA inspired British Prime Minister Thatcher and US President Reagan.

TINA holds that human fallibility vitiates any plan for an economy, suggesting that any attempt is hubristic overreach. Only the competition of the market, they say, can accomplish a desirable outcome. Paraphrasing Churchill, capitalism is the worst possible economic system, except for all the others. The view further holds that the mere attempt to plan an economy necessarily entails the destruction or severe limitation of human freedom.

In their 2019 book, Canadian socialists Leigh Phillips and Michal Rozworski counter the TINA claim by describing Walmart's operation (p. 244):

> The glimmers of hope for a different way of doing things are foreshadowed in the sophisticated economic planning and intense long-distance cooperation already happening under capitalism. If today's economic system can plan at the level of a firm larger than many national economies and produce the information that makes such planning ever more efficient, then the task for the future is obvious: we must democratize and expand this realm of planning, that is, spread it to the level of entire economies, even the entire globe.

They contrast the successful Walmart with Sears, whose CEO Edward Lampert adopted the opposite tack, instituting an "internal market" by "disaggregating the company's different divisions into competing units." The authors found that Walmart's relation to its suppliers is so integrated and coordinated that its "suppliers cannot really be considered external entities," and its "vast network of global suppliers, warehouses and retail stores is regularly described by business analysts as more akin to *behaving like a single firm*" (pp. 31 and 37, emphasis theirs). The integration, cooperation, and coordination of far-flung suppliers are all the more interesting, since the cooperation seems to be not merely possible but actually necessary to compete in a global neoliberal economy. In contrast, Lampert's "grand free market experiment to show that the invisible hand would outperform the central planning typical of any firm" (p. 41) ended up destroying Sears. In essence, Lampert believed that the market could override the fallibility of human judgment. However, his experiment failed, and did so, say the authors, "for one reason above all: the model kills cooperation" (p. 44).[12]

---

[12] We return to the issue of the market versus human judgment in our final chapter.

The charge of "hubris" appears in both the free-market school's rejection of planning and in the Green rejection of geoengineering. As a corollary, the free-market ideologues—both libertarian and neoliberal—insist that any plan that invokes cooperation and planning instead of markets and competition requires totalitarian overreach. The Green narrative's attitudes toward nuclear power, biotechnology, and geoengineering form a shadow image of this view—a hubristic devil's bargain with the uncontrollable. The Green left's aversion to *geo*engineering is an echo of the libertarian aversion to *social* engineering. "Don't mess with nature" is an echo of the call, "don't mess with the market." Both paradigms exhibit the recurrent epistemic error—incompleteness (one-sidedness) of analysis.

These examples of incomplete analysis—common to both the political right and left wings—center around opposition to the establishment. That is, whatever we think the establishment favors, we're against it, regardless of what it is and regardless of whether the dominant portion of the establishment really holds such a view.

As alluded to in Chapter 9, anti-establishmentarianism has led part of the egalitarian leftwing, without their intention or realization, directly into a default Malthusianism. Ironically and tragically, egalitarians oppose the very technologies that the exploited majority needs to overcome the profit system and flourish. In our final chapter we elaborate further on the ways this default anti-establishment foundation fosters epistemic dysfunction.

Summarizing

We have tried to clarify several reasons why wind/solar cannot possibly replace fossil fuels. Chief among those reasons are wind/solar's low EROI and intermittency, both of which make them the most inefficient and costly sources of energy and the most destructive to our common environment, among other ways, through resource depletion and waste production. We have indicated the incomplete character of all analyses that favor these renewables and dismiss nuclear.

In a worldwide economic system based on market competition, any country (aside from the few with abundant hydro or geothermal) that opted to build an all-RE (wind/solar) energy system would be economically crushed by the competition and would render itself more, not less, dependent upon its rivals for energy.

As the most popular example of incomplete energy analysis, Jacobson and his colleagues claim that the energy transition to an all-RE world would be relatively painless, despite an estimated global cost of $62 trillion (their figure), almost two-thirds of the global GDP (https://tinyurl.com/8shv9ska). They further predict that this huge global investment would pay for itself in just one to five years.[13]

This wildly over-optimistic estimate exemplifies the epistemic dysfunction that is rampant in a competitive global economic system. Competition, just as it generates the clashing of economic units and therefore produces conflicting material interests, tends to generate clashing worldviews. While failing to converge, opposing worldviews often facilitate the detection of shortcomings in rival views. But the rivalry tends to blind advocates to invalid elements in their own narratives, which, along with the fear of admitting their rivals were correct, impedes revision.

The table that initiates Chapter 9 summarizes a number of views that seek to halt AGW but that differ with respect to two categories—energy and economic system. Our analysis, represented by the lower right cell, rejects that in the upper left and agrees in part and differs in part with those in the other two cells. Some of the best energy analyses in our view come from those with a libertarian or conservative bent that endorses free-market capitalism, and also nuclear energy that is plentiful, while sometimes exhibiting a degree of skepticism if not denialism about climate change. Meanwhile, some of the best analyses of climate change in our view come from those who are anti-capitalist but at the same time antinuclear, believing either that wind/solar can power complex societies or that humanity can get by with minimal societal energy, or both.

---

[13] The subheadline to this article says "six years." In contrast to Jacobson, Smil says (2024, p. 25):

> Nobody can offer a reliable estimate of the eventual cost of a worldwide energy transition by 2050 though a recent (and almost certainly highly conservative) total suggested by McKinsey's Global Institute makes it clear that comparing this effort to any former dedicated government-funded projects is another serious category mistake. Their estimate of $275 trillion between 2021 and 2050 prorates to $9.2 trillion a year. Compared to the 2022 global GDP of $101 trillion, this implies an annual expenditure on the order of 10 percent of the total worldwide economic product for three decades, rather than 0.2 or 0.3 percent for a few years (McKinsey and Company, 2022; World Bank, 2023).
>
> In reality, the real burden would be far higher for two reasons. First, it cannot be expected that low-income countries could sustain such a diversion of their limited resources and hence this global endeavor could not succeed unless the world's high-income nations annually spend sums equal to 15 to 20 percent of their GDP.

Segue to Final Chapter

Though this isn't a book about climate change, one major motivation for writing it is our understanding that AGW is an increasingly serious problem. Our sense is that the effects of AGW may go on to kill hundreds of thousands of people, or even more, over the coming decades—affecting the poorest and most defenseless among us to a greater extent—but will not drive the human species to extinction, even though it is rapidly exterminating numerous other species.

However, in the course of our writing this book we have come to recognize that no matter how much we may agree with an assessment of just about anything, there will virtually always be some element of blindness about some aspects, and conversely no matter how much we may disagree, there will virtually always be some valuable insight in it. So we pay attention to arguments, for example, that deny the reality of AGW or minimize its effects.

One such argument comes from US physicist and Princeton professor emeritus William Happer, who grants, or at least does not dispute, that the current level of CO_2 in the atmosphere has caused some warming. But he nevertheless claims that further carbon emissions will play no part in advancing AGW because, he says, the current 425 parts per million (ppm) of CO_2 already captures as much heat emitted from the ground as it can, and even doubling the CO_2 concentration to 850 ppm will have little effect in raising the Earth's temperature any further. With this claim, Happer lets the continued reliance on fossil fuels off scot-free.

Before we cite a particular response that supplies a missing part of the story, it should be noted that even if the concentration were to remain at 425 ppm, with no further emissions, the temperature would not stabilize immediately. Rather it would continue to rise to a new equilibrium temperature, until the reradiation of energy into outer space matches the incoming radiation energy from the sun, despite the trapping of part of the heat.

This would mean that recent heat waves over 49°C (120°F) would become even hotter. At those temperatures people, and many animals, are simply incapable of shedding their heat to the environment, and many more people (and animals) would perish in summers, particularly in the earth's temperate and equatorial zones. Among other effects, glaciers would melt faster than they are melting now, causing sea levels to rise faster, obliterating many coastal cities. It's also true that without GHGs the earth's temperature would be too cold to accommodate life, so the Earth's temperature is in a Goldilocks zone of GHG concentrations for human (and other animal) life, but we're now tending to rise beyond the upper threshold (again).

Now for the promised response to Happer's argument, which offers another example of the combination of insight and blindness that characterizes epistemic dysfunction. It is given by British geophysicist and climate analyst R.S. D'Arcy in a short video (https://tinyurl.com/2p9s5dft). During his exposé of Happer's incomplete analysis, D'Arcy puts Happer's claim in the context of denialist narratives.[14] In particular, he links to a talk by libertarian guru Jordan Peterson (https://tinyurl.com/59kkxzkn), to which D'Arcy in turn responds. Peterson suggests correctly that a Green economy will condemn the world to energy poverty. However, he attributes Green motivation to the "climate con," and believes that it derives from a hatred of humanity. That is, Greens, he says, hold that the human impact on the planet is inherently destructive, that humans necessarily pollute and sicken the earth—a tangled mixture of valid and invalid beliefs and assertions.

D'Arcy, who, like Peterson, makes clear his commitment to fighting poverty in general, including energy poverty, responds to Peterson's accusations against the Greens by laughing off the critique as a strawman. However, even as he rightly rejects Peterson's claim that climate concerns are a "con," D'Arcy asserts that wind/solar is now cheaper than fossil fuels and can power the planet while producing plentiful energy. This is precisely the claim that this book (and Peterson as well) rejects as blatantly false. The debate exemplifies insights combined with blindness on the part of all adversaries—particularly with regard to the problem of AGW as well as to the possible solution.

Among D'Arcy's references is the online blog "Our World in Data" (https://tinyurl.com/mvsy3hf6), with which he bolsters his view that the progressive cheapening of wind/solar benefits from learning curves. However, he implies that fossil fuels and nuclear are immune to such curves. Schernikau and Smith (2023), cited in our Chapter 5, trenchantly criticize the optimistic Green narrative that D'Arcy accepts without question, but they, in turn, accept without question Happer's climate denialism.[15]

---

[14] We use the term "denialist" to contrast with the often-misused term "skeptic." Skepticism questions; denialism denies.

[15] For a detailed up-to-date explanation refuting the portion of Happer's claim that negates the valid portion, see https://tinyurl.com/2vp6zsuv and https://tinyurl.com/43p3jear. In addition, see Weart (2008, pp. 23–24).

The epistemological point that motivates the foregoing citations and introduces our final chapter is that both sides of the argument are situated within clashing worldviews, each of which belittles and sometimes demonizes the other while regarding their own view as scientific, but also, in many cases, appropriately anti-establishment. The valid insights in each position tend to obscure from the insiders their own respective blind spots on scientific as well as social or political questions.[16] In our final chapter we explore the deeper sources of this common epistemic breakdown.

---

[16] We make no claim to be free of this combination; we just do the best we can to be open to arguments contrary to our own.

11

Fragmentation, Competition, and Epistemic Dysfunction

Throughout this book we have explored, both implicitly and explicitly, barriers to widespread comprehension that renewables are incapable of halting AGW by themselves and are infeasible, unsustainable, and would lead to resource exhaustion and energy impoverishment. This chapter is about the Green adherence to a strict anti-establishment attitude: "Whatever the establishment wants or says must be opposed." This attitude leads to **epistemic dysfunction**, in particular to confirmation bias, and is captured in the bumper sticker "Subvert the Dominant Paradigm." This chapter offers a condensed genealogy of what we have called **anti-establishmentarianism**. We suggest that this attitude is a profoundly inadequate response to a society characterized by real competition both within and between classes, the latter known as the class struggle. In short, opposing the establishment is not necessarily the problem; letting this opposition take precedence over, or even substitute for, a multi-faceted scientific analysis is the problem.

The Historical Foundation of the Anti-Establishment Paradigm

We begin with a critique of so-called **standpoint epistemology**[1]—one central way that victims have responded to the scars inflicted

---

[1] Roughly defined, standpoint epistemology holds that oppression and discrimination confer on their victims a greater understanding of those experiences than others can possibly possess. It holds, for example, that sexism can only be comprehended by women and that racism can only be grasped by persons who are African American, Latin, Native American, or Asian. Conversely and perniciously, it denies that a sympathetic man can understand or presume to describe the experience of women or that anyone not a member of a racialized or ethnicized group has a right to speak about the experiences of the group's members.

by a competitive and exploitative class-divided society. A closely related idea, but with a somewhat different intellectual history, holds that "the margins are at the center"—that those who have been marginalized are, in fact, central and thereby possess a certain wisdom. In the philosophy of what's often called "post-structuralism" this idea takes the form of the deconstruction of binary oppositions—the concept that society is shaped by oppositions between a primary term and a secondary term. The primary term is associated with truth, presence, center, and origin while the secondary term is associated with the inferior, the derivative, the copy, and the marginal. The deconstruction consists of the recognition that the primary term is undermined or subverted or governed by the secondary term.

While Algerian-French philosopher Jacques Derrida regarded these oppositions as merely philosophical, they have come to promote an "identity politics" in which marginal identities (those who are otherized) are taken to be founts of wisdom by virtue of their positionality or standpoint. Specifically, the more exploited and oppressed, the wiser. And conversely, the less exploited and oppressed, the more delusional. The more establishment the source, the less valid are its claims; the more anti-establishment, the more valid. While the more exploited and oppressed members of society can indeed be potential sources of insight, cruder versions of standpoint theory turn these victims into near *automatic* sources of wisdom. And such wisdom is thought to be entirely internal to the group and self-validating.

Also derived from post-structuralism and reinforcing this "wisdom of the margins" is a concept introduced by the French historian of ideas Michel Foucault—that of "subjugated knowledges," a term implying subjugation of both the knowledge and the knowers.[2] However, since Foucault repudiates any normative epistemology that allows for a reliable distinction between true and false, his "subjugated knowledges" lay claim to truth only by virtue of their

---

[2] From "Two Lectures," collected in Colin Gordon, ed., *Power/Knowledge: Selected Interviews and Other Writings, 1972–1977* (1980). Focusing on penology and psychiatry, Foucault equates knowledge to power and power to domination, all possessed by the establishment. Then the "insurrection of subjugated knowledges" represents resistance to the establishment (p. 81), whereby Foucault lends undeserved authority to standpoint epistemology. In her recent book, *Left Is Not Woke* (2024), Susan Neiman does an excellent job connecting theories of power to the victimology in updated versions of standpoint theory (about which more below), both versions denying ethical universalism in favor of a dogmatic tribalism.

marginality—even as the very concept of truth is undermined by Foucault's relativism.[3]

Derrida's and Foucault's variations aside, the roots of standpoint epistemology come from the formulation, based on social/economic class, by Hungarian Marxist Georg Lukacs. It has since been updated (and oversimplified) to apply to gender, race, and, continuing the logic, intersectionality.

For Lukacs, the main source of cognitive distortion about the social totality derives from capitalism's market exchange relation—commodification, in which the value of everything is reduced to its monetary price in the market. Marx called this distorted outlook "commodity fetishism," the central function of which is to disguise the actual source of profit. As Marx showed, profit lies in the exploitation of labor, in which capitalists extract surplus value (unpaid labor) from workers, and instead pretend (and likely believe) that their profit comes from their own hard work and knowledge, and from charging more for a product in the market than it costs them to produce.[4] Such distortion constitutes a partial or fragmented cognition, which Lukacs called "reification."

Marx exposed the conflation of two different realms of an economy—production and circulation. It is in the realm of production that exploitation resides. Lukacs posits that the victims of exploitation can see through, or have the potential to see through (a crucial difference), the distorted account of their exploiters. That is, the vantage point of the working class, says Lukacs, allows it to "see the whole," to understand the social totality by coming to understand its own relation as a class to the capitalist system—in other words to become "class conscious."

The idea that the working class sees all and knows all is a significant source of dogma in leftist thinking, and a form of groupthink. Just because an individual is a victim of oppression, and therefore understands what that oppression feels like, does not

---

[3] Relativism holds that the perceptual framework of each social group is just as valid as all the others, while realism holds that there is an external reality that can be comprehended by all through emerging and evolving scientific methods, regardless of social group membership.

[4] The hidden reality is the converse. Rather than charging more than it costs to produce, capitalists pay workers less than the value of their labor time and on average, being constrained by supply and demand, are only able to charge the value of the latter. So due to constraints beyond the control of the capitalists (except for illegal collusion and price fixing), price is not set higher than cost; rather cost is held lower than price. In other words, capitalists pay workers less than the value of their labor time, while throughout the economy prices on average are pegged to the total labor (living and dead) that goes into the product (Chapter 10).

mean that the individual knows where the oppression comes from or why it exists (the upstream cause), or what to do about it (the downstream path to liberation). Nor does each individual yet have a basis for confidence that all fellow victims understand what she or he understands, until they create organization in which they openly share that understanding and act on it together.

Lukacs has mistakenly equated the "universal subjectivity" of the proletariat (its group self-knowledge) with knowledge about real social processes, so that universal subjectivity becomes in effect identical with objectivity, allowing (if not guaranteeing) knowledge of the social totality.[5] British Marxist literary theorist Terry Eagleton exposes Lukacs's error:

> If the working class is the potential bearer of such class consciousness, from what viewpoint is this judgment to be made? It cannot be made from the viewpoint of the (ideal) proletariat itself, since this simply begs the question; but if only that viewpoint is true, then it cannot be made from some standpoint external to it either. As [Indian-British Marxist political theorist] Bhikhu Parekh points out, to claim that only the proletariat allows one to grasp the truth of society as a whole already assumes that one knows what the truth is. It would seem that truth is either wholly internal to the consciousness of the working class, in which case it cannot be assessed as truth and the claim becomes simply dogmatic; or one is caught in the impossible paradox of judging the truth from outside the truth itself, in which case the claim that this form of consciousness is true simply undercuts itself.[6]

Lukacs's idea that truth can be wholly internal to an exploited and oppressed group, in this case the working class, reappears in more modern examples of standpoint epistemology with respect to categories such as gender and "race." These examples centrally feature an anti-establishmentarian attitude—an attitude prominently featured in Green misconceptions of energy (our book's main focus).

---

[5] In other words, Lukacs exaggerates the degree to which victims of exploitation and oppression understand how and why they are victimized and what efforts are necessary to eliminate their intolerable situation. If Lukacs were right, and potential knowledge were identical with actual knowledge, capitalism would have long since been abolished by the victims of its intrinsic exploitation.

[6] *Ideology: An Introduction* (1991), p. 97. Note the similarity between the conflation of "universal subjectivity" and objectivity criticized by Eagleton and the idea expressed above equating marginality with truth.

If truth is "wholly internal" to a group, even of millions, and inaccessible to those outside the group, then this truth cannot be shared across group lines, cannot be demonstrated in the ordinary sense of the term, and so cannot qualify as evidence in any scientific sense. This leaves Lukacs's idea as dogmatic. If, as he claims, the dogma of the powerful (the capitalist ruling class) is a socially produced illusion, why is not the "standpoint of the proletariat" also an illusion, though opposite?

Lukacs's outlook originated as a rebuttal to a largely "positivist" epistemology, an epistemology claiming neutrality or independence from material interest. Positivism, in turn, was a response by the Vienna circle, a group of philosophers and scientists, to the heated class struggles between the world wars. As articulated by Lukacs, the Marxist left rejected neutrality and accepted the standpoint epistemology of class (as opposed to gender or race) because it pointed to a desirable outcome. That is, it suggested that a collection of individuals might spontaneously act together just because they shared a common oppression, from which the left assumed that the proletariat also shared a common understanding of this oppression. The belief in spontaneity by some Marxists and working-class activists (though not Lukacs) led them to promote trade unionism—a limited response to exploitation, limited because it only seeks reforms that grant immediate, if temporary, relief from intense oppression but does nothing to seek permanent liberation from capitalism's exploitation and oppression.

One principal corrective to spontaneity on the revolutionary left was the Leninist Party, which held that only through the abolition of capitalism could the working class free itself from exploitation, and that deliberate organization (in the form of the Party) was crucial to achieving that end. Interestingly, in later forms of standpoint epistemology, those rooted in gender and race, this critique of spontaneity disappeared, perhaps because class and capitalism disappeared from view. We discuss one contemporary version below.[7] However, as we hope to make clear, the Leninist corrective to spontaneity is also far from adequate as it fails to confront the problem of collective illusions, or groupthink.

---

[7] We focus on racialized standpoint theory below, since its influence on current anti-establishmentarian dogma is far greater than the gendered form. The classic feminist view of gendered standpoint epistemology is expressed by Nancy Hartsock in her book *Money, Sex, and Power: Toward a Feminist Historical Materialism* (1984). While Hartsock asserts that non-females could also adopt the "feminist standpoint," such a concession, as we will see, is rejected by certain adherents of racialized standpoint theory, who claim that only members of the racially oppressed minority can understand its point of view.

While class membership neither guarantees nor precludes a grasp of truth, certain vantage points (or frameworks or paradigms), related to class or not, confer greater or lesser epistemic access to an object of investigation, make it easier or harder to grasp.[8] So when it comes to material interests, while the associated vantage point may give greater access to certain insights, it can just as easily encourage confirmation bias and blindness.[9]

By virtue of their social and economic situation, the argument goes, workers generally understand exploitation better than the exploiters. The latter are almost universally attracted to the obfuscating distortion of commodity fetishism, and therefore believe that the exchange of wages for labor time is an equal exchange. But workers' experience of exploitation does not automatically lead them to recognize the falsity of that notion or to grasp the need to abolish capitalism. Eagleton relevantly asks (1991, p. 104):

> Are the complex equations of [Marx's three-volume work] *Capital* . . . no more than a theoretical "expression" of socialist consciousness? Is not that consciousness partly constituted by such theoretical labor? And if only proletarian self-consciousness will deliver us the truth, how do we come to accept this truth as true in the first place, if not by a certain theoretical understanding that is relatively independent of it?

In other words, Eagleton makes two points: first, that the theoretical understanding by Marx and other Marxist thinkers comes primarily from intellectual labor outside the direct experience of exploitation, and second, that in general only through a standard external to the group can one judge the truth or falsity of an argument or theory.

As we suggest later in the chapter, the dangers of groupthink, not just on the left but in the entire society, have intensified with

---

[8] For example, in Chapter 8 we suggested that the LNT paradigm makes it harder for its adherents to generate questions about an adaptive protective response by the organism, since LNT either denies the existence of such a response or greatly underestimates its effect.

[9] Realist US philosopher of science Richard Boyd is known for enlarging on the concept of scientific paradigm, first elucidated by US philosopher and science historian Thomas Kuhn in his widely read book *The Structure of Scientific Revolutions* (1962). Boyd refers to successful scientific method in a mature scientific discipline as a "paradigm-dependent paradigm-modification" method. Such a method might more successfully avoid confirmation bias. Boyd's understanding of knowledge production is far more sophisticated than that offered by the versions of standpoint epistemology explored here.

the dominance of social media. Whatever epistemic advantage is afforded by "social location" has become at least partly compromised by discourses and ideologies that "contaminate" direct experience. Social media contaminates that exaggerated advantage to the point where anything like a scientific approach to bringing about socialism demands organizations whose emergent properties make its members smarter not, as groupthink does, dumber.

Recent Formulations of Standpoint Epistemology

The fallacies of *class* standpoint theory are magnified when applied to "racial" categories. The latter proposes that knowledge of systemic racism—including its sources and solutions—is afforded by the "black standpoint," or black collective self-consciousness.

A rational modification of standpoint theory would grant that oppressed people *have an interest* in resisting their oppression and so have an interest in *discovering* the source of that oppression. Inhibiting that discovery are numerous alternative descriptions and explanations, promulgated by the beneficiaries and promoters of exploitation and oppression. Coming to understand complex social relationships takes work, the right kind of work, as Eagleton's reference to Marx's theoretical labor suggests.

In structurally antagonistic societies, confirmation bias and cherry-picking can unwittingly pervade our own paradigms. We can be quite perceptive about the flaws in our opponents' positions but less so about our own. At the current historic moment in the US, liberals can easily see the gaslighting by rightwingers in the cascade of lies around January 6th and the myth of the stolen 2020 election. Those on the right, in response, have waged culture wars that focus on "wokeness," which originally meant "awoken to the injustices of the system." While the alleged sins of wokeness are often wholly fabricated by the right, wokeness contains some actual inaccuracies, though they are no match for the distortions of Trumpism.

One conservative criticism of wokeness (though the author describes himself as a liberal) correctly captures some of the flaws in standpoint epistemology even as the criticism exhibits its own areas of blindness. African-American linguist John McWhorter (2021) combines a penetrating critique of one form of woke dogma with his promotion of a culture-of-poverty analysis (the claim that poverty is self-generating and self-perpetuating, rather than being imposed by an exploitative system). McWhorter thereby displays his blindness to the inequality-producing features of capitalist competition and engages in a classic blame-the-victim narrative.

He targets popular authors like Ta-Nehisi Coates, Ibram Kendi, and Robin DiAngelo and addresses how their racialized standpoint epistemology enters into popular culture. He correctly criticizes the woke claim that the feelings of the essentialized oppressed (all black people, in this case, irrespective of class) cannot be wrong and, equally importantly, must not be questioned by those outside the group. To exemplify the latter stance, McWhorter quotes an anonymous response on Facebook to a white commenter who expressed agreement with BLM (p. 55):

> Wait a minute! You "agree" with them? That implies you get to dis-agree with them! That's like saying you agree with the law of grav-ity! You as a white person don't get to "agree" or "disagree" when black people assert something! Saying you "agree" with them is EVERY bit as arrogant as disputing them! This isn't an intellectual exercise! This is THEIR lives on the line.

This "no right to speak" dogma of racialized standpoint episte-mology has been termed "epistemic deference" by African-American philosopher Olúfẹ́mi Táíwò. Táíwò notes that epistemic deference essentializes blackness, associating it with an authentic experience of suffering, and puts white and other non-black antiracist sympathizers in an impossible position—namely, to which black authors should they defer, those who support or those who criticize standpoint theory?[10]

Táíwò mentions a white colleague who is researching some unspecified racial issue. Offering her notes to Táíwò, the colleague says, "I don't think I'm the right person to write this story—I have no idea what it's like to be Black." Táíwò says, "I flinched inwardly." Despite differences, he shares McWhorter's view that racialized standpoint epistemology essentializes blackness. It erases class difference and confers authenticity on the collective "race."

Táíwò notes that "epistemic deference" assumes that suffer-ing confers wisdom. But, while potentially *aided* by a particular experiential standpoint, epistemic privilege "is achieved only through deliberate, concerted struggle from that position." Knowledge can only be gained by study and hard work. Feeling

---

[10] "Being-in-the-Room Privilege: Elite Capture and Epistemic Deference," *The Philosopher* 108: 4 (https://tinyurl.com/ucw6hjhb).

is not knowledge, and trauma "can corrupt as readily as it can ennoble." He says:

> Contra the old expression, pain—whether borne of oppression or not—is a poor teacher. Suffering is partial, short-sighted, and self-absorbed. We shouldn't have a politics that expects different: oppression is not a prep school.

McWhorter's assertion that critical race theory (CRT) is the discourse of a virtuous "elect," in the Puritan religious sense, needs some unpacking: the religious or "revelatory" character of woke dogma can be disguised since the superior perspective of the ethnicized standpoint is presumably rooted in group experience. Because experience is assumed to carry with it the scent or feel of the real in comparison to individual (and supernatural) revelation, defined in opposition to ordinary experience, the religious character of the discourse is thereby cloaked—religious in the sense of dogmatic and irrational, both of which are evident in the quote above claiming that "whites" or other "outsiders" have no right to speak about the "other" yet are told that "silence is violence."

McWhorter's insights, however, accompany, and perhaps enable, a lack of judgment. He elevates wokeness to a greater level of threat than the tsunami of Trumpist lies, ending his first chapter with the warning: "Make no mistake. These [woke] people are coming after your kids." The pervasive entanglement of insight and blindness is also illustrated by CRT itself, which often makes factually accurate claims, denials by the rightwing notwithstanding. For example, institutions like policing, prisons, and the judiciary system are, demonstrably, systemically racist. However, mainstream CRT tends to imply that these institutions either have no effect on working-class whites or actually benefit them, if they are not also directly enforced by these white workers to enhance black oppression. This attribution lets off the hook the capitalist class, whose interests rule the society, and it turns fellow victims of exploitation into mutual enemies—the very functions of the racism that CRT purports to oppose. The factual inaccuracies of CRT percolate up into popular culture.[11]

---

[11] An excellent non-Marxist critique of the epistemology underlying mainstream CRT is a book by Helen Pluckrose (a liberal) and James Lindsay (a conservative)—an interesting collaboration—*Cynical Theories: How Activist Scholarship Made Everything about Race, Gender, and Identity—and Why This Harms Everybody* (2020). Valuable Marxist critiques include Walter Benn Michaels and Adolph Reed Jr.'s book *No Politics but Class Politics*, edited and with a foreword by Anton Jäger and Daniel Zamora (2022), and *The New York*

A rightwing partner in distortion is the popular country western song, "Rich Men North of Richmond." Despite its promising title, its singer/songwriter, Oliver Anthony, blames overweight welfare recipients (the ever-present dog whistle implies black, all the more effectively when unstated) for the plight of the white working-class. The argument over which sections of the working class oppress the others leaves the capitalists' role in the shadows, while the right and left distortions meet in the epistemically dysfunctional corridor.

A rich mixture of insight and blindness can be found in the paradigms of many who join us in opposing the LNT outlook (Chapter 8). Paradigms contain both scientific and extra-scientific components. Many scientists oppose LNT primarily on evidential grounds, but some also oppose it because it impedes expansion of nuclear power and some because it reinforces destructive radiophobia. While for some, nuclear energy is central to halting AGW, for others in the anti-LNT corner, climate change is a hoax motivated by Green advocacy for government intervention in the economy, or worse. For such libertarians, LNT serves to invite despised government intervention (misregulation), though as nuclear energy advocates, they eschew consistency and call for government intervention to fund its development.

Green *defenders* of LNT, on the other hand, assume that the formula is scientifically accurate, so they oppose nuclear energy and support aggressive government regulation of radiation. As mentioned (Chapter 9), many of the best energy analysts in our view are libertarians and favor the capitalist free market, while those Greens who recognize the contradictions and oppressions of capitalism are blind to both the incapacitating disadvantages of renewables and the compelling advantages of nuclear. Both resort to the manipulation of fear—of either capitalism's opponents or of radiation. Since paradigms generally harbor both insights and blindnesses, it obliges us to seek our own areas of blindness and be especially open to their being pointed out by others.

Anti-establishmentarianism operates through confirmation bias whereby confirmed insights *enable* the blindness instead of producing a heightened awareness of it. More precisely, it appears that science enables the extra-scientific components to function subconsciously as background ideology. While realist philosopher Richard Boyd characterizes mature science as a paradigm-depen-

---

Times's 1619 Project and the Racialist Falsification of History, eds. David North and Thomas Mackaman (2021).

dent paradigm-modification method, anti-establishmentarianism may usefully be characterized as a paradigm-dependent bias-confirmation approach. For good science, insight raises the probability of further insight, whereas anti-establishmentarianism not only *lowers* the probability of insight, it likely prohibits insight altogether.

Anti-Establishmentarianism and GMOs

Green outlooks share with standpoint epistemology the commitment to anti-establishmentarianism. Greens like Ajl (Chapter 9) romanticize marginalized indigenous populations, assuming their wisdom to be "natural" and therefore authentic. This default anti-establishment attitude enables certain Green NGOs and their well-meaning activists forcibly to impose, in good conscience, their favored technology on Indian and African farmers who struggle to survive. For these NGOs, acceptable technology is confined to organic and traditional farming, assumed to be natural, but rules out biotechnology, assumed to be unnatural, polluting, contaminating, and, the anti-establishment element, manipulable by big capital.[12]

Mark Lynas's book *Seeds of Science* (cited in Chapter 9) describes controversies around GMOs. Arguments against nuclear energy are transferred to GMOs almost without alteration, both exhibiting the cognitive distortions spawned by anti-establishmentarianism. GMO science is ignored in favor of attacks on the establishment, big capital. The formula "GMO equals Monsanto, and Monsanto is the devil" is the agricultural equivalent of "radiation equals cancer, and cancer is the devil." Demonization of GMOs is used to validate organic farming, which pretends to obey the small-is-beautiful mantra, though organic farming is also associated with large corporations.

Lynas traces to US activist Jeremy Rifkin the conflation of GMOs with Monsanto. Monsanto only entered the game after government-funded research had established the safety and effectiveness of GMOs. Rifkin's anti-GMO career followed his opposition to fascism and Naziism and was boosted by protests against the Vietnam War and against concentrated corporate power. He demonized GMOs by equating this crop science with the pseudoscience of eugenics, which originated mainly in England and the

---

[12] The celebrated association of wind and solar with nature similarly obfuscates their limitations.

US, and was later taken up by the Nazis. This guilt-by-association schema is the same whether applied to biotechnology or to nuclear energy.

A cliché of social psychology calls for wariness of thinking that features "either/or" or "us/them," an echo of post-structuralist "binary oppositions." While such oppositions can be fairly accurate about power relations in competitive societies like ours, they are at the same time oversimplified. This makes them susceptible to grotesquely distorted formulations, as the term "demonization" suggests.

The liberal hope is that mere recognition of cognitive distortions will lead to their rejection and thereby provide a foundation for the emergence of a unifying "democracy." Thwarting this hope, the truth content entangled in cognitive distortions is often substantial. Capitalism generates tremendous global inequality and poverty even as it generates enormous wealth. While most wealth is held by a few, capitalism raises many others out of poverty (particularly in China), partly inadvertently and partly to form a large consuming population. No institution under such a profit-making, inequality-producing order can be entirely independent of these contradictions and this can lead to vastly oversimplified understandings of the power dynamics involved. Oversimplification is exemplified by the demonization of corporations as well as the science they both enable and appropriate in distorted forms. Lynas's description of anti-GMO campaigns in India and Africa sheds light on the demonization that inhabits Green anti-establishment discourse.

The campaign against GMOs in India is largely led by Indian author and environmental activist Vandana Shiva. She claims that big agricultural corporations have produced a "suicide seed" to assure profits—a seed producing sterile offspring that forces poor farmers to buy new and nearly unaffordable seeds for each planting. Extracting double value from her nomenclature, she claims that the affected farmers have been driven by these seeds to an epidemic of suicides.

Worse yet, Shiva maintains that the sterile offspring of genetically engineered (GE) seeds would spread and contaminate organic crops in neighboring fields (https://tinyurl.com/bdd9m4hu). Never mind that sterile seeds cannot reproduce, so their features, including sterility, can't contaminate. She adds that the innocent neighboring organic farmers could be hit with patent infringement lawsuits for stealing intellectual property.

As discussed in Chapter 9, technologies cannot in general be reduced to their corporate uses; almost all can be repurposed. Nor is the corporate use of GMOs as nefarious as claimed. Shiva's target is a bacterium, *Bacillus thuringiensis* (*Bt*), that happens to be a natural pesticide and the first choice of even organic farmers. *Bt* genes introduced, for example, into brinjal (eggplant) very precisely target the crop's pest infestation. But because it is genetically engineered into the plants, many Greens demonize it, regardless of its proven safety and effectiveness. *Bt* is sought after by the very eggplant farmers Shiva claims to champion.

As Lynas notes, something akin to a suicide seed is familiar to many of us as hybrid seed that fails to breed true, but Shiva's "suicide seed" does not in fact exist. Nor, Lynas points out, do her imagined mass suicides. Shiva claims that GE cotton seeds increased the price of cotton by 80,000 percent [*sic*], as a consequence of which "300,000 Indian farmers have committed suicide, trapped in vicious cycles of debt and crop failures." Echoing Jacobson's resort to unwarranted precision (Chapter 3), Shiva attributes precisely 84 percent of these suicides "directly to Monsanto's *Bt* cotton," in effect accusing Monsanto of negligent homicide, if not mass murder. However, while India's cotton farmers—"40 million in the cotton growing states alone," says Lynas—may account for a significant number of suicides, one detailed study found that their suicide *rate* is lower than that of non-farmers across the cotton growing region and is similar to that of French and Scottish farmers, where no GE crops are grown.

Finally, British social statistician Ian Plewis, who studied suicide rates among farmers in the cotton growing regions of India before and after the introduction of *Bt* cotton, found, according to Lynas (p. 117):

> not only is the ubiquitous *Bt* cotton-suicides story incorrect but [says Plewis] "there is evidence to support the hypothesis that the reverse is true: male farmer suicide rates have actually declined after 2005 having been increasing before then." The Indian farmer suicide story is a myth built on tragic individual anecdotes and extrapolated to a whole country by those like Vandana Shiva with an ideological axe to grind and little concern about the true facts.

Lynas asks why these farmers would keep purchasing this seed if it were trapping them in debt and inducing them to end their lives. The answer, researchers found, is that the widespread adoption of *Bt* products (cotton in this case) led to a 24 percent increase in yields due to reduced pest damage and an increase in profits of 50 percent, along with a 50 percent drop in pesticide use in the area studied. The researchers, says Lynas, "estimated that if the benefits of reduced pesticides are extrapolated to India as a whole, '*Bt* cotton now helps to avoid at least 2.4 million cases of pesticide poisoning every year'" (p. 115).

Moving from India to Africa further illuminates the seriously destructive side of Green anti-establishment attitudes. Lynas describes the science-denying efforts by European Green NGOs in Africa to block GMO technology by erecting both legal and ideological obstacles to its employment. Their efforts range from Tanzania to Uganda, from Mozambique to Kenya, and elsewhere on the continent. In particular, European NGO delegates strongly influenced the United Nations Environment Programme under the Cartagena Protocol on Biosafety to promote laws making GMOs illegal in Africa. Under such a law, Tanzania charges large fines, and anyone involved in the production and testing of GMOs is threatened with imprisonment. Worse yet, substantial damages can be awarded "to any anti-GMO group claiming harm" (p. 135).

African scientists whom Lynas interviewed argued that organic farming and agroecology produce yields inferior to those of GMO methods, leaving farmers "trapped in a cycle of subsistence." They add that crop diseases avoided by GMOs—including cassava brown streak, banana bacterial wilt, maize lethal necrosis, and certain cotton crop diseases—cannot be prevented by organic and agroecological methods.[13]

Anti-GMO ideology is sometimes bizarre. During a talk on his changing views of GMOs that Lynas gave in Tanzania, an organic farming activist told the audience in Swahili that genes implanted in maize will turn boys into homosexuals—though Lynas only learned what had been claimed when it was later translated for him (p. 140). In Uganda, anti-GMO campaigners fought to sabotage any legislation allowing such crops to "be distributed to farmers with proper testing and safeguards."

Fearmongering was key to this campaign of sabotage. For example, an evocative warning was issued that bananas could be longer if farmers were to "pick a gene from a snake and put it into the banana so that the banana becomes the length of the snake." Muslim audiences, some of whom regard pigs as unclean, are told that scientists insert pig genes into maize to make it "fat as a pig" (p. 141). Or GMOs "are here to reduce our life expectancy" (p. 142). The cam-

---

[13] Organic and agroecological methods have their place. Plant geneticist Pamela Ronald and her husband, organic farmer Raoul Adamchak, both at the University of California, Davis, argue in their book (2018) that organic farming and GMOs can work in concert. But, they say, were organic agriculture to involve the vast extents of land required by wind and solar farms, its disadvantages might become excessive.

paigners create fake photos showing children's heads growing out of corn plants (p. 143). NGO fearmongering in Kenya blocked the introduction of disease-resistant GE maize and sweet potato.

A discredited study by Algerian-French molecular biologist Gilles-Éric Séralini, purporting to show that GE corn causes tumors in rats, has been relied upon by NGOs to scare the Kenyan health minister, Beth Mugo, herself a breast cancer patient, into banning GMO imports. In Ghana, public sector-generated GMO crops like *Bt* cowpea, cotton, and drought-resistant rice were blocked by an NGO called Food Sovereignty Ghana (FSG). The organization achieved its goal by stressing the association of *Bt* with Monsanto and demanding a "total ban on everything GMO." To ensure its success, FSG pointed to a host of ailments that it falsely laid at GMO's door—from birth defects and autism to infertility and Parkinson's (and more). "Needless to say, no genuine scientific evidence supports any of these assertions" (p. 152). Similar story in Zimbabwe, asserting associations between GMOs and a plethora of diseases, culminating with the claim that sexual dysfunction due to GMOs is a "huge problem in the USA, where males become impotent around the age of 24, at the prime of life" (p. 153).

Lynas notes, "Everywhere I went in Africa it was the same story. Foreign-funded NGOs, supported mainly by donors in Europe, were delaying or blocking the development not just of biotechnology but of modern agriculture generally across the continent" (p. 150).

Meanwhile, Ugandan scientists scramble to save banana crops, a staple food, from a bacterium that causes Banana Xanthomonas Wilt (BXW, same as banana bacterial wilt). International scientists, partnering with government-run institutions in Africa, have developed BXW-resistant bananas by inserting a sweet pepper gene. "First results were promising," as the plants with the resistance gene were thriving while the controls died (p. 146). They obtained similar results with cassava plants, fending off a virus that yellowed and shriveled the plants.

We bother to enumerate these incidents to give a full picture of the way that well-intentioned Greens can unwittingly end up killing people when they refuse to let science overrule prejudice, and even resort to fear-promoting lies in service to what they regard as a higher cause (shades of Muller's use of LNT to counter nuclear weapons testing).[14] Lynas is particularly adept at describ-

---

[14] The epistemic dysfunction of the antinuclear position depends for its persistence on default anti-establishmentarianism.

ing such Green-associated devastation as he used to be one of the activists. But he finally became aware of the negative effects of their efforts and the absence of science behind their dogma. To many of his former friends he is now an enemy.[15]

Lynas experienced first-hand the effect of aggressive Green rhetoric when he participated with Cornell's Alliance for Science, helping introduce *Bt* eggplant to Indian farmers. This research was sponsored by a public university and largely funded by the government and was not connected to Monsanto or any other big agricultural corporation. The research and seeds they produced were made available to poor farmers patent free. The *Bt* genes were introduced into seven local varieties of eggplant to reduce the use of pesticides, whose negative health effects have been found "to include non-Hodgkin's lymphoma, leukemia, birth defects and cancer" (p. 119). The use of *Bt* in eggplant was spectacularly successful, reducing pesticide use in some cases down to zero, and the plants were notably healthier than controls. Yet while this work was going on, anti-GMO activists were trying to convince the recipient farmers that the seeds were poison, and if eaten would paralyze their children.

No shill for big capital, Lynas studies the Green denunciations of Monsanto focused largely on the herbicide glyphosate and finds their evidence quite thin.[16] He is aware of a possible charge that he is letting Monsanto off the hook, but his point is worth repeating. The Green NGOs' default is not a search for truth but rather a decision to treat testimony in favor of their position at face value, "making the procedural assumption at the outset that all Monsanto critics would be telling the truth" (p. 128). The flipside is that anyone defending a product manufactured by a corporation, or by universities linked to corporations, is necessarily lying in the interest of profit. Neither, of course, need be true.

In other words, Lynas's critics are bathed in anti-establishment confirmation bias. Lynas, in contrast, distinguishes sharply between the technology and the objectionable power of corporations. Iron-

---

[15] In his initial days in the medical device evaluation section of the FDA, Sacks approached his job with the intention of protecting the public from medical devices. He assumed that most devices were harmful and designed by companies purely for their own profit interests. His anti-establishmentarian bias was soon eroded and replaced with the realization that most medical devices brought to the FDA for approval were actually beneficial—the profit motivation of their sponsors aside, whether they were small start-up firms or major corporations. His mission then became one of helping to approve these devices and hastening their availability. He also discovered that virtually all the scientists who worked at the FDA were honestly engaged in promoting the health of the public.

[16] Glyphosate is an herbicide invented in 1970 by Monsanto chemist John Franz. Designed to go with herbicide-tolerant crops, it is marketed as 'Roundup'.

ically, it turns out, however, that Monsanto's market power is actually quite modest (p. 128):

> any multinational in the world would have been judged harshly in such a process [the International Monsanto Tribunal]. Would Google have fared any better? Or Apple? Or even Amazon-owned Whole Foods, which has a similar annual turnover to Monsanto . . . ? All large companies . . . wield power that can outstrip that of elected governments, and without sufficient scrutiny and accountability this power will end up being abused. Monsanto, however, isn't even in the top 50 big corporations, appearing at 189 in the 2016 Fortune 500 list.[17]

Epistemic Dysfunction and the Proposed Remedy of Transparency and Free Speech

In our review of various irrational and destructive dogmas of leftist Green ideology, and some of their genealogies, we do not intend to let rightist dogma off the hook, which is at least as irrational and destructive as Green dogma. Rather, we concentrate on the misconceptions of subscribers to a sociopolitical stance that aims to eliminate exploitation and oppression and achieve equality for all. That is, we are trying to get our own house in order so that the project of ending exploitation and oppression can operate on sound procedural assumptions amidst a world of conflicting and confusing claims.

Liberal pluralism and **free speech** are cardinal values in most western societies. They have become the go-to values to guide us in discerning and extracting valid conclusions from various forms of groupthink. Liberals and conservatives, as noted, are positioned to see each other's blind spots with far more clarity than they see their own.

While Supreme Court Justice Louis Brandeis said, "Sunlight is said to be the best of disinfectants," or as the liberal *Washington Post's* masthead declares, "Democracy Dies in Darkness," it seems that light is necessary but not sufficient to solve the many manifestations of epistemic dysfunction. And aiming at complete absence of constraints on speech often becomes part of the problem. That is, to pursue the metaphor, sunlight produces glare just as easily as it produces illumination, and glare can be just as blinding as darkness. TMI (too much information, not the nuclear power plant)

---

[17] Monsanto has since been bought out by Bayer, which clocked in at 137 in Forbes's Global 2000 list of biggest companies (https://tinyurl.com/4rzk57ax).

can as easily hinder as enable the extraction of the valid from the invalid. Finding a needle of truth is more difficult when buried in a haystack of falsity or, for that matter, in a haystack of other valid propositions, but ones with less relevance to the matter at hand. To refer to our recurrent theme, there are Goldilocks zones for illumination and information—darkness is too little and glare is too much. Either one obscures.

The proclaimed healing powers of free speech are usefully deconstructed by legal theorist Stanley Fish in two of his books on the First Amendment (1994, 2019). The fallacies associated with the concept of competition in a free market (Chapter 10) have their correlates in the concept of free speech as a "marketplace of competing ideas." This marketplace is said to lead to the truth just as the economic marketplace is said to lead to the best products for consumers. The marketplace of ideas is promoted as the best means "of winnowing . . . the true from the false in the course of a free and open competition" (1994, p. 122). But while the censorship practiced by many governments imposes darkness, those that tout the free marketplace of ideas impose glare—even as they denounce the censorship by others and hypocritically use this denunciation to cloak their own withholding of information from the public and their suppression of many types of speech that oppose capitalism or threaten "national security."

The First Amendment of the US Constitution, in essence, prioritizes glare over censorship, either of which can obscure truth. Nor is the glare unregulated, considering the consolidation, concentration, and hence power of the mass media. Their output is confined to positions within establishment discourse, even if such positions (such as woke liberalism versus fascism) create a situation verging now on civil war. Or at least, with Trump's dog-whistle encouragement, his followers increasingly resort to threats and violence that, like the antinuclear and anti-GMO movements, aim at producing fear.

In his analysis of incoherent court findings in two First Amendment cases, the Hudnut pornography case and the Flynt/ Hustler case, Fish comments that "it is hard . . . not to feel that the entire enterprise [First Amendment jurisprudential philosophy] has gone off the rails and that you are in the hands either of charlatans or idiots." But in fact, he goes on, "you are in the hands of persons who are captive to a faulty theory, a theory that has produced among other philosophical curiosities this oft repeated dictum: 'Under the First Amendment, there is no such thing as a false idea' (Gertz v Robert Welch, 1974)" (1994, p. 124). One consequence of this "philosophical curiosit[y]" is to give equal legal status to both false ideas and true, which makes the discernment of truth more difficult and hinders broad agreement on its identity. This facilitates gaslighting, which now rages out of control in our political and social-media landscapes.

Fish points out that at the foundation of this position—which in effect rejects the idea that in order for truth to flourish false ideas must be regulated and denied the freedom to spread—lies a deep skepticism that what we take to be true today (including judgments about dangerous ideas) will remain true for all time. Since a presumed truth today may turn out to be considered false tomorrow, it, as Fish paraphrases the argument, "would be unwise to institutionalize beliefs we may not hold at a later date" (1994, p. 118). While such reversals have happened more times than we could count, the position leaves us without enough stability to guide today's actions. The need for such stability forces us to regard as true—for all practical purposes for the time being (FAPPFTTB)—propositions favored by the preponderance of evidence today. The logical conclusion of regarding all propositions as precariously perched on the verge of being overthrown is that when the day of the new truth seems to have arrived, it can be anticipated that it, too, will itself be subsequently negated, so that truth never, in fact, arrives in the present but only beckons from an ever-receding future horizon.[18]

In his updated discussion a quarter century later (2019), Fish notes that the metaphorical marketplace of ideas is intended as a haven safe from the fallibility of human judgment. This echoes the way standard free-market rhetoric opposes a planned economy in general, as well as a government role in choosing particular sources of energy in particular. Since governmental favoring of particular

---

[18] It is striking how similar both free speech epistemology and its deconstruction are to the deconstruction of a certain kind of skepticism by realist philosopher Alvin Goldman in *Epistemology and Cognition* (1986). Goldman says (p. 158):

> Suppose scientists decide that . . . in the science of 50 years ago . . . no theory of that vintage was true. If this keeps happening, shouldn't we be led to the meta-induction that . . . no present or future theory will be true? . . . won't it follow that we are not, and could not be, justified in believing any theoretical statement?
>
> No. There are several lacunae in this line of reasoning. Let us look first at the "disastrous meta-induction" itself. Upon inspection, the meta-induction is really self-undermining. We judge past theories to be false only in the light of our present theory. If we abandon our present theory, we are no longer in a position to judge past theories false. So if we use the meta-induction to conclude that no present scientific theory is true, we thereby eliminate all grounds for believing that past theories were false. But then we are no longer entitled to believe the premises of the meta-induction.

energy sources (subsidizing here, price-boosting there) promotes both winners and losers, the government, according to free market ideology, should bow out, because the judgment of government officials is fallible. It follows that both ideas and energy sources should be left to the market, which, it is assumed, will somehow produce truth and efficiency, respectively.

Here is Fish's devastating response to this argument (2019, p. 60):

> If we need the marketplace because as beings too much in love with our own views we cannot be trusted to make good judgments, won't that same fallibility (no more removable than original sin; it *is* original sin) prevent us from being able to determine when the marketplace has finally done its work and we can rest securely in what it has delivered? If no government or court can be trusted to make the necessary distinctions, neither can the marketplace be trusted to do the job, for its revolutions mark nothing more than the temporary ascendancy of someone's or some group's point of view . . . Is the Marketplace of Ideas[,] the free-speech offshoot of capitalism (as the very phrase suggests), asking us to be confident in its benign workings even though much of the evidence is to the contrary?

What unites the ideologies of the economic and philosophical marketplaces is the idea of transparency, an idea that is fundamental to libertarian views of capitalism, specifically those of F.A. Hayek—more recently reenergized in debates about Internet regulation. In this context, the notion of transparency is based on "neutrality" or the "God's Eye point of view," which fancifully regards the marketplace of ideas as capable of revealing the transparent truth.

In response, Fish expounds on the imaginary utopia of the fully transparent free market of ideas (2019, p. 153):

> transparency: the less the information we receive is filtered or curated, the closer we come to the Edenic condition of ever proliferating data accessible to more and more persons through an ever expanding network of communication channels.

The standard story that truth dies in darkness and flourishes in light, then, ignores the fact that falsehood also flourishes in light. In fact, falsehood can be the major occupant of the marketplace of ideas, since conceivable false propositions greatly outnumber the true. Neither truth nor falsehood arrives at the door sporting a label.

Fish points out that in our current capitalist social media environment, transparency (unfiltered and voluminous) like censorship (filtered and restricted) can eliminate the rational basis for good judgment. This results in largely siloed populations, drowning in TMI, accusing each other of fake news. This can lead us to throw up our hands, feeling we might as well go with our default and give free rein to confirmation bias.

US developmental psychologist Todd Rose (2022) examines the impact of TMI in terms of the way our brains process information. First, in contrast to the fantasy of transparency, allegedly characterized by an absence of filters, our brains impose the filtration. Rose notes (p. 124) that while your brain can capture

> the equivalent of eleven megabytes of information per second from your eyes, you can only "upload" about sixty bits per second into the picture you consciously "see." This is the equivalent of facing the entire population of Paris, France, but actually seeing only eight people . . .
>
> The sheer effort of being 100 percent accurate about everything would be an enormous waste of cognitive power.

A fictional representation of what would happen to someone with no filter can be found in the title character in Jorge Luis Borges's short story, "Funes the Memorious" (*Labyrinths*, 1964, pp. 59–66). Because Funes's perception and memory are infallible, he is paralyzed by TMI. In our media world of TMI, we are approaching the situation of Funes, with the difference that in an antagonistic, competitive information environment, we respond to TMI, Rose says, so that "most of the information directed at us has been tailored, personalized by us or by algorithms. In other words, we now only see the information that we want to see" (p. 135). He continues, "the combination of our brains and the internet has produced not just greater connection but also an unprecedented explosion of misunderstanding that threatens to envelop us all" (p. 136). Later in the chapter he adds that on the net, the loudest voices dominate, that such voices represent "fringe opinions," and this turns the web "into a carnival room full of distorting mirrors" so that "it's almost impossible to sort truth from falsehood, perception from reality" (p. 141).

While Rose's work is extremely enlightening about how we process information and can become overloaded, in our view he fails to give sufficient emphasis to the social environment within which our brains and the internet operate. It is good news, sort of, that in our personal lives, where we know the difference between truth and lie, brain research reveals that we prefer truth, having a "natural craving for congruence, trust, and sharing rooted in our need to survive"

(p. 162). The key question then becomes what kind of social organization can bring into line our need for truth with large-scale institutions that permit and reinforce that need, instead of undermining it.

The late US paleontologist Stephen Jay Gould, in an essay on the relationship between observation and theory, implicitly undercuts the claim that light, transparency, and the marketplace of ideas are the formula for finding truth. These positivist ideas, he says, affirm that objectivity is achieved "by clearing the mind of all preconceptions and then simply seeing, in a pure and unfettered way, what nature presents."[19] But as an observer and assimilator of nature, the brain is incapable of seeing, in a "pure and unfettered way." It necessarily sees through the lens of a paradigm, even if unaware of that paradigm. In fact, the concept of "pure and unfettered" seeing (the denial of paradigm) is itself a filter, a kind of magical thinking that interferes with insight.

Fish observes that "those who proclaim this theology [that information and transparency are all we need] think that only the non-method of having no routes, no boundaries, no categories, no silos can bring us to the river Jordan and beyond" (2019, p. 160). Blind faith in transparency is the faulty view of objectivity described by Gould, updated for the social media age. It's a view that evacuates human judgment, the only means by which the true can be distinguished from the false and information from mis/disinformation. As Fish points out, transparency becomes just another rhetorical tool in the culture wars.

It's interesting to note that the flipside of pure transparency is the epistemological anarchism of a relativist thinker like Austrian philosopher Paul Feyerabend. He proclaimed himself "against method," and his Dadaist[20] attitude that "anything goes" in the philosophy of science forswears the ability to judge creationism as any more or less scientific than the theory of evolution. But one can no more be coherently "against method" than one can be against filters. In short, neither the filter of "transparency" nor the filter of Feyerabend's "anything goes" can distinguish reliably between the true and the false.[21]

---

[19] Stephen Jay Gould, *Dinosaur in a Haystack: Reflections in Natural History* (1995, p. 148).

[20] Dadaism, a movement among artists originating during and in reaction to the horrors of World War I, places heavy emphasis on nonsense to satirize the social order.

[21] According no special place to science, Feyerabend says "basic beliefs" like evolution should be put to a vote. He later contradicts his libertarianism by approving the Chinese Communist Party's directive to "hospitals and medical schools to teach the ideas and methods contained in the Yellow Emperor's Textbook of Internal Medicine." The CCP may have

Just as some propositions, we contend, contain more truth than others, so are some paradigms better than others and that a meta-paradigm that is committed to cooperation for the common good is better than one that prizes cut-throat competition. Among other undesirable features, the latter fosters epistemic dysfunction. Transparency and the marketplace of ideas, like the concept of free speech, all contain their own self-contradictions. They are all futile attempts to substitute for judgment instead of working to strengthen our judgments through recognition of our limits and learning to compensate for them. In this regard, the insights of authors like Todd Rose (2022) and Israeli-American psychologist Daniel Kahneman (2013)[22] are indispensable. It is our belief that only a genuinely cooperative social order can fully capture and utilize these insights.[23]

Modern defenders of free speech at all costs, in the era of social media, call for "more speech" to counter disinformation. They also express skepticism about the very labeling of something as disinformation, as though it were just an excuse for government censorship. But they might as well argue that there is no such thing as

---

had good reasons to defend elements of traditional medicine, but our point is that Feyerabend's Dadaism allows him to be libertarian and anti-libertarian in the same breath and to claim that such behavior allows science to progress. His equal liking for coercion and choice allows him to grind his axe against what he sees as science worship. See Paul Feyerabend, *Against Method* (1978), and for a realist rebuttal, Richard Boyd, "How to be a Moral Realist," in G. Sayre-McCord, ed., *Essays on Moral Realism* (1988), pp. 181–228.

[22] Nobelist Kahneman (2013) explains how intuition (fast thinking) is sometimes correct but unreliably so, while investigation and reasoning (slow thinking) can act as a corrective when intuition is wrong. We always respond immediately with our intuition, based on personal experience and what we've been taught, but as Kahneman warns, without investigation and reasoning we can easily mislead ourselves without our realizing it.

[23] The present authors are moral realists. For us, moral realism is a branch of scientific realism because there are such things as moral facts: for example, slavery is wrong, which is clearly a moral position but also becomes a fact by virtue of its resting on the scientifically demonstrable fact that humans in general have the capacity to regard each other with respect for our individuality and the further fact that human flourishing is best achieved by our adhering to that respect for each other. Facts become facts by virtue of their requiring evidence for their establishment as true and by virtue of their being able themselves to serve as evidence for further gains in factual knowledge. As US philosopher Hilary Putnam has argued, the idea of a solid wall between fact and value is illusory. Many excellent moral philosophers, however, are not moral realists, but we think they should be. Susan Neiman, mentioned above, who considers herself a Kantian, says, "I have argued that hope for progress is never a matter of evidence" (2024, p. 117). But evidence for moral progress in the past as evidence that it could continue in the future should be no more controversial than making the same argument for scientific progress. Given progress in areas of scientific investigation—say, about how viruses work and how to combat them—we can expect more progress in the future. In the moral sphere, for example, the US has never had a female president, yet the likelihood of one in the future is increased by worldwide moral progress related to gender (including, of course, in the US).

disinformation, or that disinformation doesn't matter, that the value of speech per se outweighs its truth content no matter what. This is a view that inevitably downgrades truth production, which requires the winnowing of true from false, something essentially and ironically thwarted by the First Amendment.

In short, the concept of free speech enthrones the right to lie in public alongside the right to tell the truth. Though in a class-divided society dominated by one class, the right to tell *certain* truths is dethroned, while the right to lie remains unimpeded. Proponents of the marketplace of ideas may see it as the remedy to woke groupthink, or any groupthink, but what we get instead is a proliferation of mutually exclusive woke groupthinks, even if one mode of being woke is to be anti-woke. All this throws up barriers to rationality including, most importantly for our present purposes, rational energy policy.

It seems clear at this point in our story that being anti-establishment is nowhere near enough to move us either in the direction of a rational energy policy or a more cooperative society, one whose goal is human flourishing in the absence of exploitation. Moreover, these days those who occupy both the right and left wings of the establishment-approved political spectrum as a rule claim to be anti-establishment. Trump had no trouble picking up Bernie Sanders's "rigged system" to justify overturning an election. Climate deniers portray themselves as Galileo versus the climatology church, or as David versus Goliath. Anti-establishment discourses on both left and right mobilize their rhetoric to lend an aura of virtue to their fear-inducing dogmas.

In doing so, both right and left draw (however unknowingly) on a long and (we argue) flawed tradition, from Lukacs on down, which holds that membership among the exploited/oppressed grants an epistemological advantage. Applied to energy this anti-establishment ideology leads much of the egalitarian left to an unintended default Malthusian position (Chapter 9), as it rejects the very technologies required to overcome the damage and dead ends fostered by the fragmented competitive profit system. And it rejects the evidence-based arguments that might actually convince enough of those oppressed by that system to establish a successful alternative to it. This we term "epistemic dysfunction."

Acknowledgments

This project began some eight years ago when our friend Barbara Foley from the editorial board of the left-wing journal *Science and Society* asked us to write a rebuttal to an antinuclear article that the journal had published. But the research and writing for our project soon exceeded its word limit, and this book is the result.

Along the way we have been in touch with nuclear engineers and scientists, physicists, climatologists, radiobiologists, radiologists, radiation oncologists, other physicians, health physicists, science writers, philosophers, Marxist theoreticians, energy experts, electricians, electrical engineers, electrical grid experts, metallurgists, and critical theorists.

Among those who have offered the most consistent help with difficult concepts are fellow members of the international organization called Scientists for Accurate Radiation Information (SARI). SARI is a decade-old group of more than one hundred scientists, engineers, physicians, health physicists, and others from countries all around the world. SARI members exchange emails on a daily basis and continually pose and answer each others' questions—an indispensable resource.

It would be difficult to name specific individuals without unduly inflating the length of this book. However, among those whom we cannot conscionably omit are, in alphabetical order, Rod Adams, Wade Allison, Meredith Angwin, Chris Bachelder, Joe Bevelacqua, Tom Blees, Ed Calabrese, John Cardarelli, Yoon Chang, Bernard Cohen, Jim Conca, Mike Conley, Les Corrice, Jerry Cuttler, Jack Devanney, Ludwik Dobrzynski, Mohan Doss, Ludwig Feinendegen, Chris Feltham, Stephen Ferguson, Marty Goodman, Rod Green, Jim Hansen, Bob Hargraves, Gary Hoe,

Lars Jorgensen, Glenn Kissack, Len Koch, Don Luckey, Jeff Mahn, Tim Maloney, Javad Mortazavi, Mark Nelson, Jim Neilson, Leo Parascondola, Joe Ramsey, Chary Rangacharyulu, Michael Roberto, Ted Rockwell, Dave Rossin, John Sackett, Charles Sanders, Bobby Scott, Jeff Siegel, Andrzej Strupczewski, Shizuyo Sutou, Brant Ulsh, Mike Waligórski, Alan Waltar, David Walters, and Jim Welsh.

We also owe thanks to Carus Books's in-house editor David Ramsay Steele, who has steered us through multiple rounds of proofreading. Having gone through this process, we have a much fuller understanding of the usual disclaimer that any remaining errors are wholly owned by the two of us.

Both of us owe a debt of gratitude to Miriam Sacks, who read portions of the manuscript, offered suggestions for improvement, and who herself organized in years past two forums on climate and nuclear energy as well as one on radiobiology, each filling the hall with some 200 persons. It goes without saying, but Bill is saying it anyway, without her love and even-tempered guidance through many a tension-laden strait, life would have been far less tolerable during, as well as before and since, the writing of this book.

And Greg owes a debt to Sheila Wells, who trained herself to leave the room at any mention of grams of CO_2/kWh. Such discipline is undoubtedly the secret to their more than four decades of marriage.

Glossary

AGW. Anthropogenic global warming.

ALARA. As low as reasonably achievable.

Alpha particle. A helium nucleus, comprising two protons and two neutrons.

ANS. American Nuclear Society.

Becquerel (Bq). One radioactive decay (ejection of particle and energy) per second.

Beta decay. Ejection of an electron or positron.

BLM. Black Lives Matter.

Breeder reactor. Splits fissile isotopes while rapidly breeding new ones from fertile isotopes.

BWRX-300. 300 MW boiling water reactor (SMR).

C-14. Radioactive isotope of carbon with atomic weight 14.

Capacity factor (CF). Actual power over nameplate (maximum).

Chain reaction. Each event causes one further event. If more than one, it's a "branching" reaction.

Clathrates. Methane trapped in undersea ice crystals (or any other similar combination of two materials, one trapped in a crystal of the other).

Conversion device. Turns raw energy into electricity: wind turbines, solar photovoltaic (PV) panels, combustion chambers, and nuclear reactors.

Coolant. Water, gas, or liquid metal that carries heat from reactor core to make steam to drive the electric generator.

CRT. Critical race theory.

CSP. Concentrated solar power.

Demand. See "Load."

Depleted uranium (DU). See "Enrichment."

Deuterium. See "Tritium."

Dispatchable. Available on demand, where and when needed.

DOE. US Department of Energy.

DSB. Double strand break of DNA.

$E = mc^2$. More precisely the sum of energy (E) and mass (m) of an isolated system never changes, but energy and mass can exchange between the two at an exchange rate given by Einstein's famous formula. While the exchange between energy and mass is relevant to fission or fusion, most applications encountered in daily life undergo no such exchange, and E and m remain independently constant.

EBR II. Second experimental fast breeder reactor.

Ecomodernists. Favor nuclear energy and take capitalism for granted.

Ecosocialists. Reject capitalism and nuclear energy.

EIA. US DOE's Energy Information Administration.

EJ. Exajoule, or 10^{18} J.

Energy density. The standard definition is energy output per mass of fuel (MJ/kg). However, in order to include wind and solar energy, which are not fuel-based, we have extended the definition to include energy per mass of structural material plus fuel (if any) over the lifetime of the conversion apparatus.

Enrichment. Removal from raw uranium of most of the U-238, which is non-fissile. This process inevitably takes with it a small amount of U-235, but it leaves behind a higher *concentration* of U-235 than the original 0.7 percent. U-235 is fissile and constitutes the main portion of uranium used as fuel in LWRs. The removed portion is called "depleted uranium" (DU), meaning depleted with regard to the concentration of U-235, since it is almost entirely U-238. The term "enrichment" is somewhat misleading because it suggests that something is *added* to the raw uranium to "enrich" it, whereas the portion destined to serve as fuel is the result of a *removal* rather than an addition.

EROI. Energy returned (ER) over invested (EI).

Fertile. Nonfissile but transmutable to fissile.

Fissile. Fissionable by collision with a neutron.

Fission. Splitting of heavy nucleus into two smaller nuclei (fission products). Releases energy plus much smaller particles.

Fusion. Combining of light nuclei into one larger nucleus. Releases energy plus smaller particles.

GHG. Greenhouse gas.

GMO. Genetically modified organism.

GND. Green New Deal.

Goldilocks zone. Dose range that enhances health, between two ranges that diminish health.

GW. Gigawatt = billion watts.

GWC. Global warming contribution, our own definition, needed to indicate the change in CO_2's impact on AGW over time, since by definition GWP for CO_2 is always 1.

GWP. Global warming potential: ratio of heat energy absorbed per time to that absorbed by equal mass of CO_2.

HALEU. High-assay low-enriched uranium—enriched to between 5 and 20 percent. As opposed to slightly enriched uranium (SEU, up to about 1.2 percent), low-enriched uranium (LEU, between 1.2 percent and 5 percent), and highly enriched uranium (HEU, greater than 20 percent).

Half-life. Time required to halve a radioactive substance through decay, applicable to any exponential disappearance.

Hormesis. Stimulation of evolved protective biological response to repair or remove damage by an agent plus damage by endogenous processes. Leaves organism healthier.

HPS. Health Physics Society.

IAEA. International Atomic Energy Agency.

IARC. International Agency for Research on Cancer.

ICE. Internal combustion engine.

ICRP. International Commission on Radiological Protection.

IEA. International Energy Agency.

IPCC. Intergovernmental Panel on Climate Change.

Isotope. Forms of an element's nucleus differing only in number of neutrons.

Jevons's paradox. Self-negating phenomenon. increase in efficiency results in greater use, canceling inputs saved.

Joule. Unit of energy equivalent to a watt-second.

K-40. Radioactive isotope of potassium with atomic weight 40.

Load. Instantaneous electrical demand.

LCOE. Levelized cost of energy, or electricity.

LNT. Linear no-threshold: entails permanent radiogenic harm at any dose, cumulative throughout life.

Luddism. A nineteenth-century movement that sought to destroy machines, which, rather than exploitation, were wrongly thought to be the source of unemployment. Today it means opposition to technology in general.

LWR. Light water reactor.

Malthusianism. Named for the eighteenth-century clergyman and economist Thomas Malthus, who wrongly opined that population inevitably out-stripped food supply unless something drastic were done to decrease the population in the short term, such as war, genocide, disease, or starvation.

Meltdown. Melting of solid metal fuel in reactor core.

Moderator. Substance that slows neutrons through collisions, absent in fast breeder reactors.

MW. Megawatts = million watts.

MWe. MW of electricity, as opposed to MWt, meaning MW of thermal power, or total power, of which about one-third is electrical and the rest heat. Thus, the electrical efficiency of a nuclear reactor is about one-third.

Nameplate capacity. Maximum power of conversion device under ideal conditions, reduced by CF.

NCRP. National Council on Radiation Protection and Measurements.

NEA. Nuclear Energy Agency, an intergovernmental agency of the OECD.

Net energy. Energy return minus energy input (EN = ER – EI).

Neutron. Nuclear component with zero electrical charge, holds protons together.

NG. Natural gas, mainly methane (CH_4).

NGO. Non-governmental organization.

NRC. Nuclear Regulatory Commission.

Power. Energy produced or consumed per unit time. Power is measured in watts (W), energy in watt-hours (Wh).

Power density. In this book, power per land area occupied by renewables farm or power plant (watts per square meter = W/m^2).

Proton. Electrically charged nuclear component; number determines which element.

PV panels. Photovoltaic panels.

Quad. 1 quadrillion British Thermal Units (10^{15} BTU).

Radioactivity. Series of events in which unstable nuclei spontaneously emit small particles and release energy to attain stable state.

Rare earth metals. Though their ore is not rare in the ground, they are energetically and monetarily expensive to extract and process. It is for that reason they have been dubbed with their somewhat misleading name. Critical particularly in batteries and permanent magnets, though small proportion of their materials.

RBMK. Initials of Russian for high-power channel reactor.

ROS. Reactive oxygen species. Produced continuously by oxygen metabolism in cells' mitochondria or by radiation and avid in seizing electrons from other atoms and compounds (oxidation).

SLAPP. Strategic lawsuit against public participation, an illegal method of attempting to silence opposition through the courts.

SMR. Small modular reactor.

Solar PV. Photovoltaic panels, which convert sun's radiation into electricity.

SSB. Single strand break of DNA.

Tritium. Hydrogen isotope with one proton and two neutrons. Deuterium has one proton and one neutron. Isotope with one proton and no neutrons simply called hydrogen.

TW. Terawatts = trillion watts.

Uranium. U-235 and U-238 are the two naturally occurring uranium isotopes, with 92 protons and either 143 (92 + 143 = 235) or 146 neutrons (92 + 146 = 238). The number of protons defines which element it is.

UNSCEAR. United Nations Scientific Committee on the Effects of Atomic Radiation.

Watt. Unit of power equivalent to a joule per second.

Bibliography

Ajl, Max. *A People's Green New Deal* (2021).

Angell, Marcia. *The Truth About the Drug Companies: How They Deceive Us and What to Do about It* (2005).

Angus, Ian. *Facing the Anthropocene: Fossil Capitalism and the Crisis of the Earth System* (2016).

Angwin, Meredith. *Shorting the Grid: The Hidden Fragility of our Electric Grid* (2020).

Aronoff, Kate, Alyssa Battistoni, et al. *A Planet to Win: Why We Need a Green New Deal* (2019).

Ausubel, Jesse. http://phe.rockefeller.edu/docs/Density.pdf (2017).

Bakke, Gretchen. *The Grid: The Fraying Wires between Americans and Our Energy Future* (2016).

Baran, Paul A., and Paul M. Sweezy. *Monopoly Capital: An Essay on the American Economic and Social Order* (1966).

Black, Edwin. *Nazi Nexus: America's Corporate Connections to Hitler's Holocaust* (2009).

Blees, Tom. *Prescription for the Planet: The Painless Remedy for Our Energy and Environmental Crises* (2008).

Bodansky, David. *Nuclear Energy: Principles, Practices, and Prospects, 2nd Edition* (2004).

Boice, John D., Michael Mumma, and William J. Blot. Cancer Incidence and Mortality in Populations Living Near Uranium Milling and Mining Operations in Grants, New Mexico, 1950–2004. *Radiat Res* 174: 624–636 (2010). https://pubmed.ncbi.nlm.nih.gov/20954862/.

Borrego-Soto, Gissela, Rocio Ortiz-López, and Augusto Rojas-Martinez. Ionizing Radiation-Induced DNA Injury and Damage Detection in Patients with Breast Cancer. *Genet Mol Biol* 38(4): 420–32 (2015). https://doi.org/10.1590/S1415-475738420150019.

Brill, Steven. *America's Bitter Pill: Money, Politics, Backroom Deals, and the Fight to Fix Our Broken Healthcare System* (2015).

Broad, William, and Nicholas Wade. *Betrayers of the Truth: Fraud and Deceit in the Halls of Science* (1982).

Brooks, Antone L. *Low Dose Radiation: The History of the U.S. Department of Energy Research Program* (2018).

Bryce, Robert. *Power Hungry: The Myths of "Green" Energy and the Real Fuels of the Future* (2010).

———. *Smaller Faster Lighter Denser Cheaper: How Innovation Keeps Proving the Catastrophists Wrong* (2014).

Bryce, Robert. *A Question of Power: Electricity and the Wealth of Nations* (2020).

Calabrese, Edward J. Interview by the Health Physics Society, Divided into Twenty-Two Relatively Short Sessions (2022). http://hps.org/hpspublications/historylnt/episodeguide.html.

———. Podcast: The Historical Foundations of the Linear Non-Threshold (LNT) Dose Response Model for Cancer Risk Assessment (2023a). https://nam10.safelinks.protection.outlook.com/.

Calabrese, Edward J. The Gofman-Tamplin Cancer Risk Controversy and Its Impact on the Creation of BEIR I and the Acceptance of LNT. *La Medicina del Lavoro* 114(1): e2023007 (2023b). http://doi.org/10.23749/mdl.v114i1.14006.

Calabrese, Edward J., Evgenios Agathokleous, James Giordano, and Paul B. Selby. Manhattan Project Genetic Studies: Flawed Research Discredits LNT Recommendations. *Environ Pollut* 319: 120902 (2023). https://doi.org/10.1016/j.envpol.2022.120902.

Calabrese, Edward J., and James Giordano. How Hermann J. Muller Viewed the Ernest Sternglass Contributions to Hereditary and Cancer Risk Assessment. *Health Phys* 126(3): 151–55 (2024).

Calabrese, Edward J., Marc Nascarella, et al. Hormesis Determines Lifespan. *Ageing Res Rev* 94: 102181 (2024). https://doi.org/10.1016/j.arr.2023.102181.

Carchedi, Guglielmo, and Michael Roberts, eds. *World in Crisis: A Global Analysis of Marx's Law of Profitability* (2018).

Cardarelli, John J., and Brant A. Ulsh. It Is Time to Move Beyond the Linear No-Threshold Theory for Low-Dose Radiation Protection. *Dose-Response* 16(3): 1–24 (2018). https://doi.org/10.1177/1559325818779651.

Cardis, Elisabeth, Geoffrey Howe, et al. Cancer Consequences of the Chernobyl Accident: 20 Years On. *J. Radiol. Prot.* 26(2): 127–140 (2006). https://pubmed.ncbi.nlm.nih.gov/16738412/

Chang, Iris. *The Rape of Nanking: The Forgotten Holocaust of World War II* (1997).

Chen, W.L., Y.C. Luan, M.C. Shieh, et al. Effects of Cobalt-60 Exposure on Health of Taiwan Residents Suggest New Approach Needed in Radiation Protection. *Dose-Response* 5(1): 63–75 (2007). http://doi.org/10.2203/dose-response.06-105.Chen.

Clack, Christopher T.M., Staffan A. Qvist, et al. Evaluation of a Proposal for Reliable Low-Cost Grid Power with 100 Percent Wind, Water, and Solar. *Proc Natl Acad Sci USA* 114 (26): 6722–6727 (2017). https://doi.org/10.1073/pnas.1610381114.

Cohen, Bernard. *The Nuclear Energy Option* (1990).

———. Test of the Linear-No Threshold Theory of Radiation Carcinogenesis for Inhaled Radon Decay Products. *Health Phys* 68(2): 157–174 (1995). http://www.phyast.pitt.edu/~blc/LNT-1995.PDF.

————. The Linear No-Threshold Theory of Radiation Carcinogenesis Should Be Rejected. *J Am Phys Surg* 13(3):70–76 (2008). https://www.jpands.org/vol13no3/cohen.pdf.

Cohen, Mervyn. CT radiation dose reduction: can we do harm by doing good? *Pediatr Radiol* 42: 397–98 (2012). https://doi.org/10.1007/s00247-011-2315-9.

Conley, Mike. *The LNT Report* (2025).

Conley, Mike, and Tim Maloney. *Roadmap to Nowhere* (2017). Soon to be published in print, no longer available online.

————. *Earth Is a Nuclear Planet: The Environmental Case for Nuclear Power* (2024).

Cox, Stan. *The Green New Deal and Beyond: Ending the Climate Emergency While We Still Can* (2020).

Cravens, Gwyneth. *Power to Save the World: The Truth About Nuclear Energy* (2007).

Cuttler, Jerry M. Applications of Low Doses of Ionizing Radiation in Medical Therapies. *Dose-Response* 18(1): January–March (2020). https://www.ncbi.nlm.nih.gov/pmc/articles/PMC6945458/.

Cuttler, Jerry M., and James S. Welsh. Leukemia and Ionizing Radiation Revisited. *J Leukemia* 3(4):1000202 (2015). https://www.researchgate.net/publication/338819152_Leukemia_and_Ionizing_Radiation_Revisited.

David, Elroei, Marina Wolfson, and Vadim E. Fraifeld. Background Radiation Impacts Human Longevity and Cancer Mortality: Reconsidering the Linear No-Threshold Paradigm. *Biogerontology* 22:189–195 (2021). https://doi.org/10.1007/s10522-020-09909-4. Also available at https://sci-hub.wf/10.1007/s10522-020-09909-4.

Devanney, Jack. *Why Nuclear Power Has Been a Flop at Solving the Gordian Knot of Electricity, Poverty, and Global Warming, Third Edition* (2023)— available free at https://gordianknotbook.com/.

Doss, Mohan. Evidence Supporting Radiation Hormesis in Atomic Bomb Survivor Cancer Mortality Data. *Dose-Response* 10: 584–592 (2012). https://journals.sagepub.com/doi/10.2203/dose-response.12-023.Doss.

————. Low Dose Radiation Adaptive Protection to Control Neurodegenerative Diseases. *Dose-Response* 12:277-287 (2013). https://journals.sagepub.com/doi/10.2203/dose-response.13-030.Doss.

Edesess, Michael. We Need to Get Serious about the Renewable Energy Revolution—by Including Nuclear Power. *Bulletin of the Atomic Scientists* (May 5, 2022). https://thebulletin.org/2022/05/we-need-to-get-serious-about-the-renewable-energy-revolution-by-including-nuclear-power/.

Feinendegen, Ludwig E., Myron Pollycove, and Charles A. Sondhaus. Responses to Low Doses of Ionizing Radiation in Biological Systems. *Nonlinearity in Biology, Toxicology, and Medicine* 2: 143–171 (2004).

Fish, Stanley. *There's No Such Thing as Free Speech . . . and It's a Good Thing, Too* (1994).

————. *The First: How to Think about Hate Speech, Campus Speech, Religious Speech, Fake News, Post-Truth, and Donald Trump* (2019).

Floyd, Josh. Beyond This Brief Anomaly (2014). https://tinyurl.com/ns964ahp.

Fornalski, Krzysztof W., and Ludwik Dobrzy ski. The Healthy Worker Effect and Nuclear Industry Workers. *Dose-Response* 8(2): 125–147 (2010). https://scholarworks.umass.edu/dose_response/vol8/iss2/4.

Foster, John Bellamy, and Brett Clark. *The Robbery of Nature: Capitalism and the Ecological Rift* (2020).

Fox, Michael H. *Why We Need Nuclear Power: The Environmental Case* (2014).

Fuller, Gary. *The Invisible Killer: The Rising Global Threat of Air Pollution—and How We Can Fight Back* (2018).

Gale, Robert P., and Eric Lax. *Radiation: What It Is, What You Need to Know* (2013).

Gibbs, Jeff, and Ozzie Zehner. *Planet of the Humans* (documentary film, 2019). https://www.youtube.com/watch?v=Zk11vI-7czE.

Godfrey-Smith, Peter. *Other Minds: The Octopus, The Sea, and the Deep Origins of Consciousness* (2016).

Goldacre, Ben. *Bad Pharma: How Drug Companies Mislead Doctors and Harm Patients* (2012).

Goldstein, Joshua S., and Staffan A. Qvist. *A Bright Future: How Some Countries Have Solved Climate Change and the Rest Can Follow* (2019).

Greider, William. *One World Ready or Not: The Manic Logic of Global Capitalism* (1997).

Halm, B.M., A.A. Franke, J.F. Lai, D. Brenner, et al. g-H2AX Foci Are Increased in Lymphocytes in Vivo in Young Children 1 h after Very Low-dose Xirradiation: A Pilot Study. *Pediatr Radiol* 44(10):1310-1317 (2014).

Hanekamp, Yannic, James Giordano, et al. Immunomodulation Through Low-Dose Radiation for Severe COVID-19: Lessons from the Past and New Developments. *Dose-Response* 18(3): July–September (2020). https://www.ncbi.nlm.nih.gov/pmc/articles/PMC7513398/.

Hansen, James. *Storms of My Grandchildren: The Truth about the Coming Climate Catastrophe and Our Last Chance to Save Humanity* (2009).

Hargraves, Robert. *Thorium: Energy Cheaper than Coal* (2012).

Harman, Chris. *Zombie Capitalism: Global Crisis and the Relevance of Marx* (2009).

Harvey, David. *The Limits to Capital* (1982).

Heard, Benjamin, Barry Brook, et al. Burden of Proof: A Comprehensive Review of the Feasibility of 100 percent Renewable Electricity Systems. *Renewable and Sustainable Energy Reviews* 76:1122-33 (2017).

Hecht, Gabrielle. *The Radiance of France: Nuclear Power and National Identity after World War II* (2009).

Henriksen, Thormod, et al. *Radiation and Health* (2015). http://www.mn.uio.no/fysikk/tjenester/kunnskap/straling/radiation-and-health-2015.pdf.

Henriksen, Thormod, et al. *Radon, Lung Cancer, and the LNT model* (2016). https://www.mn.uio.no/fysikk/tjenester/kunnskap/straling/radon-and-lung-cancer.pdf.

Hirsh, Richard F. *Power Loss: The Origins of Deregulation and Restructuring in the American Electric Utility System* (1999).

Huber, Peter W., and Mark P. Mills. *The Bottomless Well: The Twilight of Fuel, the Virtue of Waste, and Why We Will Never Run Out of Energy* (2006).

Huke, A., G. Ruprecht, D. Weißbach, et al. The Dual Fluid Reactor—a New Concept for a Highly Effective Fast Reactor. The 19th Pacific Basin Nuclear Conference (PBNC 2014). www.researchgate.net/publication/265297594.

Jacobson, Mark Z. Review of Solutions to Global Warming, Air Pollution, and Energy Security. *Energy Environ. Sci.* 2: 148–173 (2009).

Jacobson, Mark Z., and Mark A. Delucchi. A Path to Sustainable Energy by 2030. *Scientific American* (November 2009).

———. Response to "A Critique of Jacobson and Delucchi's Proposals for a World Renewable Energy Supply" by Ted Trainer. *Energy Policy* 44: 482–84 (2012).

Jacobson, Mark Z., Mark A. Delucchi, Mary A. Cameron, and Bethany A. Frew. Low-cost Solution to the Grid Reliability Problem with 100 Percent Penetration of Intermittent Wind, Water, and Solar for All Purposes. *PNAS* 112 (49): 15060–15065 (2015). www.pnas.org/cgi/doi/10.1073/pnas.1510028112.

Jacobson, Mark Z., Mark A. Delucchi, Guillaume Bazouin, et al. 100 Percent Clean and Renewable Wind, Water, and Sunlight (WWS) All-Sector Energy Roadmaps for the 50 United States. *Energy Environ. Sci.* 8: 2093–2117 (2015). http://doi.org/10.1039/c5ee01283j.

Jacobson, Mark Z., Anna-Katharina von Krauland, et al. Zero Air Pollution and Zero Carbon from All Energy at Low Cost and Without Blackouts in Variable Weather throughout the U.S. with 100 Percent Wind-Water-Solar and Storage. *Renewable Energy* 184: 430–442 (2022). https://doi.org/10.1016/j.renene.2021.11.067.

Janiak, Marek, and Michael Waligórski. Can Low-Level Ionizing Radiation Do Us Any Harm? *Dose-Response* January–March: 1–15 (2023). http://doi.org/10.1177/15593258221148013.

Jaworowski, Zbigniew. Observations on the Chernobyl Disaster and LNT. *Dose-Response* 8:148–171 (2010).

Jones, Van. *The Green Collar Economy: How One Solution Can Fix Our Two Biggest Problems* (2008).

Kabat, Geoffrey. *Getting Risk Right: Understanding the Science of Elusive Health Risks* (2017).

Kahneman, Daniel. *Thinking, Fast and Slow* (2013).

Kalantzakos, Sophia. *China and the Geopolitics of Rare Earths* (2018).

Kara, Siddarth. *Cobalt Red: How the Blood of the Congo Powers Our Lives* (2023).

Klein, Naomi. *This Changes Everything: Capitalism vs The Climate* (2014).

———. *On Fire: The (Burning) Case for a Green New Deal* (2019).

Kliman, Andrew. *The Failure of Capitalist Production: Underlying Causes of the Great Recession* (2012).

Kovel, Joel. *The Enemy of Nature: The End of Capitalism or the End of the World?* (2007).

Li, Minqi. *The Rise of China and the Demise of the Capitalist World Economy* (2008).

Liu, Jifeng, Tengfei Ma, et al. History, Advancements, and Perspective of Biological Research in Deep Underground Laboratories: A Brief Review. *Environment International*, Volume 120 (2018).

Löbrich, M., N. Rief, M. Kühne, et al. In Vivo Formation and Repair of DNA Double-strand Breaks after Computed Tomography Examinations. *Proc Natl Acad Sci USA* 102(5): 8984–8989 (2005).

Lovering, Jessica, Arthur Yip, and Ted Nordhaus. Historical Construction Costs of Global Nuclear Reactors. *Energy Policy* 91: 371–382 (2016). https://tinyurl.com/46hkp7jd.

Luckey, T.D. *Radiation Hormesis* (1991).

Lynas, Mark. *Seeds of Science: Why We Got It So Wrong on GMOs* (2020).

Mahaffey, James. *Atomic Awakening* (2009).

———. *Atomic Accidents* (2014).

———. *Atomic Adventures* (2017).

Maloney, Tim. http://www.timothymaloney.net/Critique_of_100_WWS_Plan.html (2016).

Mann, Michael. *Our Fragile Moment* (2023)

Mann, Michael, and Tom Toles. *The Madhouse Effect* (2016).

McWhorter, John. *Woke Racism: How a New Religion Has Betrayed Black America* (2021).

Meadows, Donella H., Dennis L. Meadows, et al. *The Limits to Growth* (1972).

Mifune, M., T. Sobue, et al. Cancer Mortality Survey in a Spa Area (Misasa, Japan) with a High Radon Background. *Jpn J Cancer Res* 83(1): 1–5 (1992). http://doi.org/10.1111/j.1349-7006.1992.tb02342.x.

Mighton, John. *The Myth of Ability: Nurturing Mathematical Talent in Every Child* (2004).

Miles, Daniel. *The Phantom Fallout-Induced Cancer Epidemic in Southwestern Utah* (2008).

Miller, Lee M., and David W. Keith. Observation-based Solar and Wind Power Capacity Factors and Power Densities. *Environ. Res. Lett.* (2018). https://iopscience.iop.org/article/10.1088/1748-9326/aae102/meta.

Mitchel, R.E.J., P. Burchart, and H. Wyatt. A Lower Dose Threshold for the *In Vivo* Protective Adaptive Response to Radiation. Tumorigenesis in Chronically Exposed Normal and *Trp53* Heterozygous C57BL/6 Mice. *Radiation Research* 170: 765–775 (2008).

Montalbano, Sarah. Shattered Green Dreams: The environmental costs of wind and solar. https://www.americanexperiment.org/reports/shattered-green-dreams.

Morgan, John. https://bravenewclimate.com/2014/08/22/catch-22-of-energy-storage/#more-6460 (2014).

Mycio, Mary. *Wormwood Forest: A Natural History of Chernobyl* (2005).

Neiman, Susan. *Left Is Not Woke* (2024).

Neumaier, T., Joel Swenson, et al. Evidence for formation of DNA Repair Centers and Dose-Response Nonlinearity in Human Cells. *Proc Natl Acad Sci USA* 109(2): 443–48 (2012). https://doi.org/10.1073/pnas.1117849108.

Nuclear Energy Agency. *Status Report on Spent Fuel Pools under Loss-of-Cooling and Loss-of-Coolant Accident Conditions* (2015). https://tinyurl.com/pt8vrny2.

Paarlberg, Robert. *Resetting the Table: Straight Talk about the Food We Grow and Eat* (2021).

Palley, Reese. *The Answer: Why Only Inherently Safe, Mini Nuclear Power Plants Can Save Our World* (2011).

Palmer, Graham. *Energy in Australia: Peak Oil, Solar Power, and Asia's Economic Growth* (2014).

Palmer, Graham, and Josh Floyd. *Energy Storage and Civilization: A Systems Approach* (2020).

Petroski, Henry. *Pushing the Limits: New Adventures in Engineering* (2004).

Phillips, Leigh. *Austerity Ecology and the Collapse-Porn Addicts* (2015).

Phillips, Leigh, and Michal Rozworski. *The People's Republic of Walmart: How the World's Biggest Corporations Are Laying the Foundation for Socialism* (2019).

Pollan, Michael. *The Omnivore's Dilemma: A Natural History of Four Meals* (2006).

Powers, Richard. *The Overstory* (2018).

Prieto, Pedro and Charles Hall. *Spain's Photovoltaic Revolution: The Energy Return on Investment* (2013).

Qvist, Staffan A., and Barry Brook. Potential for Worldwide Displacement of Fossil-Fuel Electricity by Nuclear Energy in Three Decades Based on Extrapolation of Regional Deployment Data. *PLOS One* (2015) https://journals.plos.org/plosone/article?id=10.1371/journal.pone.0124074.

Ridley, Matt. *How Innovation Works: And Why It Flourishes in Freedom* (2020).

Roberto, Michael Joseph, Gregory Meyerson, et al. Moment of Transition. *Works and Days* 30(59/60):51–118 (2012).

Roberts, Michael. *Marx 200—a Review of Marx's Economics 200 Years after His Birth* (2018).

Robinson, Kim Stanley. *New York 2140* (2018).

———. *The Ministry for the Future* (2020).

Ronald, Pamela C., and Raoul W. Adamchak. *Tomorrow's Table: Organic Farming, Genetics, and the Future of Food* (2018).

Rose, Todd. *Collective Illusions: Conformity, Complicity and the Science of Why We Make Bad Decisions* (2022).

Ruhnau, Oliver, and Staffan A. Qvist. Storage Requirements in a 100 Percent Renewable Electricity System: Extreme Events and Inter-annual Variability. *Leibniz Information Centre for Economics* (2021). http://hdl.handle.net/10419/236723.

Russell, Geoff. *Greenjacked!: The Derailing of Environmental Action on Climate Change* (2016).

Sacks, Bill, and Greg Meyerson. The Nuclear Energy Solution. *Brave New Climate.com* (2012). https://bravenewclimate.files.wordpress.com/2012/04/nuclear_energy_solution_4-6-12.pdf.

———. The Left Needs to Reconsider Its Automatic Position Against Nuclear Energy. *Atomic Insights* August 7 (2015). http://tinyurl.com 3ehw95tj

Sacks, Bill, and Jeffry A. Siegel. Preserving the Anti-Scientific Linear No-Threshold Myth: Authority, Agnosticism, Transparency, and the Standard of Care. *Dose-Response* July–September, 1–4 (2017). https://journals.sagepub.com/doi/10.1177/1559325817717839.

Sacks, Bill, Gregory Meyerson, and Jeffry A. Siegel. Epidemiology without Biology: False Paradigms, Unfounded Assumptions, and Specious Statistics in Radiation Science (with Commentaries by Inge Schmitz-Feuerhake and Christopher Busby and a Reply by the Authors). *Biol Theory* 11(2): 69–101 (2016). http://link.springer.com/article/10.1007/s13752-016-0244-4.

Saez, Emmanuel, and Gabriel Zucman. *The Triumph of Injustice: How the Rich Dodge Taxes and How to Make Them Pay* (2019).

Sanders, Charles L. *Radiation Hormesis and the Linear-No-Threshold Assumption* (2010).

Schernikau, Lars, and William H. Smith. *The Unpopular Truth about Electricity and the Future of Energy* (2023).

Shaikh, Anwar. *Capitalism: Competition, Conflict, Crisis* (2016).

Shellenberger, Michael. *Apocalypse Never: Why Environmental Alarmism Hurts Us All* (2020).

Siegel, Jeffrey A., Bill Sacks, Yehoshua Socol. The LSS Cohort of Atomic Bomb Survivors and LNT. Comments on "Solid Cancer Incidence among the Life Span Study of Atomic Bomb Survivors: 1958–2009" (*Radiat Res* 187: 513–537, 2017) and "Reply to the Comments by Mortazavi and Doss" (*Radiat Res* 188: 369–371, 2017). *Radiat Res* 188: 463–64 (2017).

Siegel Jeffrey A., Bill Sacks, Bennett S. Greenspan. NRC Rejects Petitions to End Reliance on LNT Model. *J Nucl Med* 62(11): 17N–22N (2021). https://tinyurl.com/mwdwbs7u.

Slotkin, Richard. *A Great Disorder: National Myth and the Battle for America* (2024).

Smil, Vaclav. *Energy Transitions: History, Requirements, Prospects* (2010).

———. *Power Density: A Key to Understanding Energy Sources and Uses* (2016).

———. *Halfway Between Kyoto and 2050: Zero Carbon Is a Highly Unlikely Outcome* (2024). https://tinyurl.com/ycy2yt3e.

Sponsler, Ruth, and John R. Cameron. Nuclear Shipyard Worker Study (1980–1988): A Large Cohort Exposed to Low-Dose-Rate Gamma Radiation. *Int J Low Radiat* 1(4): 463–478 (2005).

Stone, Robert. *Pandora's Promise*, a documentary film (2019). www.youtube.com/watch?v=KMutoR8YTlQ.

Sutou, Shizuyo. Black Rain in Hiroshima: A Critique to the Life Span Study of A-bomb Survivors, Basis of the Linear No-Threshold Model. *Genes and Environment* 42:1 (2020). https://doi.org/10.1186/s41021-019-0141-8.

Thompson, William L. *Living on the Grid: The Fundamentals of the North American Electric Grids in Simple Language* (2016).

Trainer, Ted. 100 Percent Renewable Supply? Comments on the Reply by Jacobson and Delucchi to the Critique by Trainer. *Energy Policy* 57: 634–640 (2013).

Tubiana, Maurice. Dose-Effect Relationship and Estimation of the Carcinogenic Effects of Low Doses of Ionizing Radiation: The Joint Report of the Académie des Sciences (Paris) and of the Académie Nationale de Médecine. *Int J Radiat Onc Biol Phys* 63(2): 317–19 (2005). https://doi.org/10.1016/j.ijrobp.2005.06.013.

Tucker, Colin. *How to Drive a Nuclear Reactor* (2019).

Visscher, Marco. *The Power of Nuclear: The Rise, Fall, and Return of Our Mightiest Energy Source* (2025).

Walker, J. Samuel. *Three Mile Island: A Nuclear Crisis in Historical Perspective* (2004).

Walinder, Gunnar. *Has Radiation Protection Become a Health Hazard?* (1995).

Waltar, Alan, and Ludwig Feinendegen. The Double Threshold: Consequences for Identifying Low-Dose Radiation Effects. *Dose-Response* July–September 1–5 (2020).

http://doi.org/10.1177/1559325820949729.

Weart, Spencer. *The Discovery of Global Warming* (2003).

———. *The Rise of Nuclear Fear* (2012).

Weißbach, D., G. Ruprecht, et al. Energy Intensities, EROIs, and Energy Payback Times of Electricity Generating Power Plants. *Energy* Volume 52, April, 210–221 (2013).

doi.org/10.1016/j.energy.2013.01.029.

Wood, E.M. *The Ellen Meiksins Wood Reader* (2012).

Index